A·N·N·U·A·L E·D·I·

MW00569714

Physical Anthropology

00/01

Ninth Edition

TXAPF 1 - $2695

EDITOR

Elvio Angeloni
Pasadena City College

Elvio Angeloni received his B.A. from UCLA in 1963, his M.A. in anthropology from UCLA in 1965, and his M.A. in communication arts from Loyola Marymount University in 1976. He has produced several films, including *Little Warrior,* winner of the Cinemedia VI Best Bicentennial Theme, and *Broken Bottles,* shown on PBS. He most recently served as an academic adviser on the instructional television series, *Faces of Culture.*

Dushkin/McGraw-Hill
Sluice Dock, Guilford, Connecticut 06437

Visit us on the Internet
http://www.dushkin.com/annualeditions/

Credits

1. Natural Selection
Unit photo—courtesy of New York Public Library.
2. Primates
Unit photo—United Nations photo by George Love.
3. Sex and Society
Unit photo—courtesy Baron Hugo van Lawick; © National Geographic Society.
4. The Fossil Evidence
Unit photo—courtesy of the American Museum of Natural History.
5. Late Hominid Evolution
Unit photo—AP Photo of prehistoric cave painting by Jean Clottes.
6. Human Diversity
Unit photo—United Nations photo by Doranne Jacobson.
7. Living with the Past
Unit photo—© Napoleon Chagnon.

Copyright

Cataloging in Publication Data
Main entry under title: Annual Editions: Physical Anthropology. 2000/2001.
   1. Physical anthropology—Periodicals. I. Angeloni, Elvio, *comp.* II. Title: Physical anthropology.
ISBN 0–07–236398–3   573'.05   ISSN 1074–1844

Ninth Edition

Cover image © 2000 PhotoDisc, Inc.

Printed in the United States of America   1234567890BAHBAH543210   Printed on Recycled Paper

Editors/Advisory Board

Staff

To the Reader

In publishing ANNUAL EDITIONS we recognize the enormous role played by the magazines, newspapers, and journals of the public press in providing current, first-rate educational information in a broad spectrum of interest areas. Many of these articles are appropriate for students, researchers, and professionals seeking accurate, current material to help bridge the gap between principles and theories and the real world. These articles, however, become more useful for study when those of lasting value are carefully collected, organized, indexed, and reproduced in a low-cost format, which provides easy and permanent access when the material is needed. That is the role played by ANNUAL EDITIONS.

New to ANNUAL EDITIONS is the inclusion of related World Wide Web sites. These sites have been selected by our editorial staff to represent some of the best resources found on the World Wide Web today. Through our carefully developed topic guide, we have linked these Web resources to the articles covered in this ANNUAL EDITIONS reader. We think that you will find this volume useful, and we hope that you will take a moment to visit us on the Web at **http://www.dushkin.com** to tell us what you think.

This ninth edition of *Annual Editions: Physical Anthropology* contains a variety of articles relating to human evolution. The writings were selected for their timeliness, relevance to issues not easily treated in the standard physical anthropology textbook, and clarity of presentation.

Whereas textbooks tend to reflect the consensus within the field, *Annual Editions: Physical Anthropology 00/01* provides a forum for the controversial. We do this in order to convey to the student the sense that the study of human development is an evolving entity in which each discovery encourages further research and each added piece of the puzzle raises new questions about the total picture.

Our final criterion for selecting articles is readability. All too often, the excitement of a new discovery or a fresh idea is deadened by the weight of a ponderous presentation. We seek to avoid that by incorporating essays written with enthusiasm and with the desire to communicate some very special ideas to the general public.

Included in this volume are a number of features designed to be useful for students, researchers, and professionals in the field of anthropology. While the articles are arranged along the lines of broadly unifying subject areas, the *topic guide* can be used to establish specific reading assignments tailored to the needs of a particular course of study. Other useful features include the *table of contents* abstracts, which summarize each article and present key concepts in bold italics, and a comprehensive *index*. In addition, each unit is preceded by an overview that provides a background for informed reading of the articles, emphasizes critical issues, and presents *challenge questions*. Also included are *World Wide Web* sites that can be used to further explore the topics. These sites are cross-referenced by number in the topic guide.

In contrast to the usual textbook, which by its nature cannot be easily revised, this book will be continually updated to reflect the dynamic, changing character of its subject. Those involved in producing *Annual Editions: Physical Anthropology 00/01* wish to make the next one as useful and effective as possible. Your criticism and advice are welcomed. Please complete and return the postage-paid *article rating form* on the last page of the book and let us know your opinions. Any anthology can be improved, and this one will continue to be.

Elvio Angeloni
Editor

(E-mail address: *evangeloni@paccd.cc.ca.us*)

Contents

UNIT 1

Natural Selection

Four articles examine the link
between genetics and the
process of natural selection.

UNIT 2

Primates

Eight selections examine some of the
social relationships in the primate world
and how they mirror human society.

The concepts in bold italics are developed in the article. For further expansion please refer to the Topic Guide and the Index.

The concepts in bold italics are developed in the article. For further expansion please refer to the Topic Guide and the Index.

Overview   **70**

13. **These Are Real Swinging Primates,** Shannon Brownlee, *Discover*, April 1987.   **72**
Although the muriqui monkeys of Brazil are heavily invested in ***reproductive competition,*** they seem to get along just fine without a ***dominance hierarchy*** and fighting over females. In other words, their mission in life seems geared to making love, not war.

14. **The Myth of the Coy Female,** Carol Tavris, from *The Mismeasure of Woman*, Simon & Schuster, 1992.   **77**
It may be impossible for us to observe the behavior of other species in a way that does not mirror the assumptions of our own way of life. In this light, ***primate behavior*** and the ***theories*** it generates need to be handled with care.

15. **First, Kill the Babies,** Carl Zimmer, *Discover*, September 1996.   **81**
In the early 1970s, Sarah Hrdy proposed a new explanatory model for ***infanticide.*** Its evolutionary logic has since generated a wealth of research and continuing controversy.

16. **A Woman's Curse?** Meredith F. Small, *The Sciences*, January/February 1999.   **86**
An anthropologist's study of the ***ritual of seclusion*** surrounding ***women's menstrual cycle*** has some rather profound implications regarding human evolution, certain cultural practices, and ***women's health.***

17. **What's Love Got to Do with It?** Meredith F. Small, *Discover*, June 1992.   **91**
The ***bonobos'*** use of sex to reduce tension and to form ***alliances*** is raising some interesting questions regarding human evolution. Does this behavior help to explain the origins of our ***sexuality,*** or should we see it as just another primate aberration that occurred after humans and primates split from their common lineage?

18. **Apes of Wrath,** Barbara Smuts, *Discover*, August 1995.   **95**
Whether or not males beat up females in a particular species seems to have a great deal to do with who is forming ***alliances*** with whom. This, in turn, has powerful implications as to what can be done about ***sexual coercion*** in the human species.

Overview   **98**

19. **Sunset on the Savanna,** James Shreeve, *Discover*, July 1996.   **100**
The long-held belief that ***hominid bipedalism*** owes its origin to a shift from life in the forest to life in a more open habitat is being challenged by new evidence regarding fossils found in the wrong place at the wrong time.

20. **Early Hominid Fossils from Africa,** Meave Leakey and Alan Walker, *Scientific American*, June 1997.   **106**
Fossil finds in Africa display similarities to modern African apes. Not only do these discoveries hint at our ***common ancestry*** with apes, but the ***environmental context*** is raising questions about whether early hominids evolved on the savanna or in a more wooded setting.

UNIT 3

Sex and Society

Six articles discuss the relationship between the sexes and the evolution of a social structure.

UNIT 4

The Fossil Evidence

Six selections discuss some of the fossil evidence for hominid evolution.

UNIT 5

Late Hominid Evolution

Eight articles examine archaeological evidence of human evolution.

The concepts in bold italics are developed in the article. For further expansion please refer to the Topic Guide and the Index.

UNIT 6

Human Diversity

Five articles examine human
racial evolution and diversity.

The concepts in bold italics are developed in the article. For further expansion please refer to the Topic Guide and the Index.

ix

The concepts in bold italics are developed in the article. For further expansion please refer to the Topic Guide and the Index.

The concepts in bold italics are developed in the article. For further expansion please refer to the Topic Guide and the Index.

This topic guide suggests how the selections and World Wide Web sites found in the next section of this book relate to topics of traditional concern to physical anthropology students and professionals. It is useful for locating interrelated articles and Web sites for reading and research. The guide is arranged alphabetically according to topic.

The relevant Web sites, which are numbered and annotated on pages 4 and 5, are easily identified by the Web icon (☺) under the topic articles. By linking the articles and the Web sites by topic, this ANNUAL EDITIONS reader becomes a powerful learning and research tool.

TOPIC AREA	TREATED IN	TOPIC AREA	TREATED IN
Aggression	5. Machiavellian Monkeys 6. What Are Friends For? 9. Dim Forest, Bright Chimps 10. To Catch a Colobus 13. These Are Real Swinging Primates 15. First, Kill the Babies 18. Apes of Wrath 26. Hard Times 32. Archeologists Rediscover Cannibals 42. Exploring Our Basic Human Nature ☺ **15, 16, 18, 26, 28, 31, 34**	**Cro-Magons**	27. Old Masters 29. Dating Game 30. Neanderthal Peace ☺ **4, 5, 6, 7, 25, 26**
		Disease	2. Curse and Blessing of the Ghetto 3. Saltshaker's Curse 35. Racial Odyssey 40. HIV 1998: The Global Picture 41. Viral Superhighway 43. Dr. Darwin ☺ **6, 13, 28, 30, 32, 33, 34**
Anatomy	17. What's Love Got to Do with It? 19. Sunset on the Savannah 21. New Human Ancestor? 22. Asian Hominids 24. New Clues 25. *Erectus* Rising 26. Hard Times 28. Gift of Gab 30. Neanderthal Peace 31. Learning to Love Neanderthals 33. Lost Man 38. Profile of an Anthropologist ☺ **20, 21, 22, 23, 24, 25, 26**	**DNA (Deoxyribonucleic Acid)**	2. Curse and Blessing of the Ghetto 4. Gene for Nothing 34. Black ☺ **29, 30, 31, 33, 34**
		Dominance Hierarchy	6. What Are Friends For? 13. These Are Real Swinging Primates 17. What's Love Got to Do with It? 18. Apes of Wrath ☺ **15, 16, 18, 19, 20, 21**
Archeology	25. *Erectus* Rising 27. Old Masters 29. Dating Game 30. Neanderthal Peace 31. Learning to Love Neanderthals 32. Archeologists Rediscover Cannibals 33. Lost Man 38. Profile of an Anthropologist ☺ **22, 23, 24, 25, 26**	**Forensic Anthropology**	38. Profile of an Anthropologist ☺ **32, 33, 34**
		Genes	2. Curse and Blessing of the Ghetto 3. Saltshaker's Curse 4. Gene for Nothing 34. Black, White, Other 43. Dr. Darwin ☺ **28, 29, 30, 31**
Australopithecines	19. Sunset on the Savannah 20. Early Hominid Fossils 21. New Human Ancestor? 24. New Clues 42. Exploring Our Basic Human Nature ☺ **22, 23, 24**	**Genetic Drift**	1. Growth of Evolutionary Science 2. Curse and Blessing of the Ghetto ☺ **11, 12, 13, 14**
Bipedalism	17. What's Love Got to Do With It? 19. Sunset on the Savannah 20. Early Hominid Fossils 21. New Human Ancestor? 23. Scavenger Hunt ☺ **19, 20, 21, 22, 23, 24**	**Genetic Testing**	2. Curse and Blessing of the Ghetto 4. Gene for Nothing ☺ **11, 12, 13, 14, 28, 29**
		Homo erectus	22. Asian Hominids 25. *Erectus* Rising 32. Archeologists Rediscover Cannibals ☺ **22, 23, 24, 25, 26, 28**
Blood Groups	34. Black, White, Other 35. Racial Odyssey ☺ **29, 31, 32, 33, 34**	**Homo sapiens**	29. Dating Game ☺ **25, 26, 28**
Catastrophism	1. Growth of Evolutionary Science ☺ **11, 12, 13, 14**	**Hunting and Gathering**	9. Dim Forest, Bright Chimps 10. To Catch a Colobus 23. Scavenger Hunt 26. Hard Times 27. Old Masters 30. Neanderthal Peace ☺ **16, 18, 22, 23, 24, 25, 26**
Chain of Being	1. Growth of Evolutionary Science ☺ **11, 12, 13, 14**		

3

● AE: Physical Anthropology

The following World Wide Web sites have been carefully researched and selected to support the articles found in this reader. If you are interested in learning more about specific topics found in this book, these Web sites are a good place to start. The sites are cross-referenced by number and appear in the topic guide on the previous two pages. Also, you can link to these Web sites through our DUSHKIN ONLINE support site at *http://www.dushkin.com/online/*.

The following sites were available at the time of publication. Visit our Web site—we update DUSHKIN ONLINE regularly to reflect any changes.

General Sources

1. American Anthropological Association (AAA)
http://www.ameranthassn.org/index.htm
Maintained by the AAA, this site links to AAA's publications (including tables of contents of recent issues, style guides, and others) and links to other anthropology sites.

2. Anthromorphemics
http://www.anth.ucsb.edu/glossary/index2.html
Access anthropological glossary terms at this site.

3. Anthropology in the News
http://www.tamu.edu/anthropology/news.html
Texas A&M provides data on interesting news articles that relate to anthropology, including biopsychology and sociocultural anthropology news.

4. Anthropology on the Internet
http://www.as.ua.edu/ant/libguide.htm
This indispensable site provides excellent Web addresses and tips on acquiring links to regional studies, maps, anthropology tutorials, and other data.

5. Anthropology 1101 Human Origins Website
http://www.geocities.com/Athens/Acropolis/ 5579/TA.html
Exploring this fun site provided by the University of Minnesota will lead to information about our ancient ancestors and other topics of interest to physical anthropologists.

6. Anthropology Resources on the Internet
http://www.socscisearch.com/r7.html
This site provides extensive links to Internet resources of anthropological relevance, including Web servers in different fields. *The Education Index* rated it "one of the best education-related sites on the Web."

7. Anthropology Resources Page
http://www.usd.edu/anth/
Many topics can be accessed from this University of South Dakota site. South Dakota archaeology, American Indian issues, and paleopathology resources are just a few examples.

8. Library of Congress
http://www.loc.gov
Examine this extensive Web site to learn about resource tools, library services/resources, exhibitions, and databases in many different subfields of anthropology.

9. The New York Times
http://www.nytimes.com
Browsing through the archives of *The New York Times* will provide a wide array of articles and information related to the different subfields of anthropology.

10. The PaleoAnthro Lists Home Page
http://www.pitt.edu/~mattf/PalAntList.html
Spend time at this site and the related PaleoChat site, at *http://www.pitt.edu/~mattf/PaleoChat.html*, to exchange information related to physical anthropology.

Natural Selection

11. Charles Darwin on Human Origins
http://www.literature.org/Works/Charles-Darwin/
This Web site contains the text of Charles Darwin's classic writing, *The Origins of Species*, which presents his scientific theory of human origins.

12. Enter Evolution: Theory and History
http://www.ucmp.berkeley.edu/history/evolution.html
Find information related to Charles Darwin and other important scientists at this Web site. It addresses preludes to evolution, natural selection, and more. Topics cover systematics, dinosaur discoveries, and vertebrate flight.

13. Fossil Hominids FAQ
http://www.talkorigins.org/faqs/fossil-hominids.html
Some links to materials related to hominid species and hominid fossils are provided on this site. The purpose of the site is to refute creationist claims that there is no evidence for human evolution.

14. Harvard Dept. of MCB—Biology Links
http://mcb.harvard.edu/BioLinks.html
This site features sources on evolution and links to anthropology departments and laboratories, taxonomy, paleontology, natural history, journals, books, museums, meetings, and many other related areas.

Primates

15. African Primates at Home
http://www.indiana.edu/~primate/primates.html
Don't miss this unusual and compelling site describing African primates on their home turf. "See" and "Hear" features provide samples of vocalizations and beautiful photographs of various types of primates.

16. Chimpanzee and Great Ape Language Resources—Anthropology
http://www.brown.edu/Departments/Anthropology/ apelang.html
This series of Web sites on primates includes the Primate home page and the Gorilla home page. It provides links to the entire text of Darwin's *Origin of Species* and more.

17. Electronic Zoo/NetVet-Primate Page
http://netvet.wustl.edu/primates.htm
This site touches on every kind of primate from A to Z and related information. The long list includes Darwinian theories and the *Descent of Man*, the Ebola virus, fossil hominids, the nonhuman Primate Genetics Lab, the Simian Retrovirus Laboratory, and zoonotic diseases, with many links in between.

18. Jane Goodall Research Center
http://www.usc.edu/dept/elab/anth/goodall.html
The Jane Goodall Research Center, a program of the University of Southern California's Anthropology Department, is a repository for data gathered over more than 30 years at Gombe National Park, Tanzania. Search this site for information about primate research.

Sex and Society

19. American Anthropologist
http://www.ameranthassn.org/ameranth.htm
Check out this site—the home page of *American Anthropologist* for general information about anthropology as well as articles relating to such topics as biological research.

20. American Scientist
http://www.amsci.org/amsci/amsci.html
Investigating this site will help students of physical anthropology to explore issues related to sex and society.

21. Bonobo Sex and Society
http://songweaver.com/info/bonobos.html
Accessed through Carnegie Mellon University, this site includes a *Scientific American* article discussing a primate's behavior that challenges traditional assumptions about male supremacy in human evolution.

The Fossil Evidence

22. The African Emergence and Early Asian Dispersals of the Genus *Homo*
http://www.sigmaxi.org/amsci/subject/EvoBio.html
Explore this site and click on this title to learn about what the Rift Valley in East Africa has to tell us about early hominid species. An excellent bibliography is included.

23. Anthropology, Archaeology, and American Indian Sites on the Internet
http://dizzy.library.arizona.edu/users/jlcox/first.html
This Web page points out a number of Internet sites of interest to different kinds of anthropologists, including physical and biological anthropologists. Visit this page for links to electronic journals and more.

24. Long Foreground: Human Prehistory
http://www.wsu.edu:8001/vwsu/gened/learn-modules/top_longfor/lfopen-index.html
This Washington State University site presents a learning module covering three major topics in human evolution: Overview, Hominid Species Timeline, and Human Physical Characteristics. It also provides a helpful glossary of terms and links to other Web sites.

Late Hominid Evolution

25. Archaeology Links (NC)
http://www.arch.dcr.state.nc.us/links.htm#stuff/
North Carolina Archaeology provides this site, which has many links to physical anthropology sites, such as the paleolithic painted cave at Vallon-Pont-d'Arc (Ardeche).

26. Human Prehistory
http://users.hol.gr/~dilos/prehis.htm
The evolution of the human species, beginning with the *Australopethicus* and continuing with *Homo Habilis, Homo erectus,* and *Homo sapiens,* is examined on this site. Also included are data on the people who lived in the Palaeolithic and Neolithic Age and are the immediate ancestors of modern man.

Human Diversity

27. Cult Archaeology Topics
http://www.usd.edu/anth/cultarch/culttopics.html
This fun site provides information on interesting pseudoscientific theories that have attracted scholarly attention. The Lost Tribes and the Moundbuilder Myth; and Cryptozoology: Bigfoot, and Nessie are among the many myths debunked here.

28. Hominid Evolution Survey
http://www.geocities.com/SoHo/Atrium/1381/index.html
This survey of the Hominid family, categorizes known hominids by genus and species. Beginning with the oldest known species, data include locations and environments, physical characteristics, any technology processed, and social behaviors. Includes charts and citations.

29. Human Genome Project Information
http://www.ornl.gov/TechResources/Human_Genome/home.html
Obtain answers about the U.S. Human Genome Project from this site, which details progress, goals, support groups, ethical, legal, and social issues, and genetics information.

30. OMIM Home Page-Online Mendelian Inheritance in Man
http://www3.ncbi.nlm.nih.gov/omim/
This database from the National Center for Biotechnology Information is a catalog of human genes and genetic disorders. It contains text, pictures, and reference information of great interest to students of physical anthropology.

31. Patterns of Human Variability: The Concept of Race
http://www.as.ua.edu/ant/bindon/ant101/syllabus/race/race1.htm
This site provides a handy, at-a-glance reference to the prevailing concepts of race and the causes of human variability since ancient times. It can serve as a starting point for research and understanding into the concept of race.

Living with the Past

32. Ancestral Passions
http://www.canoe.ca/JamBooksReviewsA/ancestral_morell.html
This review of Virginia Morell's book, *Ancestral Passions,* a biography of the famously dysfunctional Leakey family, will likely spur you to the bookstore in order to learn more about the history of paleontology and the thrill and trials of the hunt for human origins. It is this evolutionary detective story that is the book's true drama. Jump over to *http://url.co.nz/african_trip/tanzania.html* to read an individual's account of a recent trip "In the Cradle of Humankind."

33. Forensic Science Reference Page
http://www.lab.fws.gov
Look over this site from the U.S. Fish and Wildlife Forensics Lab to explore topics related to forensic anthropology.

34. Zeno's Forensic Page
http://users.bart.nl/~geradts/forensic.html
A complete list of resources on forensics is presented on this Web site. It includes general information sources, DNA/serology sources and databases, forensic medicine anthropology sites, and related areas.

We highly recommend that you review our Web site for expanded information and our other product lines. We are continually updating and adding links to our Web site in order to offer you the most usable and useful information that will support and expand the value of your Annual Editions. You can reach us at: http://www.dushkin.com/annualeditions/.

www.dushkin.com/online/

Unit Selections

1. **The Growth of Evolutionary Science,** Douglas J. Futuyma
2. **Curse and Blessing of the Ghetto,** Jared Diamond
3. **The Saltshaker's Curse,** Jared Diamond
4. **A Gene for Nothing,** Robert Sapolsky

Key Points to Consider

❖ In nature, how is it that design can occur without a designer, orderliness without purpose?

❖ What is "natural selection"? How does Gregor Mendel's work relate to Charles Darwin's theory?

❖ Why is Tay-Sachs disease so common among Eastern European Jews?

❖ What is the "saltshaker's curse," and why are some people more affected by it than others?

❖ To what extent do identical twins actually differ from one another and what implication does this have for cloning humans?

❖ What do genes actually do and how predictive are they of human social behavior?

 **Links**   **www.dushkin.com/online/**

These sites are annotated on pages 4 and 5.

As the twentieth century draws to a close and we reflect upon where science has taken us over the past 100 years, it should come as no surprise that the field of genetics has swept us along a path of insight into the human condition as well as heightened controversy as to how to handle this potentially dangerous knowledge of ourselves.

Certainly, Gregor Mendel, in the late nineteenth century, could not have anticipated that his study of pea plants would ultimately lead to the better understanding of over 3,000 genetically caused diseases, such as sickle-cell anemia, Huntington's chorea, and Tay-Sachs. Nor could he have foreseen the present-day controversies over such matters as surrogate motherhood, cloning, and genetic engineering. See "A Gene for Nothing" by Robert Sapolsky.

The significance of Mendel's work, of course, was his discovery that hereditary traits are conveyed by particular units that we now call "genes," a then-revolutionary notion that has been followed by a better understanding of how and why such units change. It is knowledge of the process of "mutation," or alteration of the chemical structure of the gene, that is now providing us with the potential to control the genetic fate of individuals.

The other side of the evolutionary coin, as discussed in the unit's first article, "The Growth of Evolutionary Science," is natural selection, a concept provided by Charles Darwin and Alfred Wallace. Natural selection refers to the "weeding out" of unfavorable mutations and the perpetuation of favorable ones. The reproductive aspects of this process are also a matter of continuing investigation.

It seems that as we gain a better understanding of both of these processes, mutation and natural selection, and grasp their relevance to human beings, we draw nearer to that time when we may even control the evolutionary direction of our species. Knowledge itself, of course, is neutral—its potential for good or ill being determined by those who happen to be in a position to use it. Consider the possibility of eliminating some of the harmful hereditary traits discussed in "Curse and Blessing of the Ghetto" and "The Saltshaker's Curse," both by Jared Diamond. While it is true that many deleterious genes do get weeded out of the population by means of natural selection, there are other harmful ones, Diamond points out, that may actually have a good side to them and will therefore be perpetuated. It may be, for example, that some men are dying from a genetically caused overabundance of iron in their blood systems in a trade-off that allows some women to absorb

sufficient amounts of the element to guarantee their own survival. The question of whether we should eliminate such a gene would seem to depend on which sex we decide should reap the benefit.

The issue of just what is a beneficial application of scientific knowledge is a matter for debate. Who will have the final word as to how these technological breakthroughs will be employed in the future? Even with the best of intentions, how can we be certain of the long-range consequences of our actions in such a complicated field? Note, for example, the sweeping effects of ecological change upon the viruses of the world, which in turn seem to be paving the way for new waves of human epidemics. Generally speaking, there is an element of purpose and design in our machinations. Yet, even with this clearly in mind, the whole process seems to be escalating out of human control. It seems that the whole world has become an experimental laboratory in which we know not what we do until we have already done it.

As we read the essays in this unit and contemplate the significance of genetic diseases for human evolution, we can hope that a better understanding of congenital diseases will lead to a reduction of human suffering. At the same time, we must remain aware that someone, at some time, may actually use the same knowledge to increase rather than reduce the misery that exists in the world.

The Growth of Evolutionary Science

Douglas J. Futuyma

Today, the theory of evolution is an accepted fact for everyone but a fundamentalist minority, whose objections are based not on reasoning but on doctrinaire adherence to religious principles.

—James D. Watson, 1965*

In 1615, Galileo was summoned before the Inquisition in Rome. The guardians of the faith had found that his "proposition that the sun is the center [of the solar system] and does not revolve about the earth is foolish, absurd, false in theology, and heretical, because expressly contrary to Holy Scripture." In the next century, John Wesley declared that "before the sin of Adam there were no agitations within the bowels of the earth, no violent convulsions, no concussions of the earth, no earthquakes, but all was unmoved as the pillars of heaven." Until the seventeenth century, fossils were interpreted as "stones of a peculiar sort, hidden by the Author of Nature for his own pleasure." Later they were seen as remnants of the Biblical deluge. In the middle of the eighteenth century, the great French naturalist Buffon speculated on the possibility of cosmic and organic evolution and was forced by the clergy to recant: "I abandon everything in my book respecting the formation of the earth, and generally all of which may be contrary to the narrative of Moses." For had not St. Augustine written, "Nothing is to be ac-

*James D. Watson, a molecularbiologist, shared the Nobel Prize for his work in discovering the structure of DNA.

cepted save on the authority of Scripture, since greater is that authority than all the powers of the human mind"?

When Darwin published *The Origin of Species,* it was predictably met by a chorus of theological protest. Darwin's theory, said Bishop Wilberforce, "contradicts the revealed relations of creation to its Creator." "If the Darwinian theory is true," wrote another clergyman, "Genesis is a lie, the whole framework of the book of life falls to pieces, and the revelation of God to man, as we Christians know it, is a delusion and a snare." When *The Descent of Man* appeared, Pope Pius IX was moved to write that Darwinism is "a system which is so repugnant at once to history, to the tradition of all peoples, to exact science, to observed facts, and even to Reason herself, [that it] would seem to need no refutation, did not alienation from God and the leaning toward materialism, due to depravity, eagerly seek a support in all this tissue of fables."[1] Twentieth-century creationism continues this battle of medieval theology against science.

One of the most pervasive concepts in medieval and post-medieval thought was the "great chain of being," or *scala naturae.*[2] Minerals, plants, and animals, according to his concept, formed a gradation, from the lowliest and most material to the most complex and spiritual, ending in man, who links the animal series to the world of intelligence and spirit. This "scale of nature" was the manifestation of God's infinite benevolence. In his goodness, he had conferred

existence on all beings of which he could conceive, and so created a complete chain of being, in which there were no gaps. All his creatures must have been created at once, and none could ever cease to exist, for then the perfection of his divine plan would have been violated. Alexander Pope expressed the concept best:

> Vast chain of being! which from God
> began,
> Natures aethereal, human, angel, man,
> Beast, bird, fish, insect, what no eye
> can see,
> No glass can reach; from Infinite to
> thee,
> From thee to nothing.—On superior
> pow'rs
> Were we to press, inferior might on
> ours;
> Or in the full creation leave a void,
> Where, one step broken, the great
> scale's destroy'd;
> From Nature's chain whatever link you
> strike,
> Tenth, or ten thousandth, breaks the
> chain alike.

Coexisting with this notion that all of which God could conceive existed so as to complete his creation was the idea that all things existed for man. As the philosopher Francis Bacon put it, "Man, if we look to final causes, may be regarded as the centre of the world... for the whole world works together in the service of man... all things seem to be going about man's business and not their own."

"Final causes" was another fundamental concept of medieval and post-medieval thought. Aristotle had distinguished final causes from efficient causes, and the

Western world saw no reason to doubt the reality of both. The "efficient cause" of an event is the mechanism responsible for its occurrence: the cause of a ball's movement on a pool table, for example, is the impact of the cue or another ball. The "final cause," however, is the goal, or purpose for its occurrence: the pool ball moves because I wish it to go into the corner pocket. In post-medieval thought there was a final cause—a purpose—for everything; but purpose implies intention, or foreknowledge, by an intellect. Thus the existence of the world, and of all the creatures in it, had a purpose; and that purpose was God's design. This was self-evident, since it was possible to look about the world and see the palpable evidence of God's design everywhere. The heavenly bodies moved in harmonious orbits, evincing the intelligence and harmony of the divine mind; the adaptations of animals and plants to their habitats likewise reflected the devine intelligence, which had fitted all creatures perfectly for their roles in the harmonious economy of nature.

Before the rise of science, then, the causes of events were sought not in natural mechanisms but in the purposes they were meant to serve, and order in nature was evidence of divine intelligence. Since St. Ambrose had declared that "Moses opened his mouth and poured forth what God had said to him," the Bible was seen as the literal word of God, and according to St. Thomas Aquinas, "Nothing was made by God, after the six days of creation, absolutely new." Taking Genesis literally, Archbishop Ussher was able to calculate that the earth was created in 4004 B.C. The earth and the heavens were immutable, changeless. As John Ray put it in 1701 in *The Wisdom of God Manifested in the Works of the Creation,* all living and nonliving things were "created by God at first, and by Him conserved to this Day in the same State and Condition in which they were first made."[3]

The evolutionary challenge to this view began in astronomy. Tycho Brahe found that the heavens were not immutable when a new star appeared in the constellation Cassiopeia in 1572. Copernicus displaced the earth from the center of the universe, and Galileo found that the perfect heavenly bodies weren't so perfect: the sun had spots that changed from time to time, and the moon had craters that strongly implied alterations of its surface. Galileo, and after him Buffon, Kant, and many others, concluded that change was natural to all things.

A flood of mechanistic thinking ensued. Descartes, Kant, and Buffon concluded that the causes of natural phenomena should be sought in natural laws. By 1755, Kant was arguing that the laws of matter in motion discovered by Newton and other physicists were sufficient to explain natural order. Gravitation, for example, could aggregate chaotically dispersed matter into stars and planets. These would join with one another until the only ones left were those that cycled in orbits far enough from each other to resist gravitational collapse. Thus order might arise from natural processes rather than from the direct intervention of a supernatural mind. The "argument from design"—the claim that natural order is evidence of a designer—had been directly challenged. So had the universal belief in final causes. If the arrangement of the planets could arise merely by the laws of Newtonian physics, if the planets could be born, as Buffon suggested, by a collision between a comet and the sun, then they did not exist for any purpose. They merely came into being through impersonal physical forces.

From the mutability of the heavens, it was a short step to the mutability of the earth, for which the evidence was far more direct. Earthquakes and volcanoes showed how unstable terra firma really is. Sedimentary rocks showed that materials eroded from mountains could be compacted over the ages. Fossils of marine shells on mountain-tops proved that the land must once have been under the sea. As early as 1718, the Abbé Moro and the French academician Bernard de Fontenelle had concluded that the Biblical deluge could not explain the fossilized oyster beds and tropical plants that were found in France. And what of the great, unbroken chain of being if the rocks were full of extinct species?

To explain the facts of geology, some authors—the "catastrophists"—supposed that the earth had gone through a series of great floods and other catastrophes that successively extinguished different groups of animals. Only this, they felt, could account for the discovery that higher and lower geological strata had different fossils. Buffon, however, held that to explain nature we should look to the natural causes we see operating around us: the gradual action of erosion and the slow buildup of land during volcanic eruptions. Buffon thus proposed what came to be the foundation of geology, and indeed of all science, the principle of uniformitarianism, which holds that the same causes that operate now have always operated. By 1795, the Scottish geologist James Hutton had suggested that "in examining things present we have data from which to reason with regard to what has been." His conclusion was that since "rest exists not anywhere," and the forces that change the face of the earth move with ponderous slowness, the mountains and canyons of the world must have come into existence over countless aeons.

If the entire nonliving world was in constant turmoil, could it not be that living things themselves changed? Buffon came close to saying so. He realized that the earth had seen the extinction of countless species, and supposed that those that perished had been the weaker ones. He recognized that domestication and the forces of the environment could modify the variability of many species. And he even mused, in 1766, that species might have developed from common ancestors:

If it were admitted that the ass is of the family of the horse, and different from the horse only because it has varied from the original form, one could equally well say that the ape is of the family of man, that he is a degenerate man, that man and ape have a common origin; that, in fact, all the families among plants as well as animals have come from a single stock, and that all animals are descended from a single animal, from which have sprung in the course of time, as a result of process or of degeneration, all the other races of animals. For if it were once shown that we are justified in establishing these families; if it were granted among animals and plants there has been (I do not say several species) but even a single one, which has been produced in the course of direct descent from another species . . . then there would no longer be any limit to the power of nature, and we should not be wrong in supposing that, with sufficient time, she has been able from a single being to derive all the other organized beings.[4]

This, however, was too heretical a thought; and in any case, Buffon thought the weight of evidence was against common descent. No new species had been observed to arise within recorded history, Buffon wrote; the sterility of hybrids between species appeared an impossible barrier to such a conclusion; and if species had emerged gradually, there should have been innumerable intermediate variations between the horse and ass, or any other species. So Buffon concluded: "But this [idea of a common ancestor] is by no means a proper representation of nature. We are assured by the authority of revelation that all animals have participated equally in the grace of direct Creation and that the first pair of every species issued fully formed from the hands of the Creator."

Buffon's friend and protégé, Jean Baptiste de Monet, the Chevalier de Lamarck, was the first scientist to take the big step. It is not clear what led Lamarck to his uncompromising belief in evolution; perhaps it was his studies of fossil molluscs, which he came to believe were the ancestors of similar species living today. Whatever the explanation, from 1800 on he developed the notion that fossils were not evidence of extinct species but of ones that had gradually been transformed into living species. To be sure, he wrote, "an enormous time and wide variation in successive conditions must doubtless have been required to enable nature to bring the organization of animals to that degree of complexity and development in which we see it at its perfection"; but "time has no limits and can be drawn upon to any extent."

Lamarck believed that various lineages of animals and plants arose by a continual process of spontaneous generation from inanimate matter, and were transformed from very simple to more complex forms by an innate natural tendency toward complexity caused by "powers conferred by the supreme author of all things." Various specialized adaptations of species are consequences of the fact that animals must always change in response to the needs imposed on them by a continually changing environment. When the needs of a species change, so does its behavior. The animal then uses certain organs more frequently than before, and these organs, in turn, become more highly developed by such use, or else "by virtue of the operations of their own inner senses." The classic example of Lamarckism is the giraffe: by straining upward for foliage, it was thought, the animal had acquired a longer neck, which was then inherited by its off-spring.

In the nineteenth century it was widely believed that "acquired" characteristics—alterations brought about by use or disuse, or by the direct influence of the environment—could be inherited. Thus it was perfectly reasonable for Lamarck to base his theory of evolutionary change partly on this idea. Indeed, Darwin also allowed for this possibility, and the inheritance of acquired characteristics was not finally prove impossible until the 1890s.

Lamarck's ideas had a wide influence; but in the end did not convince many scientists of the reality of evolution. In France, Georges Cuvier, the foremost paleontologist and anatomist of his time, was an influential opponent of evolution. He rejected Lamarck's notion of the spontaneous generation of life, found it inconceivable that changes in behavior could produce the exquisite adaptations that almost every species shows, and emphasized that in both the fossil record and among living animals there were numerous "gaps" rather than intermediate forms between species. In England, the philosophy of "natural theology" held sway in science, and the best-known naturalists continued to believe firmly that the features of animals and plants were evidence of God's design. These devout Christians included the foremost geologist of the day, Charles Lyell, whose *Principles of Geology* established uniformitarianism once and for all as a guiding principle. But Lyell was such a thorough uniformitarian that he believed in a steady-state world, a world that was always in balance between forces such as erosion and mountain building, and so was forever the same. There was no room for evolution, with its concept of steady change, in Lyell's world view, though he nonetheless had an enormous impact on evolutionary thought, through his influence on Charles Darwin.

Darwin (1809–1882) himself, unquestionably one of the greatest scientists of all time, came only slowly to an evolutionary position. The son of a successful physician, he showed little interest in the life of the mind in his early years. After unsuccessfully studying medicine at Edinburgh, he was sent to Cambridge to prepare for the ministry, but he had only a half-hearted interest in his studies and spent most of his time hunting, collecting beetles, and becoming an accomplished amateur naturalist. Though he received his B.A. in 1831, his future was quite uncertain until, in December of that year, he was enlisted as a naturalist aboard *H.M.S. Beagle,* with his father's very reluctant agreement. For five years (from December 27, 1831, to October 2, 1836) the *Beagle* carried him about the world, chiefly along the coast of South America, which it was the *Beagle's* mission to survey. For five years Darwin collected geological and biological specimens, made geological observations, absorbed Lyell's *Principles of Geology,* took voluminous notes, and speculated about everything from geology to anthropology. He sent such massive collections of specimens back to England that by the time he returned he had already gained a substantial reputation as a naturalist.

Shortly after his return, Darwin married and settled into an estate at Down where he remained, hardly traveling even to London, for the rest of his life. Despite continual ill health, he pursued an extraordinary range of biological studies: classifying barnacles, breeding pigeons, experimenting with plant growth, and much more. He wrote no fewer than sixteen books and many papers, read voraciously, corresponded extensively with everyone, from pigeon breeders to the most eminent scientists, whose ideas or information might bear on his theories, and kept detailed notes on an amazing variety of subjects. Few people have written authoritatively on so many different topics: his books include not only *The Voyage of the Beagle, The Origin of Species,* and *The Descent of Man,* but also *The Structure and Distribution of Coral Reefs* (containing a novel theory of the formation of coral atolls which is still regarded as correct), *A Monograph on the Sub-class Cirripedia* (the definitive study of barnacle classification), *The Various Contrivances by Which Orchids are Fertilised by Insects, The Variation of Animals and Plants Under Domestication* (an exhaustive summary of information on variation, so crucial to his evolutionary theory), *The Effects of Cross and Self Fertilisation in the Vegetable Kingdom* (an analysis of sexual reproduction and the sterility of hybrids between species), *The Expression of the Emotions in Man and Animals* (on the evolution of human behavior from animal behavior), and *The Formation of Vegetable Mould Through the Action of Worms.* There is every reason to believe that almost all these books bear, in one way or another, on the principles and ideas that were inherent in Darwin's theory of evolution. The worm book, for example, is devoted to showing how great the impact of a seemingly trivial process like worm burrowing may be on ecology and geology if it persists for a long time. The idea of such cumulative slight effects is, of course, inherent in Darwin's view of evolution: successive slight modifications of a species, if continued long enough, can transform it radically.

When Darwin embarked on his voyage, he was a devout Christian who did not doubt the literal truth of the Bible, and did not believe in evolution any more than did Lyell and the other English scientists he had met or whose books he had read. By the time he returned to England in 1836 he had made numerous observations that would later convince him of evolution. It seems likely, however, that the idea itself did not occur to him until the spring of 1837, when the ornithologist John Gould, who was working on some of Darwin's collections, pointed out to him

that each of the Galápagos Islands, off the coast of Ecuador, had a different kind of mockingbird. It was quite unclear whether they were different varieties of the same species, or different species. From this, Darwin quickly realized that species are not the discrete, clear-cut entities everyone seemed to imagine. The possibility of transformation entered his mind, and it applied to more than the mockingbirds: "When comparing . . . the birds from the separate islands of the Galápagos archipelago, both with one another and with those from the American mainland, I was much struck how entirely vague and arbitrary is the distinction between species and varieties."

In July 1837 he began his first notebook on the "Transmutation of Species." He later said that the Galápagos species and the similarity between South American fossils and living species were at the origin of all his views.

> During the voyage of the *Beagle* I had been deeply impressed by discovering in the Pampean formation great fossil animals covered with armour like that on the existing armadillos; secondly, by the manner in which closely allied animals replace one another in proceeding southward over the continent; and thirdly, by the South American character of most of the productions of the Galápagos archipelago, and more especially by the manner in which they differ slightly on each island of the group; none of these islands appearing to be very ancient in a geological sense. It was evident that such facts as these, as well as many others, could be explained on the supposition that species gradually become modified; and the subject has haunted me.

The first great step in Darwin's thought was the realization that evolution had occurred. The second was his brilliant insight into the possible cause of evolutionary change. Lamarck's theory of "felt needs" had not been convincing. A better one was required. It came on September 18, 1838, when after grappling with the problem for fifteen months, "I happened to read for amusement Malthus on Population, and being well prepared to appreciate the struggle for existence which everywhere goes on from long- continued observa-

tion of the habits of animals and plants, it at once struck me that under these circumstances favorable variations would tend to be preserved, and unfavorable ones to be destroyed. The result of this would be the formation of new species. Here, then, I had at last got a theory by which to work."

Malthus, an economist, had developed the pessimistic thesis that the exponential growth of human populations must inevitably lead to famine, unless it were checked by war, disease, or "moral restraint." This emphasis on exponential population growth was apparently the catalyst for Darwin, who then realized that since most natural populations of animals and plants remain fairly stable in numbers, many more individuals are born than survive. Because individuals vary in their characteristics, the struggle to survive must favor some variant individuals over others. These survivors would then pass on their characteristics to future generations. Repetition of this process generation after generation would gradually transform the species.

Darwin clearly knew that he could not afford to publish a rash speculation on so important a subject without developing the best possible case. The world of science was not hospitable to speculation, and besides, Darwin was dealing with a highly volatile issue. Not only was he affirming that evolution had occurred, he was proposing a purely material explanation for it, one that demolished the argument from design in a single thrust. Instead of publishing his theory, he patiently amassed a mountain of evidence, and finally, in 1844, collected his thoughts in an essay on natural selection. But he still didn't publish. Not until 1856, almost twenty years after he became an evolutionist, did he begin what he planned to be a massive work on the subject, tentatively titled *Natural Selection.*

Then, in June 1858, the unthinkable happened. Alfred Russel Wallace (1823–1913), a young naturalist who had traveled in the Amazon Basin and in the Malay Archipelago, had also become interested in evolution. Like Darwin, he was struck by the fact that "the most closely allied species are found in the same locality or in closely adjoining lo-

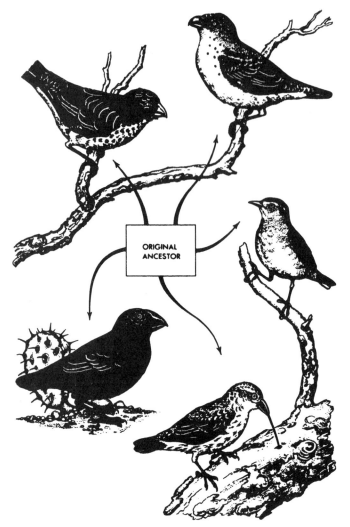

Figure 1. *Some species of Galápagos finches. Several of the most different species are represented here; intermediate species also exist. Clockwise from lower left are a male ground-finch (the plumage of the female resembles that of the tree-finches); the vegetarian tree-finch; the insectivorous tree-finch; the warbler-finch; and the woodpecker-finch, which uses a cactus spine to extricate insects from crevices. The slight differences among these species, and among species in other groups of Galápagos animals such as giant tortoises, were one of the observations that led Darwin to formulate his hypothesis of evolution. (From D. Lack, Darwin's Finches [Oxford: Oxford University Press, 1944].)*

calities and . . . therefore the natural sequence of the species by affinity is also geographical." In the throes of a malarial fever in Malaya, Wallace conceived of the same idea of natural selection as Darwin had, and sent Darwin a manuscript "On the Tendency of Varieties to Depart Indefinitely from the Original Type." Darwin's friends Charles Lyell and Joseph Hooker, a botanist, rushed in to help Darwin establish the priority of his ideas, and on July 1, 1858, they presented to the Linnean Society of London both Wallace's paper and extracts from Darwin's 1844 essay. Darwin abandoned his big book on natural selection and condensed the argument into a 490-page "abstract" that was published on November 24, 1859, under the title *The Origin of Species by Means of Natural Selection; or, the Preservation of Favored Races in the Struggle for Life.* Because it was an abstract, he had to leave out many of the detailed observations and references to the literature that he had amassed, but these were later provided in his other books, many of which are voluminous expansions on the contents of *The Origin of Species.*

The first five chapters of the *Origin* lay out the theory that Darwin had conceived. He shows that both domesticated and wild species are variable, that much of that variation is hereditary, and that breeders, by conscious selection of desirable varieties, can develop breeds of pigeons, dogs, and other forms that are more different from each other than species or even families of wild animals and plants are from each other. The differences between related species then are no more than an exaggerated form of the kinds of variations one can find in a single species; indeed, it is often extremely difficult to tell if natural populations are distinct species or merely well-marked varieties.

Darwin then shows that in nature there is competition, predation, and a struggle for life.

> Owing to this struggle, variations, however slight and from whatever cause proceeding, if they be in any degree profitable to the individuals of a species, in their infinitely complex relations to other organic beings and to their physical conditions of life, will tend to the preservation of such individuals, and will generally be inherited by the offspring. The offspring, also, will thus have a better chance of surviving, for, of the many individuals of any species which are periodically born, but a small number can survive. I have called this principle, by which each slight variation, if useful, is preserved, by the term natural selection, in order to mark its relation to man's power of selection.

Darwin goes on to give examples of how even slight variations promote survival, and argues that when populations are exposed to different conditions, different variations will be favored, so that the descendants of a species become diversified in structure, and each ancestral species can give rise to several new ones. Although "it is probable that each form remains for long periods unaltered," successive evolutionary modifications will ultimately alter the different species so greatly that they will be classified as different genera, families, or orders.

Competition between species will impel them to become more different, for "the more diversified the descendants from any one species become in structure, constitution and habits, by so much will they be better enabled to seize on many and widely diversified places in the polity of nature, and so be enabled to increase in numbers." Thus different adaptations arise, and "the ultimate result is that each creature tends to become more and more improved in relation to its conditions. This improvement inevitably leads to the greater advancement of the organization of the greater number of living beings throughout the world." But lowly organisms continue to persist, for "natural selection, or the survival of the fittest, does not necessarily include progressive development—it only takes advantage of such variations as arise and are beneficial to each creature under its complex relations of life." Probably no organism has reached a peak of perfection, and many lowly forms of life continue to exist, for "in some cases variations or individual differences of a favorable nature may never have arisen for natural selection to act on or accumulate. In no case, probably, has time sufficed for the utmost possible amount of development. In some few cases there has been what we must call retrogression of organization. But the main cause lies in the fact that under very simple conditions of life a high organization would be of no service. . . ."

In the rest of *The Origin of Species*, Darwin considers all the objections that might be raised against his theory; discusses the evolution of a great array of phenomena—hybrid sterility, the slave-making instinct of ants, the similarity of vertebrate embryos; and presents an enormous body of evidence for evolution. He draws his evidence from comparative anatomy, embryology, behavior, geographic variation, the geographic distribution of species, the study of rudimentary organs, atavistic variations ("throwbacks"), and the geological record to show how all of biology provides testimony that species have descended with modification from common ancestors.

Darwin's triumph was in synthesizing ideas and information in ways that no one had quite imagined before. From Lyell and the geologists he learned uniformitarianism: the cause of past events must be found in natural forces that operate today; and these, in the vastness of time, can accomplish great change. From Malthus and the nineteenth-century economists he learned of competition and the struggle for existence. From

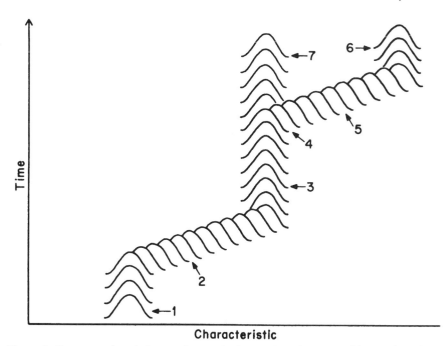

Figure 2. *Processes of evolutionary change. A characteristic that is variable (1) often shows a bell-shaped distribution—individuals vary on either side of the average. Evolutionary change (2) consists of a shift in successive generations, after which the characteristic may reach a new equilibrium (3). When the species splits into two different species (4), one of the species may undergo further evolutionary change (5) and reach a new equilibrium (6). The other may remain unchanged (7) or not. Each population usually remains variable throughout this process, but the average is shifted, ordinarily by natural selection.*

his work on barnacles, his travels, and his knowledge of domesticated varieties he learned that species do not have immutable essences but are variable in all their properties and blend into one another gradually. From his familiarity with the works of Whewell, Herschel, and other philosophers of science he developed a powerful method of pursuing science, the "hypothetico-deductive" method, which consists of formulating a hypothesis or speculation, deducing the logical predictions that must follow from the hypothesis, and then testing the hypothesis by seeing whether or not the predictions are verified. This was by no means the prevalent philosophy of science in Darwin's time.[5]

Darwin brought biology out of the Middle Ages. For divine design and unknowable supernatural forces he substituted natural material causes that could be studied by the methods of science. Instead of catastrophes unknown to physical science he invoked forces that could be studied in anyone's laboratory or garden. He replaced a young, static world by one in which there had been constant change for countless aeons. He established that life had a history, and

this proved the essential view that differentiated evolutionary thought from all that had gone before.

For the British naturalist John Ray, writing in 1701, organisms had no history—they were the same at that moment, and lived in the same places, doing the same things, as when they were first created. For Darwin, organisms spoke of historical change. If there has indeed been such a history, then fossils in the oldest rocks must differ from those in younger rocks: trilobites, dinosaurs, and mammoths will not be mixed together but will appear in some temporal sequence. If species come from common ancestors, they will have the same characteristics, modified for different functions: the same bones used by bats for flying will be used by horses for running. If species come from ancestors that lived in different environments, they will carry the evidence of their history with them in the form of similar patterns of embryonic development and in vestigial, rudimentary organs that no longer serve any function. If species have a history, their geographical distribution will reflect it: oceanic islands won't have elephants because they wouldn't have been able to get there.

Once the earth and its living inhabitants are seen as the products of historical change, the theological philosophy embodied in the great chain of being ceases to make sense; the plenitude, or fullness, of the world becomes not an eternal manifestation of God's bountiful creativity but an illusion. For most of earth's history, most of the present species have not existed; and many of those that did exist do so no longer. But the scientific challenge to medieval philosophy goes even deeper. If evolution has occurred, and if it has proceeded from the natural causes that Darwin envisioned, then the adaptations of organisms to their environment, the intricate construction of the bird's wing and the orchid's flower, are evidence not of divine design but of the struggle for existence. Moreover, and this may be the deepest implication of all, Darwin brought to biology, as his predecessors had brought to astronomy and geology, the sufficiency of efficient causes. No longer was there any reason to look for final causes or goals. To the questions "What purpose does this species serve? Why did God make tapeworms?" the answer is "To no purpose." Tapeworms were not put here to serve a purpose, nor were planets, nor plants, nor people. They came into existence not by design but by the action of impersonal natural laws.

By providing materialistic, mechanistic explanations, instead of miraculous ones, for the characteristics of plants and animals, Darwin brought biology out of the realm of theology and into the realm of science. For miraculous spiritual forces fall outside the province of science; all of science is the study of material causation.

Of course, *The Origin of Species* didn't convince everyone immediately. Evolution and its material cause, natural selection, evoked strong protests from ecclesiastical circles, and even from scientists.[6] The eminent geologist Adam Sedgwick, for example, wrote in 1860 that species must come into existence by creation,

a power I cannot imitate or comprehend; but in which I can believe, by a legitimate conclusion of sound rea-

son drawn from the laws and harmonies of Nature. For I can see in all around me a design and purpose, and a mutual adaptation of parts which I *can* comprehend, and which prove that there is exterior to, and above, the mere phenomena of Nature a great prescient and designing cause.... The pretended physical philosophy of modern days strips man of all his moral attributes, or holds them of no account in the estimate of his origin and place in the created world. A cold atheistical materialism is the tendency of the so-called material philosophy of the present day.

Among the more scientific objections were those posed by the French paleontologist François Pictet, and they were echoed by many others. Since Darwin supposes that species change gradually over the course of thousands of generations, then, asked Pictet, "Why don't we find these gradations in the fossil record ... and why, instead of collecting thousands of identical individuals, do we not find more intermediary forms? ... How is it that the most ancient fossil beds are rich in a variety of diverse forms of life, instead of the few early types Darwin's theory leads us to expect? How is it that no species has been seen to evolve during human history, and that the 4000 years which separates us from the mummies of Egypt have been insufficient to modify the crocodile and the ibis?" Pictet protested that, although slight variations might in time alter a species slightly, "all known facts demonstrate ... that the prolonged influence of modifying causes has an action which is constantly restrained within sufficiently confined limits."

The anatomist Richard Owen likewise denied "that ... variability is progressive and unlimited, so as, in the course of generations, to change the species, the genus, the order, or the class." The paleontologist Louis Agassiz insisted that organisms fall into discrete groups, based on uniquely different created plans, between which no intermediates could exist. He chose the birds as a group that showed the sharpest of boundaries. Only a few years later, in 1868, the fossil *Archaeopteryx*, an exquisite intermediate between birds and reptiles, demolished Agassiz's argument,

and he had no more to say on the unique character of the birds.

Within twelve years of *The Origin of Species,* the evidence for evolution had been so thoroughly accepted that philosopher and mathematician Chauncey Wright could point out that among the students of science, "orthodoxy has been won over to the doctrine of evolution." However, Wright continued, "While the general doctrine of evolution has thus been successfully redeemed from theological condemnation, this is not yet true of the subordinate hypothesis of Natural Selection."

Natural selection turned out to be an extraordinarily difficult concept for people to grasp. St. George Mivart, a Catholic scholar and scientist, was not unusual in equating natural selection with chance. "The theory of Natural Selection may (though it need not) be taken in such a way as to lead man to regard the present organic world as formed, so to speak, *accidentally,* beautiful and wonderful as is the confessedly haphazard result." Many like him simply refused to understand that natural selection is the antithesis of chance and consequently could not see how selection might cause adaptation or any kind of progressive evolutionary change. Even in the 1940s there were those, especially among paleontologists, who felt that the progressive evolution of groups like the horses, as revealed by the fossil record, must have had some unknown cause other than natural selection. Paradoxically, then, Darwin had convinced the scientific world of evolution where his predecessors had failed; but he had not convinced all biologists of his truly original theory, the theory of natural selection.

Natural selection fell into particular disrepute in the early part of the twentieth century because of the rise of genetics—which, as it happened, eventually became the foundation of the modern theory of evolution. Darwin's supposition that variation was unlimited, and so in time could give rise to strikingly different organisms, was not entirely convincing because he had no good idea of where variation came from. In 1865, the Austrian monk Gregor Mendel discovered, from his crosses of pea plants, that discretely different characteristics such as wrinkled versus smooth seeds were inherited from generation to generation without being altered, as if they were caused by particles that passed from parent to offspring. Mendel's work was ignored for thirty-five years, until, in 1900, three biologists discovered his paper and realized that it held the key to the mystery of heredity. One of the three, Hugo de Vries, set about to explore the problem as Mendel had, and in the course of his studies of evening primroses observed strikingly different variations arise, *de novo*. The new forms were so different that de Vries believed they represented new species, which had arisen in a single step by alteration or, as he called it, mutation, of the hereditary material.

In the next few decades, geneticists working with a great variety of organisms observed many other drastic changes arise by mutation: fruit flies (Drosophila), for example, with white instead of red eyes or curled instead of straight wings. These laboratory geneticists, especially Thomas Hunt Morgan, an outstanding geneticist at Columbia University, asserted that evolution must proceed by major mutational steps, and that mutation, not natural selection, was the cause of evolution. In their eyes, Darwin's theory was dead on two counts: evolution was not gradual, and it was not caused by natural selection. Meanwhile, naturalists, taxonomists, and breeders of domesticated plants and animals continued to believe in Darwinism, because they saw that populations and species differed quantitatively and gradually rather than in big jumps, that most variation was continuous (like height in humans) rather than discrete, and that domesticated species could be altered by artificial selection from continuous variation.

The bitter conflict between the Mendelian geneticists and the Darwinians was resolved in the 1930s in a "New Synthesis" that brought the opposing views into a "neo-Darwinian" theory of evolution.[7] Slight variations in height, wing length, and other characteristics proved, under careful genetic analysis, to be inherited as particles, in the same way as the discrete variations studied by the Mendelians. Thus a large animal

simply has inherited more particles, or genes, for large size than a smaller member of the species has. The Mendelians were simply studying particularly well marked variations, while the naturalists were studying more subtle ones. Variations could be very slight, or fairly pronounced, or very substantial, but all were inherited in the same manner. All these variations, it was shown, arose by a process of mutation of the genes.

Three mathematical theoreticians, Ronald Fisher and J. B. S. Haldane in England and Sewall Wright in the United States, proved that a newly mutated gene would not automatically form a new species. Nor would it automatically replace the preexisting form of the gene, and so transform the species. Replacement of one gene by a mutant form of the gene, they said, could happen in two ways. The mutation could enable its possessors to survive or reproduce more effectively than the old form; if so, it would increase by natural selection, just as Darwin had said. The new characteristic that evolved in this way would ordinarily be considered an improved adaptation.

Sewall Wright pointed out, however, that not all genetic changes in species need be adaptive. A new mutation might be no better or worse than the preexisting gene—it might simply be "neutral." In small populations such a mutation could replace the previous gene purely by chance—a process he called random genetic drift. The idea, put crudely, is this. Suppose there is a small population of land snails in a cow pasture, and that 5 percent of them are brown and the rest are yellow. Purely by chance, a greater percentage of yellow snails than of brown ones get crushed by cows' hooves in one generation. The snails breed, and there will now be a slightly greater percentage of yellow snails in the next generation than there had been. But in the next generation, the yellow ones may suffer more trampling, purely by chance. The proportion of yellow offspring will then be lower again. These random events cause fluctuations in the percentage of the two types. Wright proved mathematically that eventually, if no other factors intervene, these fluctuations will bring the population either to 100 percent yellow or 100 percent

brown, purely by chance. The population will have evolved, then, but not by natural selection; and there is no improvement of adaptation.

During the period of the New Synthesis, though, genetic drift was emphasized less than natural selection, for which abundant evidence was discovered. Sergei Chetverikov in Russia, and later Theodosius Dobzhansky working in the United States, showed that wild populations of fruit flies contained an immense amount of genetic variation, including the same kinds of mutations that the geneticists had found arising in their laboratories. Dobzhansky and other workers went on to show that these variations affected survival and reproduction: that natural selection was a reality. They showed, moreover, that the genetic differences among related species were indeed compounded of the same kinds of slight genetic variations that they found within species. Thus the taxonomists and the geneticists converged onto a neo-Darwinian theory of evolution: evolution is due not to mutation *or* natural selection, but to both. Random mutations provide abundant genetic variation; natural selection, the antithesis of randomness, sorts out the useful from the deleterious, and transforms the species.

In the following two decades, the paleontologist George Gaylord Simpson showed that this theory was completely adequate to explain the fossil record, and the ornithologists Bernhard Rensch and Ernst Mayr, the botanist G. Ledyard Stebbins, and many other taxonomists showed that the similarities and differences among living species could be fully explained by neo-Darwinism. They also clarified the meaning of "species." Organisms belong to different species if they do not interbreed when the opportunity presents itself, thus remaining genetically distinct. An ancestral species splits into two descendant species when different populations of the ancestor, living in different geographic regions, become so genetically different from each other that they will not or cannot interbreed when they have the chance to do so. As a result, evolution can happen without the formation of new species: a single species can be genetically transformed without splitting into several descendants. Conversely, new species can be formed without much genetic change. If one population becomes different from the rest of its species in, for example, its mating behavior, it will not interbreed with the other populations. Thus it has become a new species, even though it may be identical to its "sister species" in every respect except its behavior. Such a new species is free to follow a new path of genetic change, since it does not become homogenized with its sister species by interbreeding. With time, therefore, it can diverge and develop different adaptations.

The conflict between the geneticists and the Darwinians that was resolved in the New Synthesis was the last major conflict in evolutionary science. Since that time, an enormous amount of research has confirmed most of the major conclusions of neo-Darwinism. We now know that populations contain very extensive genetic variation that continually arises by mutation of pre-existing genes. We also know what genes are and how they become mutated. Many instances of the reality of natural selection in wild populations have been documented, and there is extensive evidence that many species form by the divergence of different populations of an ancestral species.

The major questions in evolutionary biology now tend to be of the form, "All right, factors x and y both operate in evolution, but how important is x compared to y?" For example, studies of biochemical genetic variation have raised the possibility that nonadaptive, random change (genetic drift) may be the major reason for many biochemical differences among species. How important, then, is genetic drift compared to natural selection? Another major question has to do with rates of evolution: Do species usually diverge very slowly, as Darwin thought, or does evolution consist mostly of rapid spurts, interspersed with long periods of constancy? Still another question is raised by mutations, which range all the way from gross changes of the kind Morgan studied to very slight alterations. Does evolution consist entirely of the substitution of mutations that have very slight effects, or are major mutations sometimes important too? Partisans on each side of all these questions argue vigorously for their interpretation of the evidence, but they don't doubt that the major factors of evolution are known. They simply emphasize one factor or another. Minor battles of precisely this kind go on continually in every field of science; without them there would be very little advancement in our knowledge.

Within a decade or two of *The Origin of Species,* the belief that living organisms had evolved over the ages was firmly entrenched in biology. As of 1982, the historical existence of evolution is viewed as fact by almost all biologists. To explain how the fact of evolution has been brought about, a theory of evolutionary mechanisms—mutation, natural selection, genetic drift, and isolation—has been developed.[8] But exactly what is the evidence for the fact of evolution?

Notes

1. Andrew Dickson White, *A History of the Warfare of Science with Theology in Christendom* vol. I (London: Macmillan, 1896; reprint ed., New York: Dover, 1960).
2. A. O. Lovejoy, *The Great Chain of Being* (Cambridge, Mass.: Harvard University Press, 1936).
3. Much of this history is provided by J. C. Greene, *The Death of Adam: Evolution and its Impact on Western Thought* (Ames: Iowa State University Press, 1959).
4. A detailed history of this and other developments in evolutionary biology is given by Ernst Mayr, *The Growth of Biological Thought: Diversity, Evolution, Inheritance* (Cambridge, Mass.: Harvard University Press, 1982).
5. See D. L. Hull, *Darwin and His Critics* (Cambridge, Mass.: Harvard University Press, 1973).
6. Ibid.
7. E. Mayr and W. B. Provine, *The Evolutionary Synthesis* (Cambridge, Mass.: Harvard University Press, 1980).
8. Our modern understanding of the mechanisms of evolution is described in many books. Elementary textbooks include G. L. Stebbins, *Processes of Organic Evolution,* (Englewood Cliffs, N.J.: Prentice-Hall, 1971), and J. Maynard Smith, *The Theory of Evolution* (New York: Penguin Books, 1975). More advanced textbooks include Th. Dobzhansky, F. J. Ayala, G. L. Stebbins, and J. W. Valentine, *Evolution* (San Francisco: Freeman, 1977), and D. J. Futuyma, *Evolutionary Biology* (Sunderland, Mass.: Sinauer, 1979). Unreferenced facts and theories described in the text are familiar enough to most evolutionary biologists that they will be found in most or all of the references cited above.

Curse and Blessing of the Ghetto

Tay-Sachs disease is a choosy killer, one that for centuries targeted Eastern European Jews above all others. By decoding its lethal logic, we can learn a lot about how genetic diseases evolve— and how they can be conquered.

Jared Diamond

Contributing editor Jared Diamond is a professor of physiology at the UCLA School of Medicine.

Marie and I hated her at first sight, even though she was trying hard to be helpful. As our obstetrician's genetics counselor, she was just doing her job, explaining to us the unpleasant results that might come out of the genetic tests we were about to have performed. As a scientist, though, I already knew all I wanted to know about Tay-Sachs disease, and I didn't need to be reminded that the baby sentenced to death by it could be my own.

Fortunately, the tests would reveal that my wife and I were not carriers of the Tay-Sachs gene, and our preparent-hood fears on that matter at least could be put to rest. But at the time I didn't yet know that. As I glared angrily at that poor genetics counselor, so strong was my anxiety that now, four years later, I can still clearly remember what was going through my mind: If I were an evil deity, I thought, trying to devise exquisite tortures for babies and their parents, I would be proud to have designed Tay-Sachs disease.

Tay-Sachs is completely incurable, unpreventable, and preprogrammed in the genes. A Tay-Sachs infant usually appears normal for the first few months after birth, just long enough for the parents to grow to love him. An exaggerated "startle reaction" to sounds is the first ominous sign. At about six months the baby starts to lose control of his head and can't roll over or sit without support. Later he begins to drool, breaks out into unmotivated bouts of laughter, and suffers convulsions. Then his head grows abnormally large, and he becomes blind. Perhaps what's most frightening for the parents is that their baby loses all contact with his environment and becomes virtually a vegetable. By the child's third birthday, if he's still alive, his skin will turn yellow and his hands pudgy. Most likely he will die before he's four years old.

My wife and I were tested for the Tay-Sachs gene because at the time we rated as high-risk candidates, for two reasons. First, Marie was carrying twins, so we had double the usual chance to bear a Tay-Sachs baby. Second, both she and I are of Eastern European Jewish ancestry, the population with by far the world's highest Tay-Sachs frequency.

In peoples around the world Tay-Sachs appears once in every 400,000 births. But it appears a hundred times more frequently—about once in 3,600 births—among descendants of Eastern European Jews, people known as Ashkenazim. For descendants of most other groups of Jews—Oriental Jews, chiefly from the Middle East, or Sephardic Jews, from Spain and other Mediterranean countries—the frequency of Tay-Sachs disease is no higher than in non-Jews. Faced with such a clear correlation, one cannot help but wonder: What is it about this one group of people that produces such an extraordinarily high risk of this disease?

Finding the answer to this question concerns all of us, regardless of our ancestry. Every human population is especially susceptible to certain diseases, not only because of its life-style but also because of its genetic inheritance. For example, genes put European whites at high risk for cystic fibrosis, African blacks for sickle-cell disease, Pacific Islanders for diabetes—and Eastern European Jews for ten different diseases, including Tay-Sachs. It's not that Jews

are notably susceptible to genetic diseases in general; but a combination of historical factors has led to Jews' being intensively studied, and so their susceptibilities are far better known than those of, say, Pacific Islanders.

Tay-Sachs exemplifies how we can deal with such diseases; it has been the object of the most successful screening program to date. Moreover, Tay-Sachs is helping us understand how ethnic diseases evolve. Within the past couple of years discoveries by molecular biologists have provided tantalizing clues to precisely how a deadly gene can persist and spread over the centuries. Tay-Sachs may be primarily a disease of Eastern European Jews, but through this affliction of one group of people, we gain a window on how our genes simultaneously curse and bless us all.

It seems unlikely that genetic accidents would have pumped up the frequency of the same gene not once but twice in the same population.

The disease's hyphenated name comes from the two physicians—British ophthalmologist W. Tay and New York neurologist B. Sachs—who independently first recognized the disease, in 1881 and 1887, respectively. By 1896 Sachs had

seen enough cases to realize that the disease was most common among Jewish children.

Not until 1962, however, were researchers able to trace the cause of the affliction to a single biochemical abnormality: the excessive accumulation in nerve cells of a fatty substance called G_{M2} ganglioside. Normally G_{M2} ganglioside is present at only modest levels in cell membranes, because it is constantly being broken down as well as synthesized. The breakdown depends on the enzyme hexosaminidase A, which is found in the tiny structures within our cells known as lysosomes. In the unfortunate Tay-Sachs victims this enzyme is lacking, and without it the ganglioside piles up and produces all the symptoms of the disease.

We have two copies of the gene that programs our supply of hexosaminidase A, one inherited from our father, the other from our mother; each of our parents, in turn, has two copies derived from their own parents. As long as we have one good copy of the gene, we can produce enough hexosaminidase A to prevent a buildup of G_{M2} ganglioside and we won't get Tay-Sachs. This genetic disease is of the sort termed recessive rather than dominant—meaning that to get it, a child must inherit a defective gene not just from one parent but from both of them. Clearly, each parent must have had one good copy of the gene along with the defective copy—if either had had two defective genes, he or she would have died of the disease long before reaching the age of reproduction. In genetic terms the diseased child is homozygous for the defective gene and both parents are heterozygous for it.

None of this yet gives any hint as to why the Tay-Sachs gene should be most common among Eastern European Jews. To come to grips with that question, we must take a short detour into history.

From their biblical home of ancient Israel, Jews spread peacefully to other Mediterranean lands, Yemen, and India. They were also dispersed violently through conquest by Assyrians, Babylonians, and Romans. Under the Carolingian kings of the eighth and ninth centuries Jews were invited to settle in

France and Germany as traders and financiers. In subsequent centuries, however, persecutions triggered by the Crusades gradually drove Jews out of Western Europe; the process culminated in their total expulsion from Spain in 1492. Those Spanish Jews—called Sephardim—fled to other lands around the Mediterranean. Jews of France and Germany—the Ashkenazim—fled east to Poland and from there to Lithuania and western Russia, where they settled mostly in towns, as businessmen engaged in whatever pursuit they were allowed.

There the Jews stayed for centuries, through periods of both tolerance and oppression. But toward the end of the nineteenth century and the beginning of the twentieth, waves of murderous anti-Semitic attacks drove millions of Jews out of Eastern Europe, with most of them heading for the United States. My mother's parents, for example, fled to New York from Lithuanian pogroms of the 1880s, while my father's parents fled from the Ukrainian pogroms of 1903–6. The more modern history of Jewish migration is probably well known to you all: most Jews who remained in Eastern Europe were exterminated during World War II, while most the survivors immigrated to the United States and Israel. Of the 13 million Jews alive today, more than three-quarters are Ashkenazim, the descendants of the Eastern European Jews and the people most at risk for Tay-Sachs.

Have these Jews maintained their genetic distinctness through the thousands of years of wandering? Some scholars claim that there has been so much intermarriage and conversion that Ashkenazic Jews are now just Eastern Europeans who adopted Jewish culture. However, modern genetic studies refute that speculation.

First of all, there are those ten genetic diseases that the Ashkenazim have somehow acquired, by which they differ both from other Jews and from Eastern European non-Jews. In addition, many Ashkenazic genes turn out to be ones typical of Palestinian Arabs and other peoples of the Eastern Mediterranean areas where Jews originated. (In fact, by genetic standards the current Arab-Is-raeli conflict is an internecine civil war.) Other Ashkenazic genes have indeed diverged from Mediterranean ones (including genes of Sephardic and Oriental Jews) and have evolved to converge on genes of Eastern European non-Jews subject to the same local forces of natural selection. But the degree to which Ashkenazim prove to differ genetically from Eastern European non-Jews implies an intermarriage rate of only about 15 percent.

Can history help explain why the Tay-Sachs gene in particular is so much more common in Ashkenazim than in their non-Jewish neighbors or in other Jews? At the risk of spoiling a mystery, I'll tell you now that the answer is yes, but to appreciate it, you'll have to understand the four possible explanations for the persistence of the Tay-Sachs gene.

First, new copies of the gene might be arising by mutation as fast as existing copies disappear with the death of Tay-Sachs children. That's the most likely explanation for the gene's persistence in most of the world, where the disease frequency is only one in 400,000 births—that frequency reflects a typical human mutation rate. But for this explanation to apply to the Ashkenazim would require a mutation rate of at least one per 3,600 births—far above the frequency observed for any human gene. Furthermore, there would be no precedent for one particular gene mutating so much more often in one human population than in others.

As a second possibility, the Ashkenazim might have acquired the Tay-Sachs gene from some other people who already had the gene at high frequency. Arthur Koestler's controversial book *The Thirteenth Tribe,* for example, popularized the view that the Ashkenazim are really not a Semitic people but are instead descended from the Khazar, a Turkic tribe whose rulers converted to Judaism in the eighth century. Could the Khazar have brought the Tay-Sachs gene to Eastern Europe? This speculation makes good romantic reading, but there is no good evidence to support it. Moreover, it fails to explain why deaths of Tay-Sachs children didn't eliminate the gene by natural selection in the past 1,200 years, nor how the Khazar ac-quired high frequencies of the gene in the first place.

The third hypothesis was the one preferred by a good many geneticists until recently. It invokes two genetic processes, termed the founder effect and genetic drift, that may operate in small populations. To understand these concepts, imagine that 100 couples settle in a new land and found a population that then increases. Imagine further that one parent among those original 100 couples happens to have some rare gene, one, say, that normally occurs at a frequency of one in a million. The gene's frequency in the new population will now be one in 200 as a result of the accidental presence of that rare founder.

Or suppose again that 100 couples found a population, but that one of the 100 men happens to have lots of kids by his wife or that he is exceptionally popular with other women, while the other 99 men are childless or have few kids or are simply less popular. That one man may thereby father 10 percent rather than a more representative one percent of the next generation's babies, and their genes will disproportionately reflect that man's genes. In other words, gene frequencies will have drifted between the first and second generation.

Through these two types of genetic accidents a rare gene may occur with an unusually high frequency in a small expanding population. Eventually, if the gene is harmful, natural selection will bring its frequency back to normal by killing off gene bearers. But if the resultant disease is recessive—if heterozygous individuals don't get the disease and only the rare, homozygous individuals die of it—the gene's high frequency may persist for many generations.

These accidents do in fact account for the astonishingly high Tay-Sachs gene frequency found in one group of Pennsylvania Dutch: out of the 333 people in this group, 98 proved to carry the Tay-Sachs gene. Those 333 are all descended from one couple who settled in the United States in the eighteenth century and had 13 children. Clearly, one of that founding couple must have carried the gene. A similar accident may explain why Tay-Sachs is also relatively common among French Canadians, who number 5 million today but are descended from fewer than 6,000 French immigrants who arrived in the New World between 1638 and 1759. In the two or three centuries since both these founding events, the high Tay-Sachs gene frequency among Pennsylvania Dutch and French Canadians has not yet had enough time to decline to normal levels.

The same mechanisms were one proposed to explain the high rate of Tay-Sachs disease among the Ashkenazim. Perhaps, the reasoning went, the gene just happened to be overrepresented in the founding Jewish population that settled in Germany or Eastern Europe. Perhaps the gene just happened to drift up in frequency in the Jewish populations scattered among the isolated towns of Eastern Europe.

But geneticists have long questioned whether the Ashkenazim population's history was really suitable for these genetic accidents to have been significant. Remember, the founder effect and genetic drift become significant only in small populations, and the founding populations of Ashkenazim may have been quite large. Moreover, Ashkenazic communities were considerably widespread; drift would have sent gene frequencies up in some towns but down in others. And, finally, natural selection has by now had a thousand years to restore gene frequencies to normal.

Granted, those doubts are based on historical data, which are not always as precise or reliable as one might want. But within the past several years the case against those accidental explanations for Tay-Sachs disease in the Ashkenazim has been bolstered by discoveries by molecular biologists.

Like all proteins, the enzyme absent in Tay-Sachs children is coded for by a piece of our DNA. Along that particular stretch of DNA there are thousands of different sites where a mutation could occur that would result in no enzyme and hence in the same set of symptoms. If molecular biologists had discovered that all cases of Tay-Sachs in Ashkenazim involved damage to DNA at the same site, that would have been strong evidence that in Ashkenazim the disease stems from a single mutation that has been multiplied by the founder effect or

genetic drift—in other words, the high incidence of Tay-Sachs among Eastern European Jews is accidental.

In reality, though, several different mutations along this stretch of DNA have been identified in Ashkenazim, and two of them occur much more frequently than in non-Ashkenazim populations. It seems unlikely that genetic accidents would have pumped up the frequency of the same gene not once but twice in the same population.

We're not a melting pot, and we won't be for a long time. Each ethnic group has some characteristic genes of its own, a legacy of its distinct history.

And that's not the sole unlikely coincidence arguing against accidental explanations. Recall that Tay-Sachs is caused by the excessive accumulation of one fatty substance, G_{M2} ganglioside, from a defect in one enzyme, hexosaminidase A. But Tay-Sachs is one of ten genetic diseases characteristic of Ashkenazim. Among those other nine, two—Gaucher's disease and Niemann-Pick disease—result from the accumulation of two other fatty substances similar to G_{M2} ganglioside, as a result of defects in two other enzymes similar to hexosaminidase A. Yet our bodies contain thousands of different enzymes.

It would have been an incredible roll of the genetic dice if, by nothing more than chance, Ashkenazim had independently acquired mutations in three closely related enzymes—and had acquired mutations in one of those enzymes twice.

All these facts bring us to the fourth possible explanation of why the Tay-Sachs gene is so prevalent among Ashkenazim: namely, that something about them favored accumulation of G_{M2} ganglioside and related fats.

For comparison, suppose that a friend doubles her money on one stock while you are getting wiped out with your investments. Taken alone, that could just mean she was lucky on that one occasion. But suppose that she doubles her money on each of two different stocks and at the same time rings up big profits in real estate while also making a killing in bonds. That implies more than lady luck; it suggests that something about your friend—like shrewd judgment—favors financial success.

What could be the blessings of fat accumulation in Eastern European Jews? At first this question sounds weird. After all, that fat accumulation was noticed only because of the curses it bestows: Tay-Sachs, Gaucher's, or Niemann-Pick disease. But many of our common genetic diseases may persist because they bring both blessings and curses (see "The Cruel Logic of Our Genes," *Discover,* November 1989). They kill or impair individuals who inherit two copies of the faulty gene, but they help those who receive only one defective gene by protecting them against other diseases. The best understood example is the sickle-cell gene of African blacks, which often kills homozygotes but protects heterozygotes against malaria. Natural selection sustains such genes because more heterozygotes than normal individuals survive to pass on their genes, and those extra gene copies offset the copies lost through the deaths of homozygotes.

So let us refine our question and ask, What blessing could the Tay-Sachs gene bring to those individuals who are heterozygous for it? A clue first emerged back in 1972, with the publication of the results of a questionnaire that had asked U.S. Ashkenazic parents of Tay-Sachs children what their own Eastern European-born parents had died of. Keep in mind that since these unfortunate children had to be homozygotes, with two copies of the Tay-Sachs gene, all their parents had to be heterozygotes, with one copy, and half of the parents' parents also had to be heterozygotes.

As it turned out, most of those Tay-Sachs grandparents had died of the usual causes: heart disease, stroke, cancer, and diabetes. But strikingly, only one of the 306 grandparents had died of tuberculosis, even though TB was generally one of the big killers in these grandparents' time. Indeed, among the general population of large Eastern European cities in the early twentieth century, TB caused up to 20 percent of all deaths.

This big discrepancy suggested that Tay-Sachs heterozygotes might somehow have been protected against TB. Interestingly, it was already well known that Ashkenazim in general had some such protection: even when Jews and non-Jews were compared within the same European city, class, and occupational group (for example, Warsaw garment workers), Jews had only half the TB death rate of non-Jews, despite their being equally susceptible to infection. Perhaps, one could reason, the Tay-Sachs gene furnished part of that well-established Jewish resistance.

A second clue to a heterozygote advantage conveyed by the Tay-Sachs gene emerged in 1983, with a fresh look at the data concerning the distributions of TB and the Tay-Sachs gene within Europe. The statistics showed that the Tay-Sachs gene was nearly three times more frequent among Jews originating from Austria, Hungary, and Czechoslovakia—areas where an amazing 9 to 10 percent of the population were heterozygotes—than among Jews from Poland, Russia, and Germany. At the same time records from an old Jewish TB sanatorium in Denver in 1904 showed that among patients born in Europe between 1860 and 1910, Jews from Austria and Hungary were overrepresented.

Initially, in putting together these two pieces of information, you might be tempted to conclude that because the highest frequency of the Tay-Sachs gene appeared in the same geographic region

that produced the most cases of TB, the gene in fact offers no protection whatsoever. Indeed, this was precisely the mistaken conclusion of many researchers who had looked at these data before. But you have to pay careful attention to the numbers here: even at its highest frequency the Tay-Sachs gene was carried by far fewer people than would be infected by TB. What the statistics really indicate is that where TB is the biggest threat, natural selection produces the biggest response.

Think of it this way: You arrive at an island where you find that all the inhabitants of the north end wear suits of armor, while all the inhabitants of the south end wear only cloth shirts. You'd be pretty safe in assuming that warfare is more prevalent in the north—and that war-related injuries account for far more deaths there than in the south. Thus, if the Tay-Sachs gene does indeed lend heterozygotes some protection against TB, you would expect to find the gene most often precisely where you find TB most often. Similarly, the sickle-cell gene reaches its highest frequencies in those parts of Africa where malaria is the biggest risk.

But you may believe there's still a hole in the argument: If Tay-Sachs heterozygotes are protected against TB, you may be asking, why is the gene common just in the Ashkenazim? Why did it not become common in the non-Jewish populations also exposed to TB in Austria, Hungary, and Czechoslovakia?

At this point we must recall the peculiar circumstances in which the Jews of Eastern Europe were forced to live. They were unique among the world's ethnic groups in having been virtually confined to towns for most of the past 2,000 years. Being forbidden to own land, Eastern European Jews were not peasant farmers living in the countryside, but businesspeople forced to live in crowded ghettos, in an environment where tuberculosis thrived.

Of course, until recent improvements in sanitation, these towns were not very healthy places for non-Jews either. Indeed, their populations couldn't sustain themselves: deaths exceeded births, and the number of dead had to be balanced by continued emigration from the coun-

tryside. For non-Jews, therefore, there was no genetically distinct urban population. For ghetto-bound Jews, however, there could be no emigration from the countryside; thus the Jewish population was under the strongest selection to evolve genetic resistance to TB.

Those are the conditions that probably led to Jewish TB resistance, whatever particular genetic factors prove to underlie it. I'd speculate that G_{M2} and related fats accumulate at slightly higher-than-normal levels in heterozygotes, although not at the lethal levels seen in homozygotes. (The fat accumulation in heterozygotes probably takes place in the cell membrane, the cell's "armor.") I'd also speculate that the accumulation provides heterozygotes with some protection against TB, and that that's why the genes for Tay-Sachs, Gaucher's, and Niemann-Pick disease reached high frequencies in the Ashkenazim.

Having thus stated the case, let me make clear that I don't want to overstate it. The evidence is still speculative. Depending on how you do the calculation, the low frequency of TB deaths in Tay-Sachs grandparents either barely reaches or doesn't quite reach the level of proof that statisticians require to accept an effect as real rather than as one that's arisen by chance. Moreover, we have no idea of the biochemical mechanism by which fat accumulation might confer resistance against TB. For the moment, I'd say that the evidence points to some selective advantage of Tay-Sachs heterozygotes among the Ashkenazim, and that TB resistance is the only plausible hypothesis yet proposed.

For now Tay-Sachs remains a speculative model for the evolution of ethnic diseases. But it's already a proven model of what to do about them. Twenty years ago a test was developed to identify Tay-Sachs heterozygotes, based on their lower-than-normal levels of hexosaminidase A. The test is simple, cheap, and accurate: all I did was to donate a small sample of my blood, pay $35, and wait a few days to receive the results.

If that test shows that at least one member of a couple is not a Tay-Sachs heterozygote, then any child of theirs can't be a Tay-Sachs homozygote. If both parents prove to be heterozygotes, there's a one-in-four chance of their child being a homozygote; that can then be determined by other tests performed on the mother early in pregnancy. If the results are positive, it's early enough for her to abort, should she choose to. That critical bit of knowledge has enabled parents who had gone through the agony of bearing a Tay-Sachs baby and watching him die to find the courage to try again.

The Tay-Sachs screening program launched in the United States in 1971 was targeted at the high-risk population: Ashkenazic Jewish couples of childbearing age. So successful has this approach been that the number of Tay-Sachs babies born each year in this country has declined tenfold. Today, in fact, more Tay-Sachs cases appear here in non-Jews than in Jews, because only the latter couples are routinely tested. Thus, what used to be the classic genetic disease of Jews is so no longer.

There's also a broader message to the Tay-Sachs story. We commonly refer to the United States as a melting pot, and in many ways that metaphor is apt. But in other ways we're not a melting pot, and we won't be for a long time. Each ethnic group has some characteristic genes of its own, a legacy of its distinct history. Tuberculosis and malaria are not major causes of death in the United States, but the genes that some of us evolved to protect ourselves against them are still frequent. Those genes are frequent only in certain ethnic groups, though, and they'll be slow to melt through the population.

With modern advances in molecular genetics, we can expect to see more, not less, ethnically targeted practice of medicine. Genetic screening for cystic fibrosis in European whites, for example, is one program that has been much discussed recently; when it comes, it will surely be based on the Tay-Sachs experience. Of course, what that may mean someday is more anxiety-ridden parents-to-be glowering at more dedicated genetics counselors. It will also mean fewer babies doomed to the agonies of diseases we may understand but that we'll never be able to accept.

The Saltshaker's Curse

Physiological adaptations that helped American blacks survive slavery may now be predisposing their descendants to hypertension

Jared Diamond

Jared Diamond is a professor of physiology at UCLA Medical School.

On the walls of the main corridor at UCLA Medical School hang thirty-seven photographs that tell a moving story. They are the portraits of each graduating class, from the year that the school opened (Class of 1955) to the latest crop (Class of 1991). Throughout the 1950s and early 1960s the portraits are overwhelmingly of young white men, diluted by only a few white women and Asian men. The first black student graduated in 1961, an event not repeated for several more years. When I came to UCLA in 1966, I found myself lecturing to seventy-six students, of whom seventy-four were white. Thereafter the numbers of blacks, Hispanics, and Asians exploded, until the most recent photos show the number of white medical students declining toward a minority.

In these changes of racial composition, there is of course nothing unique about UCLA Medical School. While the shifts in its student body mirror those taking place, at varying rates, in other professional groups throughout American society, we still have a long way to go before professional groups truly mirror society itself. But ethnic diversity among physicians is especially impor-

tant because of the dangers inherent in a profession composed of white practitioners for whom white biology is the norm.

Different ethnic groups face different health problems, for reasons of genes as well as of life style. Familiar examples include the prevalence of skin cancer and cystic fibrosis in whites, stomach cancer and stroke in Japanese, and diabetes in Hispanics and Pacific islanders. Each year, when I teach a seminar course in ethnically varying disease patterns, these by-now-familiar textbook facts assume a gripping reality, as my various students choose to discuss some disease that affects themselves or their relatives. To read about the molecular biology of sickle-cell anemia is one thing. It's quite another thing when one of my students, a black man homozygous for the sickle-cell gene, describes the pain of his own sickling attacks and how they have affected his life.

Sickle-cell anemia is a case in which the evolutionary origins of medically important genetic differences among peoples are well understood. (It evolved only in malarial regions because it confers resistance against malaria.) But in many other cases the evolutionary origins are not nearly so transparent. Why is it, for example, that only some human populations have a high frequency of the Tay-Sachs gene or of diabetes? . . .

Compared with American whites of the same age and sex, American blacks have, on the average, higher blood pressure, double the risk of developing hypertension, and nearly ten times the risk of dying of it. By age fifty, nearly half of U.S. black men are hypertensive. For a given age and blood pressure, hypertension more often causes heart disease and especially kidney failure and strokes in U.S. blacks than whites. Because the frequency of kidney disease in U.S. blacks is eighteen times that in whites, blacks account for about two-thirds of U.S. patients with hypertensive kidney failure, even though they make up only about one-tenth of the population. Around the world, only Japanese exceed U.S. blacks in their risk of dying from stroke. Yet it was not until 1932 that the average difference in blood pressure between U.S. blacks and whites was clearly demonstrated, thereby exposing a major health problem outside the norms of white medicine.

What is it about American blacks that makes them disproportionately likely to develop hypertension and then to die of its consequences? While this question is of course especially "interesting" to black readers, it also concerns all Americans, because other ethnic groups in the United States are not so far behind blacks in their risk of hypertension. If

Natural History readers are a cross section of the United States, then about one-quarter of you now have high blood pressure, and more than half of you will die of a heart attack or stroke to which high blood pressure predisposes. Thus, we all have valid reasons for being interested in hypertension.

First, some background on what those numbers mean when your doctor inflates a rubber cuff about your arm, listens, deflates the cuff, and finally pronounces, "Your blood pressure is 120 over 80." The cuff device is called a sphygmomanometer, and it measures the pressure in your artery in units of millimeters of mercury (that's the height to which your blood pressure would force up a column of mercury in case, God forbid, your artery were suddenly connected to a vertical mercury column). Naturally, your blood pressure varies with each stroke of your heart, so the first and second numbers refer, respectively, to the peak pressure at each heartbeat (systolic pressure) and to the minimum pressure between beats (diastolic pressure). Blood pressure varies somewhat with position, activity, and anxiety level, so the measurement is usually made while you are resting flat on your back. Under those conditions, 120 over 80 is an average reading for Americans.

There is no magic cutoff between normal blood pressure and high blood pressure. Instead, the higher your blood pressure, the more likely you are to die of a heart attack, stroke, kidney failure, or ruptured aorta. Usually, a pressure reading higher than 140 over 90 is arbitrarily defined as constituting hypertension, but some people with lower readings will die of a stroke at age fifty, while others with higher readings will die in a car accident in good health at age ninety.

Why do some of us have much higher blood pressure than others? In about 5 percent of hypertensive patients there is an identifiable single cause, such as hormonal imbalance or use of oral contraceptives. In 95 percent of such cases, though, there is no such obvious cause. The clinical euphemism for our ignorance in such cases is "essential hypertension."

Nowadays, we know that there is a big genetic component in essential hypertension, although the particular genes involved have not yet been identified. Among people living in the same household, the correlation coefficient for blood pressure is 0.63 between identical twins, who share all of their genes. (A correlation coefficient of 1.00 would mean that the twins share identical blood pressures as well and would suggest that pressure is determined entirely by genes and not at all by environment.) Fraternal twins or ordinary siblings or a parent and child, who share half their genes and whose blood pressure would therefore show a correlation coefficient of 0.5 if purely determined genetically, actually have a coefficient of about 0.25. Finally, adopted siblings or a parent and adopted child, who have no direct genetic connection, have a correlation coefficient of only 0.05. Despite the shared household environment, their blood pressures are barely more similar than those of two people pulled randomly off the street. In agreement with this evidence for genetic factors underlying blood pressure itself, your risk of actually developing hypertensive disease increases from 4 percent to 20 percent to 35 percent if, respectively, none or one or both of your parents were hypertensive.

But these same facts suggest that environmental factors also contribute to high blood pressure, since identical twins have similar but not identical blood pressures. Many environmental or life style factors contributing to the risk of hypertension have been identified by epidemiological studies that compare hypertension's frequency in groups of people living under different conditions. Such contributing factors include obesity, high intake of salt or alcohol or saturated fats, and low calcium intake. The proof of this approach is that hypertensive patients who modify their life styles so as to minimize these putative factors often succeed in reducing their blood pressure. Patients are especially advised to reduce salt intake and stress, reduce intake of cholesterol and saturated fats and alcohol, lose weight, cut out smoking, and exercise regularly.

Here are some examples of the epidemiological studies pointing to these risk factors. Around the world, comparisons within and between populations show that both blood pressure and the frequency of hypertension increase hand in hand with salt intake. At the one extreme, Brazil's Yanomamö Indians have the world's lowest-known salt consumption (somewhat above 10 milligrams per day!), lowest average blood pressure (95 over 61!), and lowest incidence of hypertension (no cases!). At the opposite extreme, doctors regard Japan as the "land of apoplexy" because of the high frequency of fatal strokes (Japan's leading cause of death, five times more frequent than in the United States), linked with high blood pressure and notoriously salty food. Within Japan itself these factors reach their extremes in Akita Prefecture, famous for its tasty rice, which Akita farmers flavor with salt, wash down with salty miso soup, and alternate with salt pickles between meals. Of 300 Akita adults studied, not one consumed less than five grams of salt daily, the average consumption was twenty-seven grams, and the most salt-loving individual consumed an incredible sixty-one grams—enough to devour the contents of the usual twenty-six-ounce supermarket salt container in a mere twelve days. The average blood pressure in Akita by age fifty is 151 over 93, making hypertension (pressure higher than 140 over 90) the norm. Not surprisingly, Akitas' frequency of death by stroke is more than double even the Japanese average, and in some Akita villages 99 percent of the population dies before age seventy.

Why salt intake often (in about 60 percent of hypertensive patients) leads to high blood pressure is not fully understood. One possible interpretation is that salt intake triggers thirst, leading to an increase in blood volume. In response, the heart increases its output and blood pressure rises, causing the kidneys to filter more salt and water under that increased pressure. The result is a new steady state, in which salt and water excretion again equals intake, but more salt and water are stored in the body and blood pressure is raised.

At this point, let's contrast hypertension with a simple genetic disease like Tay-Sachs disease. Tay-Sachs is due to a defect in a single gene; every Tay Sachs patient has a defect in that same gene. Everybody in whom that gene is defective is certain to die of Tay-Sachs, regardless of their life style or environment. In contrast, hypertension involves several different genes whose molecular products remain to be identified. Because there are many causes of raised blood pressure, different hypertensive patients may owe their condition to different gene combinations. Furthermore, whether someone genetically predisposed to hypertension actually develops symptoms depends a lot on life style. Thus, hypertension is not one of those uncommon, homogeneous, and intellectually elegant diseases that geneticists prefer to study. Instead, like diabetes and ulcers, hypertension is a shared set of symptoms produced by heterogeneous causes, all involving an interaction between environmental agents and a susceptible genetic background.

Since U.S. blacks and whites differ on the average in the conditions under which they live, could those differences account for excess hypertension in U.S. blacks? Salt intake, the dietary factor that one thinks of first, turns out on the average not to differ between U.S. blacks and whites. Blacks do consume less potassium and calcium, do experience more stress associated with more difficult socioeconomic conditions, have much less access to medical care, and are therefore much less likely to be diagnosed or treated until it is too late. Those factors surely contribute to the frequency and severity of hypertension in blacks.

However, those factors don't seem to be the whole explanation: hypertensive blacks aren't merely like severely hypertensive whites. Instead, physiological differences seem to contribute as well. On consuming salt, blacks retain it on average far longer before excreting it into the urine, and they experience a greater rise in blood pressure on a high-salt diet. Hypertension is more likely to be "salt-sensitive" in blacks than in whites, meaning that blood pressure is more likely to rise and fall with rises

and falls in dietary salt intake. By the same token, black hypertension is more likely to be treated successfully by drugs that cause the kidneys to excrete salt (the so-called thiazide diuretics) and less likely to respond to those drugs that reduce heart rate and cardiac output (so-called beta blockers, such as propanolol). These facts suggest that there are some qualitative differences between the causes of black and white hypertension, with black hypertension more likely to involve how the kidneys handle salt.

Physicians often refer to this postulated feature as a "defect": for example, "kidneys of blacks have a genetic defect in excreting sodium." As an evolutionary biologist, though, I hear warning bells going off inside me whenever a seemingly harmful trait that occurs frequently in an old and large human population is dismissed as a "defect." Given enough generations, genes that greatly impede survival are extremely unlikely to spread, unless their net effect is to increase survival and reproductive success. Human medicine has furnished the best examples of seemingly defective genes being propelled to high frequency by counterbalancing benefits. For example, sickle-cell hemoglobin protects far more people against malaria than it kills of anemia, while the Tay-Sachs gene may have protected far more Jews against tuberculosis than it killed of neurological disease. Thus, to understand why U.S. blacks now are prone to die as a result of their kidneys' retaining salt, we need to ask under what conditions people might have benefited from kidneys good at retaining salt.

That question is hard to understand from the perspective of modern Western society, where saltshakers are on every dining table, salt (sodium chloride) is cheap, and our bodies' main problem is getting rid of it. But imagine what the world used to be like before saltshakers became ubiquitous. Most plants contain very little sodium, yet animals require sodium at high concentrations in all their extracellular fluids. As a result, carnivores readily obtain their needed sodium by eating herbivores, but herbivores themselves face big problems in acquiring that sodium. That's why the

animals that one sees coming to salt licks are deer and antelope, not lions and tigers. Similarly, some human hunter-gatherers obtained enough salt from the meat that they ate. But when we began to take up farming ten thousand years ago, we either had to evolve kidneys superefficient at conserving salt or learn to extract salt at great effort or trade for it at great expense.

Examples of these various solutions abound. I already mentioned Brazil's Yanomamö Indians, whose staple food is low-sodium bananas and who excrete on the average only 10 milligrams of salt daily—barely one-thousandth the salt excretion of the typical American. A single Big Mac hamburger analyzed by *Consumer Reports* contained 1.5 grams (1,500 milligrams) of salt, representing many weeks of intake for a Yanomamö. The New Guinea highlanders with whom I work, and whose diet consists up to 90 percent of low-sodium sweet potatoes, told me of the efforts to which they went to make salt a few decades ago, before Europeans brought it as trade goods. They gathered leaves of certain plant species, burned them, scraped up the ash, percolated water through it to dissolve the solids, and finally evaporated the water to obtain small amounts of bitter salt.

Thus, salt has been in very short supply for much of recent human evolutionary history. Those of us with efficient kidneys able to retain salt even on a low-sodium diet were better able to survive our inevitable episodes of sodium loss (of which more in a moment). Those kidneys proved to be a detriment only when salt became routinely available, leading to excessive salt retention and hypertension with its fatal consequences. That's why blood pressure and the frequency of hypertension have shot up recently in so many populations around the world as they have made the transition from being self-sufficient subsistence farmers to members of the cash economy and patrons of supermarkets.

This evolutionary argument has been advanced by historian-epidemiologist Thomas Wilson and others to explain the current prevalence of hypertension in American blacks in particular. Many West African blacks, from whom most

American blacks originated via the slave trade, must have faced the chronic problem of losing salt through sweating in their hot environment. Yet in West Africa, except on the coast and certain inland areas, salt was traditionally as scarce for African farmers as it has been for Yanomamö and New Guinea farmers. (Ironically, those Africans who sold other Africans as slaves often took payment in salt traded from the Sahara.) By this argument, the genetic basis for hypertension in U.S. blacks was already widespread in many of their West African ancestors. It required only the ubiquity of saltshakers in twentieth-century America for that genetic basis to express itself as hypertension. This argument also predicts that as Africa's life style becomes increasingly Westernized, hypertension could become as prevalent in West Africa as it now is among U.S. blacks. In this view, American blacks would be no different from the many Polynesian, Melanesian, Kenyan, Zulu, and other populations that have recently developed high blood pressure under a Westernized life style.

But there's an intriguing extension to this hypothesis, proposed by Wilson and physician Clarence Grim, collaborators at the Hypertension Research Center of Drew University in Los Angeles. They suggest a scenario in which New World blacks may now be at more risk for hypertension than their African ancestors. That scenario involves very recent selection for super-efficient kidneys, driven by massive mortality of black slaves from salt loss.

Grim and Wilson's argument goes as follows. Black slavery in the Americas began about 1517, with the first imports of slaves from West Africa, and did not end until Brazil freed its slaves barely a century ago in 1888. In the course of the slave trade an estimated 12 million Africans were brought to the Americas. But those imports were winnowed by deaths at many stages, from an even larger number of captives and exports.

First, slaves captured by raids in the interior of West Africa were chained together, loaded with heavy burdens, and marched for one or two months, with little food and water, to the coast. About 25 percent of the captives died en route.

While awaiting purchase by slave traders, the survivors were held on the coast in hot, crowded buildings called barracoons, where about 12 percent of them died. The traders went up and down the coast buying and loading slaves for a few weeks or months until a ship's cargo was full (5 percent more died). The dreaded Middle Passage across the Atlantic killed 10 percent of the slaves, chained together in a hot, crowded, unventilated hold without sanitation. (Picture to yourself the result of those toilet "arrangements.") Of those who lived to land in the New World, 5 percent died while awaiting sale, and 12 percent died while being marched or shipped from the sale yard to the plantation. Finally, of those who survived, between 10 and 40 percent died during the first three years of plantation life, in a process euphemistically called seasoning. At that stage, about 70 percent of the slaves initially captured were dead, leaving 30 percent as seasoned survivors.

Even the end of seasoning, however, was not the end of excessive mortality. About half of slave infants died within a year of birth because of the poor nutrition and heavy workload of their mothers. In plantation terminology, slave women were viewed as either "breeding units" or "work units," with a built-in conflict between those uses: "These Negroes breed the best, whose labour is least," as an eighteenth-century observer put it. As a result, many New World slave populations depended on continuing slave imports and couldn't maintain their own numbers because death rates exceeded birth rates. Since buying new slaves cost less than rearing slave children for twenty years until they were adults, slave owners lacked economic incentive to change this state of affairs.

Recall that Darwin discussed natural selection and survival of the fittest with respect to animals. Since many more animals die than survive to produce offspring, each generation becomes enriched in the genes of those of the preceding generation that were among the survivors. It should now be clear that slavery represented a tragedy of unnatural selection in humans on a gigantic scale. From examining accounts of slave mortality, Grim and Wilson argue that death was indeed selective: much of it was related to unbalanced salt loss, which quickly brings on collapse. We think immediately of salt loss by sweating under hot conditions: while slaves were working, marching, or confined in unventilated barracoons or ships' holds. More body salt may have been spilled with vomiting from seasickness. But the biggest salt loss at every stage was from diarrhea due to crowding and lack of sanitation—ideal conditions for the spread of gastrointestinal infections. Cholera and other bacterial diarrheas kill us by causing sudden massive loss of salt and water. (Picture your most recent bout of *turista*, multiplied to a diarrheal fluid output of twenty quarts in one day, and you'll understand why.) All contemporary accounts of slave ships and plantation life emphasized diarrhea, or "fluxes" in eighteenth-century terminology, as one of the leading killers of slaves.

Grim and Wilson reason, then, that slavery suddenly selected for superefficient kidneys surpassing the efficient kidneys already selected by thousands of years of West African history. Only those slaves who were best able to retain salt could survive the periodic risk of high salt loss to which they were exposed. Salt supersavers would have had the further advantage of building up, under normal conditions, more of a salt reserve in their body fluids and bones, thereby enabling them to survive longer or more frequent bouts of diarrhea. Those superkidneys became a disadvantage only when modern medicine began to reduce diarrhea's lethal impact, thereby transforming a blessing into a curse.

Thus, we have two possible evolutionary explanations for salt retention by New World blacks. One involves slow selection by conditions operating in Africa for millennia; the other, rapid recent selection by slave conditions within the past few centuries. The result in either case would make New World blacks more susceptible than whites to hypertension, but the second explanation would, in addition, make them more susceptible than African blacks. At present, we don't know the relative importance of these two explanations. Grim and Wilson's provocative hypothesis is likely to stimulate medical and physiological comparisons of American blacks with African blacks and thereby to help resolve the question.

While this piece has focused on one medical problem in one human population, it has several larger morals. One, of course, is that our differing genetic heritages predispose us to different diseases, depending on the part of the world where our ancestors lived. Another is that our genetic differences reflect not only ancient conditions in different parts of the world but also recent episodes of migration and mortality. A well-established example is the decrease in the frequency of the sickle-

cell hemoglobin gene in U.S. blacks compared with African blacks, because selection for resistance to malaria is now unimportant in the United States. The example of black hypertension that Grim and Wilson discuss opens the door to considering other possible selective effects of the slave experience. They note that occasional periods of starvation might have selected slaves for superefficient sugar metabolism, leading under modern conditions to a propensity for diabetes.

Finally, consider a still more universal moral. Almost all people alive today exist under very different conditions from those under which every human lived 10,000 years ago. It's remarkable that our old genetic heritage now per-

mits us to survive at all under such different circumstances. But our heritage still catches up with most of us, who will die of life style related diseases such as cancer, heart attack, stroke, and diabetes. The risk factors for these diseases are the strange new conditions prevailing in modern Western society. One of the hardest challenges for modern medicine will be to identify for us which among all those strange new features of diet, life style, and environment are the ones getting us into trouble. For each of us, the answers will depend on our particular genes, hence on our ancestry. Only with such individually tailored advice can we hope to reap the benefits of modern living while still housed in bodies designed for life before saltshakers.

A Gene for Nothing

One to make you happy, one to make you sad, one to make you anxious,
one to make you mad—is that really the way your genes work?

By Robert Sapolsky

ROBERT SAPOLSKY is a professor of neuroscience at Stanford. He is the author of The Trouble With Testosterone and Other Essays on the Biology of the Human Predicament.

WELL, THESE LAST SIX MONTHS have been an exciting time for the sheep named Dolly, ever since it was revealed that she was the first mammal cloned from adult cells. There was the night she spent in the Lincoln bedroom and the photo op with Al Gore; the triumphant ticker-tape parade down Broadway, the billboard ads for Guess Genes. Throughout the media circus, Dolly has been poised, patient, cordial, and even-tempered—the epitome of what we look for in a celebrity and role model. But despite her charm, people keep saying mean things about Dolly. Heads of state, religious leaders, and editorialists fall over themselves in calling her an aberration of nature and an insult to the sacred biological wonder of reproduction. They thunder about the anathema of even considering applying to humans the technology that spawned her.

What's everyone so upset about? Why is cloning so disturbing? Clearly, it's not the potential for droves of clones running around with the exact same renal filtration rate that has everyone up in arms. It's probably not even the threat of winding up with a bunch of clones who look identical, creepy though that would be. No, the real horror is the prospect of having multiple copies of a sin-

gle brain, with the same neurons and the same genes directing those neurons, one multibodied consciousness among the clones, an army of photocopies of the same soul, all thinking, feeling, and acting identically.

Fortunately, that can't happen, as people have known ever since scientists discovered identical twins. Such individuals constitute genetic clones, just like Dolly and her "mother"—the sheep from which the original cell was taken. Despite all those breathless stories about identical twins separated at birth who flush the toilet before using it, twins are not melded in mind, do not behave identically. For example, if an identical twin is schizophrenic, the sibling, with the identical "schizophrenia gene(s)," has only about a 50 percent chance of having the disease.

A similar finding comes from a fascinating experiment by Dan Weinberger of the National Institute of Mental Health. Give identical twins a puzzle to solve and they might come up with closer answers than one would expect from a pair of strangers. While they're working on the puzzle, however, hook the twins up to a PET scanner, a brain-imaging instrument that visualizes metabolic demands in different regions of the brain. You'll find the pattern of activation in the pair differing considerably, despite the similarity of their solutions. Or use an MRI to get some detailed pictures of the brains of identical twins and start measuring stuff ob-

sessively—the length of this part, the width of that, the volume of another region, and the surface area of the cortex—and those identical twins with their identical genes never have identical brains. Every measure differs.

The careful editorialists have made this point. Nonetheless, that business about identical genes producing identical brains tugs at a lot of people. Gene-behavior stories are constantly getting propelled to the front pages of newspapers. One popped up shortly before Dolly, when a team of researchers reported that a single gene, called fru, determines the sexual behavior of male fruit flies. Courtship, opening lines, foreplay, who they come on to—the works. Mutate that gene and—get this—you can even change the sexual orientation of the fly. What made the story front-page news, of course, wasn't our insatiable fly voyeurism. "Could our sexual behaviors be determined by a single gene as well?" every article asked. And a bit earlier, there was the hubbub about the isolation of a gene related to anxiety in humans, and shortly before that, a gene related to novelty-seeking behavior, and a while before that, a gene whose mutation in one family was associated with violent antisocial behavior, and before that . . .

Why do these stories command attention? For many, genes and the DNA they contain represent the holy grail of biology, the code of codes (two phrases often used in lay discussions of genetics). The worship at the altar of the gene rests on two assumptions. The first concerns the autonomy of genetic regulation: it is the notion that biological information begins with genes—DNA is the commander, the epicenter from which biology emanates. Nobody tells a gene what to do; it's always the other way around. The second assumption is that when genes give a command, biological systems listen. Genes, the story goes, instruct your cells as to their structure and function. And when those cells are neurons, their functions include thought, feelings, and behavior. Thus, the gene worshipers believe, we are finally identifying the biological factors that make us do what we do.

A typical example of the code-of-codes view recently appeared in a lead *New Yorker* article by Louis Menand, an English professor at the City University of New York. Menand ruminates on anxiety genes, when "one little gene is firing off a signal to bite your fingernails" (there's that first assumption—autonomous genes firing off whenever some notion pops into their heads). He asks himself how we can reconcile societal, economic, and psychological explanations of behavior with those ironclad genes. "The view that behavior is determined by an inherited genetic package" (there's the second assumption—genes as irresistible commanders) "is not easily reconciled with the view that behavior is determined by the kinds of movies a person watches." And what is the solution? "It is like having the Greek gods and the Inca gods occupying the same pantheon. Somebody's got to go."

In other words, if you buy into the notion of genes firing off and determining our behaviors, such modern scientific findings are simply incompatible with the environment having an influence. Something's gotta go.

Now, I'm not sure what sort of genetics they teach in Menand's English department, but the something's-gotta-go loggerhead is what most behavioral biologists have been trying to unteach for decades, apparently with limited success. Which is why it's worth another try.

Okay. You've got nature—neurons, brain chemicals, hormones, and of course, at the bottom of the cereal box,

genes. And then there's nurture, all those environmental breezes gusting about. Again and again, behavioral biologists insist that you can't talk meaningfully about nature or nurture, only about their interaction. But somehow people can't seem to keep that thought in their heads. Instead, whenever a new gene is trotted out that "determines" a behavior by "firing off," they see environmental influences as the irrelevant something that has to go. Soon poor, sweet Dolly is a menace to our autonomy as individuals, and genes are understood to control who you go to bed with and whether you feel anxious about it.

A gene, a stretch of DNA, does not produce a behavior, an emotion, or even a fleeting thought. It produces a protein.

Let's try to undo the notion of genes as neurobiological and behavioral destiny by examining those two assumptions, beginning with the second one—that cells, including those in our heads, obey genetic commands. What exactly do genes do? A gene, a stretch of DNA, does not produce a behavior. A gene does not produce an emotion, or even a fleeting thought. It produces a protein. Each gene is a specific DNA sequence that codes for a specific protein. Some of these proteins certainly have lots to do with behavior and feelings and thoughts; proteins include some hormones (which carry messages between cells) and neurotransmitters (which carry messages between nerve cells); they also include receptors that receive hormonal and neurotransmitter messages, the enzymes that synthesize and degrade those messages, many of the intracellular messengers triggered by those hormones, and so on. All those proteins are vital for a brain to do its business. But only very rarely do things like hormones and neurotransmitters

cause a behavior to happen. Instead they produce tendencies to respond to the environment in certain ways.

To illustrate this critical point, let's consider anxiety. When an organism is confronted with a threat, it typically becomes vigilant, searches for information about the nature of the threat, and struggles to find an effective coping response. Once it receives a signal indicating safety—the lion has been evaded, the traffic cop buys the explanation and doesn't issue a ticket—the organism can relax. But that's not what happens with an anxious individual. Instead this person will skitter frantically among coping responses, abruptly shifting from one to another without checking whether anything has worked. He may have a hard time detecting the safety signal and knowing when to stop his restless vigilance. Moreover, the world presents a lot of triggers that not everyone reacts to. For the anxious individual, the threshold is lower, so that the mere sight of a police car in the rearview mirror can provoke the same storm of uneasiness as actually being stopped. By definition, anxiety makes little sense outside the context of what the environment is doing to an individual. In that framework, the brain chemicals and genes relevant to anxiety don't make you anxious. They make you more responsive to anxiety-provoking situations, make it harder to detect safety signals.

The same theme continues in other behaviors as well. The exciting (made-of-protein) receptor that apparently has to do with novelty-seeking behavior doesn't actually make you seek novelty. It makes you more pleasurably excited than folks without that receptor variant get when you happen to encounter a novel environment. And those (genetically influenced) neurochemical abnormalities of depression don't make you depressed. They make you more vulnerable to stressors over a one-year period instead of in the environment, to deciding that you are helpless even when you're not.

One might retort that in the long run we are all exposed to anxiety-provoking circumstances, all exposed to the depressing world around us. If we are all exposed to those same environmental factors but only the people who are genetically prone to depression get depressed, that is a pretty powerful vote for genes. In that scenario, the "genes don't cause things, they just make you more sensitive to the environment" argument becomes empty and semantic.

The problems here, however, are twofold. First, a substantial minority of people with a genetic legacy of depression do not get depressed, and not everyone who has a major depression has a genetic legacy for it. Genetic status is not all that predictive by itself. Second, we share the same environments only on a very superficial level. For example, the incidence of depression (and its probably biological underpinnings) seem to be roughly equal throughout the world. However, geriatric depression is epidemic in our society and far less prevalent in traditional societies in the developing world. Why? Different societies produce remarkably different social environments, in which old age can mean being a powerful village elder or an infantilized has-been put out to a shuffleboard pasture.

The environmental differences can be more subtle. Periods of psychological stress involving loss of control and predictability during childhood may well predispose one toward adult depression. Two children may have had similar childhood lessons in "there's bad things out there that I can't control"—both may have seen their parents divorce, lost a grandparent, tearfully buried a pet in the backyard, faced the endless menacing of a bully. Yet the temporal pattern of their experience is unlikely to be identical, and the child who experiences all those stressors over a one-year period instead of over six years is far more likely to come with the cognitive distortion, "There's bad things out there that I can't control and, in fact, I can't control anything," that sets you up for depression. The biological factors that genes code for in the nervous system typically don't determine behavior. Instead they affect how you respond to often very subtle influences in the environment. There are genetic vulnerabilities, tendencies, predispositions—but rarely genetic inevitabilities.

Now let's go back to that first assumption about behavioral genetics—that genes always have minds of their own. It takes just two startling facts about the structure of genes to blow this one out of the water.

A chromosome is made of DNA, a vastly long string of it, a long sequence of letters coding for genetic information. People used to think that Gene 1 would comprise the first eleventy letters of the DNA message. A special letter sequence would signal the end of that gene, the next eleventy and a half letters would code for Gene 2, and so on, through tens of thousands of genes. Gene 1 might specify the construction of insulin in your pancreas; Gene 2 might specify protein pigments that give eyes their color; and Gene 3, active in neurons, might make you aggressive. Ah, caught you: might make you more sensitive to aggression-provoking stimuli in the environment. Different people have different versions of Genes 1, 2, and 3, some of which work better than others. An army of biochemicals do the scut work, transcribing the genes, reading the DNA sequences, and following the instructions they contain for constructing the appropriate proteins.

As it turns out, that's not really how things work. Instead of one gene coming immediately after another, with the entire string of DNA devoted to coding for different proteins, there are long stretches of DNA that don't get transcribed. Sometimes those stretches even split up a gene into subsections. Some of the nontranscribed, noncoding DNA doesn't seem to do anything. It may have some function that we don't yet understand, or it may have none at all. But some of the noncoding DNA does something very interesting indeed. It's the instruction manual for how and when to activate genes. These stretches have many names—regulatory elements, promoters, responsive elements. Various biochemical messengers may bind to each of them, altering the activity of the gene immediately "downstream"—immediately following it in the string of DNA.

Far from being autonomous sources of information, then, genes must obey other factors that regulate when and how they function. Very often, those factors

are environmental. For example, suppose something stressful happens to a primate. A drought, say, forces it to forage miles each day for food. As a result, the animal secretes a stress hormone, cortisol, from its adrenals. Cortisol molecules enter fat cells and bind to cortisol receptors. These hormone-receptor complexes find their way to the DNA and bind to a regulatory stretch of DNA. Whereupon a gene downstream is activated, which produces a protein, which indirectly inhibits that fat cell from storing fat. It's a logical thing to do—while starving and walking the grasslands in search of a meal, the primate needs fat to fuel muscles, not to laze around in fat cells.

In effect, regulatory elements introduce the possibility of environmentally modulated if-then clauses. If the environment is tough and you're working hard to find food, then make use of your genes to divert energy to exercising muscles. The environment, of course, doesn't mean just the weather. The biology is essentially the same if a human refugee travels miles from home with insufficient food because of civil strife. The behavior of one human can change the pattern of gene activity in another.

Let's look at a fancier example of how environmental factors control the regulatory elements of DNA. Suppose that Gene 4037 (not its real name—it has one, but I'll spare you the jargon), when left to its own devices, is transcriptionally active, generating its protein. However, as long as a particular messenger binds to a regulatory element that comes just before 4037 in the DNA string, Gene 4037 shuts down. Fine. Now suppose that inhibitory messenger happens to be very sensitive to temperature. In fact, if the cell gets hot, the messenger goes to pieces and comes floating off the regulatory element. Freed from the inhibitory regulation, Gene 4037 will suddenly become active. Maybe it's a gene that works in the kidney and codes for a protein relevant to water retention. Boring—another metabolic story, this one having to do with how a warm environment triggers metabolic adaptations that stave off dehydration. But suppose, instead, Gene 4037 codes for an array of proteins that have something to do with sexual behavior. What have you just invented? Seasonal mating. Winter is waning, each day gets a little warmer, and in relevant cells in the brain, pituitary, or gonads, genes like 4037 are gradually becoming active. Finally some threshold is passed and, *wham,* everyone starts rutting and ovulating, snorting and pawing at the ground, and generally carrying on. (Actually, in most seasonal matters, the environmental signal for mating is the amount of daily light exposure, or the days are getting longer, rather than temperature, or the days are getting warmer. But the principle is the same.)

Here's a final, elegant example. Every cell in your body has a distinctive protein signature that marks it as yours. These "major histocompatability" proteins allow your immune system to tell the difference between you and some invading bacterium—that's why your body will reject a transplanted organ with a very different signature. When those signature proteins get into a mouse's urine, they help make its odor distinct. For a rodent, that's important stuff. Design receptors in olfactory cells in a rodent's nose that can distinguish signature odorant proteins similar to its own from totally novel ones. The greater the similarity, the tighter the protein will fit into the receptor. What have you just invented? A way to distinguish between the smells of relatives and strangers—something rodents do effortlessly.

Keep tinkering with this science project. Now couple those olfactory receptors to a cascade of chemical messengers inside the cell, one messenger triggering the next until you get to the DNA's regulatory elements. What might you want to construct? How about: *If* an olfactory receptor binds an odorant indicating the presence of a relative, *then* trigger a cascade that ultimately inhibits the activity of genes related to reproduction. You've just invented a mechanism by which animals could avoid mating with close relatives. Or you can construct a different cascade: if an olfactory receptor binds an odorant indicating a relative, then inhibit genes that are normally active and that regulate the synthesis of testosterone. There you have a means by which rodents get bristly and

aggressive when a strange male stinks up their burrow but not when the scent belongs to their kid brother.

IN EACH OF THESE EXAMPLES YOU CAN BEgin to see the logic, an elegance that teams of engineers couldn't do much to improve. And now for the two facts about this regulation of genes that will dramatically change your view of them. First, when it comes to mammals, by the best estimates available, more than 95 percent of DNA is noncoding. Ninety-five percent. Sure, a lot of it may have no function, but your average gene comes with a huge instruction manual for how to operate it, and the operator is very often environmental. With a percentage like that, if you think about genes and behavior, you have to think about how the environment regulates genes and behavior.

The second fact involves genetic variation between individuals. A gene's DNA sequence often varies from person to person, which often translates into proteins that differ in how well they do their job. This is the grist for natural selection: Which is the most adaptive version of some (genetically influenced) trait? Given that evolutionary change occurs at the level of DNA, "survival of the fittest" really means "reproduction of individuals whose DNA sequences make for the most adaptive collection of proteins." But—here's that startling second fact—when you examine variability in DNA sequences among individuals, the noncoding regions of DNA are considerably more variable than are the regions that code for genes. Okay, a lot of that variability is attributable to DNA that doesn't do much and so is free to drift genetically over time without consequence. But there seems to be a considerable amount of variability in regulatory regions as well.

What does this mean? By now, I hope, we've gotten past "genes determine behavior" to "genes modulate how one responds to the environment." The business about 95 percent of DNA being noncoding should send us even further, to "genes can be convenient tools used by environmental factors to influence behavior." And that second fact about

variability in noncoding regions means that it's less accurate to think "evolution is about natural selection for different assemblages of genes" than it is to think "evolution is about natural selection for different sensitivities and responses to environmental influences."

Sure, some behaviors are overwhelmingly under genetic control. Just consider all those mutant flies hopping into the sack with insects their parents disapprove of. And some mammalian behaviors, even human ones, are probably pretty heavily under genetic regulation as well. These are likely to code for behaviors that must be performed by everyone in much the same way for genes to be passed on. For example, all male primates have to go about the genetically based behavior of pelvic thrusting in fairly similar ways if they plan to reproduce successfully. But by the time you get to courtship, or emotions, or creativity, or mental illness, or any complex aspect of our lives, the intertwining of biological and environmental components utterly defeats any attempt to place them into separate categories, let alone to then decide that one of them has got to go.

I'm a bit hesitant to reveal the most telling example of how individuals with identical genes can nonetheless come up with very different behaviors, as I have it thirdhand through the science grapevine, and I'll probably get some of the details wrong. But what the hell, it's such an interesting finding. It concerns the very extensive opinion poll that was carried out among sheep throughout the British Isles. Apparently, the researchers managed to get data from both Dolly and her gene-donor mother. So get a load of this bombshell: Dolly's mother voted Tory, listed the Queen Mum as her favorite royal, worried about mad cow disease ("Is it good or bad for the sheep?"), enjoyed Gilbert and Sullivan, and endorsed the statement, "Behavior? It's all nature." And Dolly? Votes Green Party, thinks Harry and William are the cutest, worries about "the environment," listens to the Spice Girls, and endorsed the statement, "Behavior? Nature. Or nurture. Whatever." You see, there's more to behavior than just genes.

Unit 2

Unit Selections

Key Points to Consider

❖ What is the role of deception among primates, and how might it have led to greater intelligence?

❖ Why is friendship important to olive baboons? What implications does this have for the origins of pair-bonding in hominid evolution?

❖ How is it possible to objectively study and assess emotional and mental states of nonhuman primates?

❖ What are the implications for human evolution of tool use, social hunting, and food sharing among Ivory Coast chimpanzees?

❖ Why is the mountain gorilla in danger of extinction?

❖ To what extent do apes have language skills?

❖ Are we in anthropodenial? Explain your answer.

 **Links** **www.dushkin.com/online/**

These sites are annotated on pages 4 and 5.

Primates are fun. They are active, intelligent, colorful, emotionally expressive, and unpredictable. In other words, observing them is like holding up an opaque mirror to ourselves. The image may not be crystal-clear or, indeed, what some would consider flattering, but it is certainly familiar enough to be illuminating.

Primates are, of course, but one of many orders of mammals that adaptively radiated into the variety of ecological niches vacated at the end of the Age of Reptiles about 65 million years ago. Whereas some mammals took to the sea (cetaceans), and some took to the air (chiroptera, or bats), primates are characterized by an arboreal or forested adaptation. Whereas some mammals can be identified by their food-getting habits, such as the meat-eating carnivores, primates have a penchant for eating almost anything and are best described as omnivorous. In taking to the trees, primates did not simply develop a full-blown set of distinguishing characteristics that set them off easily from other orders of mammals, the way the rodent order can be readily identified by its gnawing set of front teeth. Rather, each primate seems to represent degrees of anatomical, biological, and behavioral characteristics on a continuum of progress with respect to the particular traits we humans happen to be interested in.

None of this is meant to imply, of course, that the living primates are our ancestors. Since the prosimians, monkeys, and apes are our contemporaries, they are no more our ancestors than we are theirs, and, as living end-products of evolution, we have all descended from a common stock in the distant past. So, if we are interested primarily in our own evolutionary past, why study primates at all? Because, by the criteria we have set up as significant milestones in the evolution of humanity, an inherent reflection of our own bias, primates have not evolved as far as we have. They and their environments, therefore, may represent glimmerings of the evolutionary stages and ecological circumstances through which our own ancestors may have gone. What we stand to gain, for instance, is an educated guess as to how our own ancestors might have appeared and behaved as semierect creatures before becoming bipedal. Aside from being a pleasure to observe, then, living primates can teach us something about our past.

Another reason for studying primates is that they allow us to test certain notions too often taken for granted. For instance, Barbara Smuts, in "What Are Friends For?" reveals that friendship bonds, as illustrated by the olive baboons of East Africa, have little if anything to do with a sexual division of labor or even sexual exclusivity between a pair-bonded male and female. Smuts challenges the traditional male-oriented idea that primate societies are dominated solely by males for males.

This unit demonstrates that relationships between the sexes are subject to wide variation, that the kinds of answers obtained depend upon the kinds of questions asked, and that we have to be very careful in making inferences about human beings from any one particular primate study. We may, if we are not careful, draw conclusions that say more about our own skewed perspectives than about that which we claim to understand. Still another benefit of primate field research is that it provides us with perspectives that the bones and stones of the fossil hunters will never reveal: a sense of the richness and variety of social patterns that must have existed in the primate order for many tens of millions of years. (See James Shreeve's report "Machiavellian Monkeys," Sy Montgomery's essay, "Dian Fossey and Digit," and Jane Goodall's "The Mind of the Chimpanzee.")

Even if we had the physical remains of the earliest hominids in front of us, which we do not have, there is no way such evidence could thoroughly answer the questions that physical anthropologists care most deeply about: How did these creatures move about and get their food? Did they cooperate and share? On what levels did they think and communicate? Did they have a sense of family, let alone a sense of self? In one way or another, all of the previously mentioned articles on primates relate to these issues, as do some of the subsequent ones on the fossil evidence. But what sets off this unit from the others is how some of the authors attempt to deal with these matters head-on, even in the absence of direct fossil evidence. Christophe Boesch and Hedwige Boesch-Achermann, in "Dim Forest, Bright Chimps," indicate that some aspects of "hominization" (the acquisition of such humanlike qualities as cooperative hunting and food sharing) actually may have begun in the African rain forest rather than in the dry savanna, as has usually been proposed. They base their suggestions on some remarkable first-hand observations of forest-dwelling chimpanzees.

As if to show that chimpanzee behavior may vary according to local circumstances, just as we know human behavior does, Craig Stanford, in "To Catch a Colobus," contrasts his observations of chimpanzee hunting in Gombe National Park with the findings of the Boesches.

Recent research, such as that by E. S. Savage-Rumbaugh ("Language Training of Apes") has shown some striking resemblances between apes and humans, hinting that such qualities might have been characteristic of our common ancestor. Following this line of reasoning, Frans de Waal ("Are We in Anthropodenial?") argues that we can make educated guesses as to the mental and physical processes of our hominid predecessors.

Taken collectively, the articles in this section show how far anthropologists are willing to go to construct theoretical formulations based upon limited data. Although making so much out of so little may be seen as a fault and may generate irreconcilable differences among theorists, a readiness to entertain new ideas should be welcomed for what it is—a stimulus for more intensive and meticulous research.

Machiavellian Monkeys

The sneaky skills of our primate cousins suggest that we may owe our great intelligence to an inherited need to deceive.

James Shreeve

This is a story about frauds, cheats, liars, faithless lovers, incorrigible con artists, and downright thieves. You're gonna love 'em.

Let's start with a young rascal named Paul. You'll remember his type from your days back in the playground. You're minding your own business, playing on the new swing set, when along comes Paul, such a little runt that you hardly notice him sidle up to you. All of a sudden he lets out a scream like you've run him through with a white-hot barbed harpoon or something. Of course the teacher comes running, and the next thing you know you're being whisked inside with an angry finger shaking in your face. That's the end of recess for you. But look out the window: there's Paul, having a great time on *your* swing. Cute kid.

Okay, you're a little older now and a little smarter. You've got a bag of chips stashed away in your closet, where for once your older brother won't be able to find them. You're about to open the closet door when he pokes his head in the room. Quickly you pretend to be fetching your high tops; he gives you a look but he leaves. You wait a couple of minutes, lacing up the sneakers in case he walks back in, then you dive for the chips. Before you can get the bag open, he's over your shoulder, snatching it out of your hands. "Nice try, punk," he says through a mouthful, "but I was hiding outside your room the whole time."

This sort of trickery is such a common part of human interaction that we hardly notice how much time we spend defending ourselves against it or perpetrating it ourselves. What's so special about the fakes and cheaters here, however, is that they're not human. Paul is a young baboon, and your big brother is, well, a chimpanzee. With some admittedly deceptive alterations of scenery and props, the situations have been lifted from a recent issue of *Primate Report*. The journal is the work of Richard Byrne and Andrew Whiten, two psychologists at the University of St. Andrews in Scotland, and it is devoted to cataloging the petty betrayals of monkeys and apes as witnessed by primatologists around the world. It is a testament to the evolutionary importance of what Byrne and Whiten call Machiavellian intelligence—a facility named for the famed sixteenth-century author of *The Prince,* the ultimate how-to guide to prevailing in a complex society through the judicious application of cleverness, deceit, and political acumen.

Deception is rife in the natural world. Stick bugs mimic sticks. Harmless snakes resemble deadly poisonous ones. When threatened, blowfish puff themselves up and cats arch their backs and bristle their hair to seem bigger than they really are. All these animals could be said to practice deception because they fool other animals—usually members of other species—into thinking they are something that they patently are not. Even so, it would be overreading the situation to attribute Machiavellian cunning to a blowfish, or to accuse a stick bug of being a lying scoundrel. Their deceptions, whether in their looks or in their actions, are programmed genetic responses. Biology leaves them no choice but to dissemble: they are just being true to themselves.

The kind of deception that interests Byrne and Whiten—what they call tactical deception—is a different kettle of blowfish altogether. Here an animal has the mental flexibility to take an "honest" behavior and use it in such a way that another animal—usually a member of the deceiver's own social group—is misled, thinking that a normal, familiar state of affairs is under way, while, in fact, something quite different is happening.

Take Paul, for example. The real Paul is a young chacma baboon that caught Whiten's attention in 1983, while he and Byrne were studying foraging among the chacma in the Drakensberg Mountains of southern Africa. Whiten saw a member of Paul's group, an adult female named Mel, digging in the ground, trying to extract a nutritious plant bulb. Paul approached and looked around. There were no other baboons within sight. Suddenly he let out a yell, and within seconds his mother came running, chasing the startled Mel over a small cliff. Paul then took the bulb for himself.

In this case the deceived party was Paul's mother, who was misled by his scream into believing that Paul was being attacked, when actually no such attack was taking place. As a result of her apparent misinterpretation Paul was left alone to eat the bulb that Mel had carefully extracted—a morsel, by the way, that he would not have had the strength to dig out on his own.

If Paul's ruse had been an isolated case, Whiten might have gone on with his foraging studies and never given it a second thought. But when he compared his field notes with Byrne's, he noticed that both their notebooks were sprinkled with similar incidents and had been so all summer long. After they returned home to Scotland, they boasted about their "dead smart" baboons to their colleagues in pubs after conferences, expecting them to be suitably impressed. Instead the other researchers countered with tales about their own shrewd vervets or Machiavellian macaques.

"That's when we realized that a whole phenomenon might be slipping through a sieve," says Whiten. Researchers had assumed that this sort of complex trickery was a product of the sophisticated human brain. After all, deceitful behavior seemed unique to humans, and the human brain is unusually large, even for primates—"three times as big as you would expect for a primate of our size," notes Whiten, if you're plotting brain size against body weight.

But if primates other than humans deceived one another on a regular basis, the two psychologists reasoned, then it raised the extremely provocative possibility that the primate brain, and ultimately the human brain, is an instrument crafted for social manipulation. Humans evolved from the same evolutionary stock as apes, and if tactical deception was an important part of the lives of our evolutionary ancestors, then the sneakiness and subterfuge that human beings are so manifestly capable of might not be simply a result of our great intelligence and oversize brain, but a driving force behind their development.

To Byrne and Whiten these were ideas worth pursuing. They fit in with a

theory put forth some years earlier by English psychologist Nicholas Humphrey. In 1976 Humphrey had eloquently suggested that the evolution of primate intelligence might have been spurred not by the challenges of environment, as was generally thought, but rather by the complex cognitive demands of living with one's own companions. Since then a number of primatologists had begun to flesh out his theory with field observations of politically astute monkeys and apes.

Suddenly Paul let out a yell, and his mother came running, chasing Mel over a small cliff.

Deception, however, had rarely been reported. And no wonder: If chimps, baboons, and higher primates generally are skilled deceivers, how could one ever know it? The best deceptions would by their very nature go undetected by the other members of the primate group, not to mention by a human stranger. Even those ruses that an observer could see through would have to be rare, for if used too often, they would lose their effectiveness. If Paul always cried wolf, for example, his mother would soon learn to ignore his ersatz distress. So while the monkey stories swapped over beers certainly suggested that deception was widespread among higher primates, it seemed unlikely that one or even a few researchers could observe enough instances of it to scientifically quantify how much, by whom, when, and to what effect.

Byrne and Whiten's solution was to extend their pub-derived data base with a more formal survey. In 1985 they sent a questionnaire to more than 100 primatologists working both in the field and in labs, asking them to report back any incidents in which they felt their subjects had perpetrated deception on one another. The questionnaire netted a promising assortment of deceptive tactics used by a variety of monkeys and all the great apes. Only the relatively small-brained and socially simple lemur

family, which includes bush babies and lorises, failed to elicit a single instance. This supported the notion that society, sneakiness, brain size, and intelligence are intimately bound up with one another. The sneakier the primate, it seemed, the bigger the brain.

Byrne and Whiten drew up a second, much more comprehensive questionnaire in 1989 and sent it to hundreds more primatologists and animal behaviorists, greatly increasing the data base. Once again, when the results were tallied, only the lemur family failed to register a single case of deception.

All the other species, however, represented a simian rogues' gallery of liars and frauds. Often deception was used to distract another animal's attention. In one cartoonish example, a young baboon, chased by some angry elders, suddenly stopped, stood on his hind legs, and stared at a spot on the horizon, as if he noticed the presence of a predator or a foreign troop of baboons. His pursuers braked to a halt and looked in the same direction, giving up the chase. Powerful field binoculars revealed that no predator or baboon troop was anywhere in sight.

Sometimes the deception was simply a matter of one animal hiding a choice bit of food from the awareness of those strong enough to take it away. One of Jane Goodall's chimps, for example, named Figan, was once given some bananas after the more dominant members of the troop had wandered off. In the excitement, he uttered some loud "food barks"; the others quickly returned and took the bananas away. The next day Figan again waited behind the others and got some bananas. This time, however, he kept silent, even though the human observers, Goodall reported, "could hear faint choking sounds in his throat."

Concealment was a common ruse in sexual situations as well. Male monkeys and chimpanzees in groups have fairly strict hierarchies that control their access to females. Animals at the top of the order intimidate those lower down, forcing them away from females. Yet one researcher reported seeing a male stump-tailed macaque of a middle rank leading a female out of sight of the more dominant males and then mating with her silently, his climax unaccompanied by the harsh, low-pitched grunts that the male stump-tailed normally makes. At one point during the tryst the female turned and stared into his face, then covered his mouth with her hand. In another case a subordinate chimpanzee, aroused by the presence of a female in estrus, covered his erect penis with his hand when a dominant male approached, thus avoiding a likely attack.

In one particularly provocative instance a female hamadryas baboon slowly shuffled toward a large rock, appearing to forage, all the time keeping an eye on the most dominant male in the group. After 20 minutes she ended up with her head and shoulders visible to the big, watchful male, but with her hands happily engaged in the elicit activity of grooming a favorite subordinate male, who was hidden from view behind the rock.

Baboons proved singularly adept at a form of deception that Byrne and Whiten call "using a social tool." Paul's scam is a perfect example: he fools his mother into acting as a lever to pry the plant bulb away from the adult female, Mel. But can it be said unequivocally that he intended to deceive her? Perhaps Paul had simply learned through trial and error that letting out a yell brought his mother running and left him with food, in which case there is no reason to endow his young baboon intellect with Machiavellian intent. How do we know that Mel didn't actually threaten Paul in some way that Byrne and Whiten, watching, could not comprehend? While we're at it, how do we know that any of the primate deceptions reported here were really deliberate, conscious acts?

"It has to be said that there is a whole school of psychology that would deny such behavior even to humans," says Byrne. The school in question—strict behaviorism—would seek an explanation for the baboons' behavior not by trying to crawl inside their head but by carefully analyzing observable behaviors and the stimuli that might be triggering them. Byrne and Whiten's strategy against such skepticism was to be hyperskeptical themselves. They accepted that trial-and-error learning or simple conditioning, in which an ani-

mal's actions are reinforced by a reward, might account for a majority of the incidents reported to them—even when they believed that tactical deception was really taking place. But when explaining things "simply" led to a maze of extraordinary coincidences and tortuous logic, the evidence for deliberate deception seemed hard to dismiss.

Society, sneakiness, brain size, and intelligence are intimately bound up with one another.

Paul, for instance, *might* have simply learned that screaming elicits the reward of food, via his mother's intervention. But Byrne witnessed him using the same tactic several times, and in each case his mother was out of sight, able to hear his yell but not able to see what was really going on. If Paul was simply conditioned to scream, why would he do so only when his mother could not see who was—or was not—attacking her son?

Still, it is possible that she was not intentionally deceived. But in at least one other, similar case there is virtually no doubt that the mother was responding to a bogus attack, because the alleged attacker was quite able to verbalize his innocence. A five-year-old male chimp named Katabi, in the process of weaning, had discovered that the best way to get his reluctant mother to suckle him was to convince her he needed reassurance. One day Katabi approached a human observer—Japanese primatologist Toshisada Nishida—and began to screech, circling around the researcher and waving an accusing hand at him. The chimp's mother and her escort immediately glared at Nishida, their hair erect. Only by slowly backing away from the screaming youngster did Nishida avoid a possible attack from the two adult chimps.

"In fact I did nothing to him," Nishida protested. It follows that the adults were indeed misled by Katabi's

hysterics—unless there was some threat in Nishida unknown even to himself.

"If you try hard enough," says Byrne, "you can explain every single case without endowing the animal with the ability to deceive. But if you look at the whole body of work, there comes a point where you have to strive officiously to deny it."

The cases most resistant to such officious denials are the rarest—and the most compelling. In these interactions the primate involved not only employed tactical deception but clearly understood the concept. Such comprehension would depend upon one animal's ability to "read the mind" of another: to attribute desires, intentions, or even beliefs to the other creature that do not necessarily correspond to its own view of the world. Such mind reading was clearly evident in only 16 out of 253 cases in the 1989 survey, all of them involving great apes.

For example, consider Figan again, the young chimp who suppressed his food barks in order to keep the bananas for himself. In his case, mind reading is not evident: he might simply have learned from experience that food barks in certain contexts result in a loss of food, and thus he might not understand the nature of his own ruse, even if the other chimps are in fact deceived.

But contrast Figan with come chimps observed by Dutch primatologist Frans Plooij. One of these chimps was alone in a feeding area when a metal box containing food was opened electronically. At the same moment another chimp happened to approach. (Sound familiar? It's your older brother again.) The first chimp quickly closed the metal box (that's you hiding your chips), walked away, and sat down, looking around as if nothing had happened. The second chimp departed, but after going some distance away he hid behind a tree and peeked back at the first chimp. When the first chimp thought the coast was clear, he opened the box. The second chimp ran out, pushed the other aside, and ate the bananas.

Chimp One might be a clever rogue, but Chimp Two, who counters his deception with a ruse of his own, is the true mind reader. The success of his ploy is based on his insight that Chimp One was trying to deceive *him* and on his ability to adjust his behavior accordingly. He has in fact performed a prodigious cognitive leap—proving himself capable of projecting himself into another's mental space, and becoming what Humphrey would call a natural psychologist.

Niccolò Machiavelli might have called him good raw material. It is certainly suggestive that only the great apes—our closest relatives—seem capable of deceits based on such mind reading, and chimpanzees most of all. This does not necessarily mean that chimps are inherently more intelligent: the difference may be a matter of social organization. Orangutans live most of their lives alone, and thus they would not have much reason to develop such a complex social skill. And gorillas live in close family groups, whose members would be more familiar, harder to fool, and more likely to punish an attempted swindle. Chimpanzees, on the other hand, spend their lives in a shifting swirl of friends and relations, where small groups constantly form and break apart and reform with new members.

"What an opportunity for lying and cheating!" muses Byrne. Many anthropologists now believe that the social life of early hominids—our first non-ape ancestors—was much like that of chimps today, with similar opportunities to hone their cognitive skills on one another. Byrne and Whiten stop just short of saying that mind reading is the key to understanding the growth of human intelligence. But it would be disingenuous to ignore the possibility. If you were an early hominid who could comprehend the subjective impressions of others and manipulate them to your own ends, you might well have a competitive advantage over those less psychosocially nimble, perhaps enjoying slightly easier access to food and to the mating opportunities that would ensure your genetic survival.

Consider too how much more important your social wits would be in a world where the targets of your deceptions were constantly trying to outsmart *you*. After millennia of intrigue and counterintrigue, a hominid species might well evolve a brain three times bigger than it "should" be—and capable of far more than deceiving other hominids. "The ability to attribute other intentions to other people could have been an enormous building block for many human achievements, including language," says Whiten. "That this leap seems to have been taken by chimps and possibly the other great apes puts that development in human mentality quite early."

So did our intellect rise to its present height on a tide of manipulation and deceit? Some psychologists, even those who support the notion that the evolution of intelligence was socially driven, think that Byrne and Whiten's choice of the loaded adjective *Machiavellian* might be unnecessarily harsh.

"In my opinion," says Humphrey, "the word gives too much weight to the hostile use of intelligence. One of the functions of intellect in higher primates and humans is to keep the social unit together and make it able to successfully exploit the environment. A lot of intelligence could better be seen as driven by the need for cooperation and compassion." To that, Byrne and Whiten only point out that cooperation is itself an excellent Machiavellian strategy—sometimes.

The Scottish researchers are not, of course, the first to have noticed this. "It is good to appear clement, trustworthy, humane, religious, and honest, and also to be so," Machiavelli advised his aspiring Borgia prince in 1513. "But always with the mind so disposed that, when the occasion arises not to be so, you can become the opposite."

What Are Friends For?

*Among East African baboons, friendship means companions, health, safety . . .
and, sometimes, sex*

Barbara Smuts

Virgil, a burly adult male olive baboon, closely followed Zizi, a middle-aged female easily distinguished by her grizzled coat and square muzzle. On her rump Zizi sported a bright pink swelling, indicating that she was sexually receptive and probably fertile. Virgil's extreme attentiveness to Zizi suggested to me—and all rival males in the troop—that he was her current and exclusive mate.

Zizi, however, apparently had something else in mind. She broke away from Virgil, moved rapidly through the troop, and presented her alluring sexual swelling to one male after another. Before Virgil caught up with her, she had managed to announce her receptive condition to several of his rivals. When Virgil tried to grab her, Zizi screamed and dashed into the bushes with Virgil in hot pursuit. I heard sounds of chasing and fighting coming from the thicket. Moments later Zizi emerged from the bushes with an older male named Cyclops. They remained together for several days, copulating often. In Cyclops's presence, Zizi no longer approached or even glanced at other males.

Primatologists describe Zizi and other olive baboons (*Papio cynocephalus anubis*) as promiscuous, meaning that both males and females usually mate with several members of the opposite sex within a short period of time. Promiscuous mating behavior characterizes many of the larger, more familiar primates, including chimpanzees, rhesus macaques, and gray langurs, as well as olive, yel-low, and chacma baboons, the three sub-species of savanna baboon. In colloquial usage, promiscuity often connotes wanton and random sex, and several early studies of primates supported this stereotype. However, after years of laboriously recording thousands of copulations under natural conditions, the Peeping Toms of primate fieldwork have shown that, even in promiscuous species, sexual pairings are far from random.

Some adult males, for example, typically copulate much more often than others. Primatologists have explained these differences in terms of competition: the most dominant males monopolize females and prevent lower-ranking rivals from mating. But exceptions are frequent. Among baboons, the exceptions often involve scruffy, older males who mate in full view of younger, more dominant rivals.

A clue to the reason for these puzzling exceptions emerged when primatologists began to question an implicit assumption of the dominance hypothesis—that females were merely passive objects of male competition. But what if females were active arbiters in this system? If females preferred some males over others and were able to express these preferences, then models of mating activity based on male dominance alone would be far too simple.

Once researchers recognized the possibility of female choice, evidence for it turned up in species after species. The story of Zizi, Virgil, and Cyclops is one of hundreds of examples of female pri-mates rejecting the sexual advances of particular males and enthusiastically cooperating with others. But what is the basis for female choice? Why might they prefer some males over others?

This question guided my research on the Eburru Cliffs troop of olive baboons, named after one of their favorite sleeping sites, a sheer rocky outcrop rising several hundred feet above the floor of the Great Rift Valley, about 100 miles northwest of Nairobi, Kenya. The 120 members of Eburru Cliffs spent their days wandering through open grassland studded with occasional acacia thorn trees. Each night they retired to one of a dozen sets of cliffs that provided protection from nocturnal predators such as leopards.

Most previous studies of baboon sexuality had focused on females who, like Zizi, were at the peak of sexual receptivity. A female baboon does not mate when she is pregnant or lactating, a period of abstinence lasting about eighteen months. The female then goes into estrus, and for about two weeks out of every thirty-five-day cycle, she mates. Toward the end of this two-week period she may ovulate, but usually the female undergoes four or five estrous cycles before she conceives. During pregnancy, she once again resumes a chaste existence. As a result, the typical female baboon is sexually active for less than 10 percent of her adult life. I thought that by focusing on the other 90 percent, I might learn something new. In particular, I suspected that routine, day-to-day relationships between males and pregnant or lactating (nonestrous) females might provide clues to female mating preferences.

Nearly every day for sixteen months, I joined the Eburru Cliffs baboons at their sleeping cliffs at dawn and traveled several miles with them while they foraged for roots, seeds, grass, and occasionally, small prey items, such as baby gazelles or hares (see "Predatory Baboons of Kekopey," *Natural History,* March 1976). Like all savanna baboon troops, Eburru Cliffs functioned as a cohesive unit organized around a core of related females, all of whom were born in the troop. Unlike the females, male savanna baboons leave their natal troop to join another where they may remain for many years, so most of the Eburru Cliffs adult males were immigrants. Since membership in the troop remained relatively constant during the period of my study, I learned to identify each individual. I relied on differences in size, posture, gait, and especially, facial features. To the practiced observer, baboons look as different from one another as human beings do.

As soon as I could recognize individuals, I noticed that particular females tended to turn up near particular males again and again. I came to think of these pairs as friends. Friendship among animals is not a well-documented phenomenon, so to convince skeptical colleagues that baboon friendship was real, I needed to develop objective criteria for distinguishing friendly pairs.

I began by investigating grooming, the amiable simian habit of picking through a companion's fur to remove dead skin and ectoparasites (see "Little Things That Tick Off Baboons," *Natural History,* February 1984). Baboons spend much more time grooming than is necessary for hygiene, and previous research had indicated that it is a good measure of social bonds.

Although eighteen adult males lived in the troop, each nonestrous female performed most of her grooming with just one, two, or occasionally, three males. For example, of Zizi's twenty-four grooming bouts with males, Cyclops accounted for thirteen, and a second male, Sherlock, accounted for all

the rest. Different females tended to favor different males as grooming partners.

Another measure of social bonds was simply who was observed near whom. When foraging, traveling, or resting, each pregnant or lactating female spent a lot of time near a few males and associated with the others no more often than expected by chance. When I compared the identities of favorite grooming partners and frequent companions, they overlapped almost completely. This enabled me to develop a formal definition of friendship: any male that scored high on both grooming and proximity measures was considered a friend.

Virtually all baboons made friends; only one female and three males who had most recently joined the troop lacked such companions. Out of more than 600 possible adult female-adult male pairs in the troop, however, only about one in ten qualified as friends; these really were special relationships.

Several factors seemed to influence which baboons paired up. In most cases, friends were unrelated to each other, since the male had immigrated from another troop. (Four friendships, however, involved a female and an adolescent son who had not yet emigrated. Unlike other friends, these related pairs never mated.) Older females tended to be friends with older males; younger females with younger males. I witnessed occasional May–December romances, usually involving older females and young adult males. Adolescent males and females were strongly rule-bound, and with the exception of mother-son pairs, they formed friendships only with one another.

Regardless of age or dominance rank, most females had just one or two male friends. But among males, the number of female friends varied greatly from none to eight. Although high-ranking males enjoyed priority of access to food and sometimes mates, dominant males did not have more female friends than low-ranking males. Instead it was the older males who had lived in the troop for many years who had the most friends. When a male had several female friends, the females were often closely related to one another. Since female baboons spend a lot of time near their kin, it is probably easier for a male to main-

tain bonds with several related females at once.

When collecting data, I focused on one nonestrous female at a time and kept track of her every movement toward or away from any male; similarly, I noted every male who moved toward or away from her. Whenever the female and male moved close enough to exchange intimacies, I wrote down exactly what happened. When foraging together, friends tended to remain a few yards apart. Males more often wandered away from females than the reverse, and females, more often than males, closed the gap. The female behaved as if she wanted to keep the male within calling distance, in case she needed his protection. The male, however, was more likely to make approaches that brought them within actual touching distance. Often, he would plunk himself down right next to his friend and ask her to groom him by holding a pose with exaggerated stillness. The female sometimes responded by grooming, but more often, she exhibited the most reliable sign of true intimacy: she ignored her friend and simply continued whatever she was doing.

In sharp contrast, when a male who was not a friend moved close to a female, she dared not ignore him. She stopped whatever she was doing and held still, often glancing surreptitiously at the intruder. If he did not move away, she sometimes lifted her tail and presented her rump. When a female is not in estrus, this is a gesture of appeasement, not sexual enticement. Immediately after this respectful acknowledgement of his presence, the female would slip away. But such tense interactions with nonfriend males were rare, because females usually moved away before the males came too close.

These observations suggest that females were afraid of most of the males in their troop, which is not surprising: male baboons are twice the size of females, and their canines are longer and sharper than those of a lion. All Eburru Cliffs males directed both mild and severe aggression toward females. Mild aggression, which usually involved threats and chases but no body contact, occurred most often during feeding competition or when the male redirected aggression toward a female after losing a fight with another male. Females and juveniles showed aggression toward other females and juveniles in similar circumstances and occasionally inflicted superficial wounds. Severe aggression by males, which involved body contact and sometimes biting, was less common and also more puzzling, since there was no apparent cause.

An explanation for at least some of these attacks emerged one day when I was watching Pegasus, a young adult male, and his friend Cicily, sitting together in the middle of a small clearing. Cicily moved to the edge of the clearing to feed, and a higher-ranking female, Zora, suddenly attacked her. Pegasus stood up and looked as if he were about to intervene when both females disappeared into the bushes. He sat back down, and I remained with him. A full ten minutes later, Zora appeared at the edge of the clearing; this was the first time she had come into view since her attack on Cicily. Pegasus instantly pounced on Zora, repeatedly grabbed her neck in his mouth and lifted her off the ground, shook her whole body, and then dropped her. Zora screamed continuously and tried to escape. Each time, Pegasus caught her and continued his brutal attack. When he finally released her five minutes later she had a deep canine gash on the palm of her hand that made her limp for several days.

This attack was similar in form and intensity to those I had seen before and labeled "unprovoked." Certainly, had I come upon the scene after Zora's aggression toward Cicily, I would not have understood why Pegasus attacked Zora. This suggested that some, perhaps many, severe attacks by males actually represented punishment for actions that had occurred some time before.

Whatever the reasons for male attacks on females, they represent a serious threat. Records of fresh injuries indicated that Eburru Cliffs adult females received canine slash wounds from males at the rate of one for every female each year, and during my study, one female died of her injuries. Males probably pose an even greater threat to infants. Although only one infant was

killed during my study, observers in Botswana and Tanzania have seen recent male immigrants kill several young infants.

Protection from male aggression, and from the less injurious but more frequent aggression of other females and juveniles, seems to be one of the main advantages of friendship for a female baboon. Seventy times I observed an adult male defend a female or her offspring against aggression by another troop member, not infrequently a high-ranking male. In all but six of these cases, the defender was a friend. Very few of these confrontations involved actual fighting; no male baboon, subordinate or dominant, is anxious to risk injury by the sharp canines of another.

Males are particularly solicitous guardians of their friends' youngest infants. If another male gets too close to an infant or if a juvenile female plays with it too roughly, the friend may intervene. Other troop members soon learn to be cautious when the mother's friend is nearby, and his presence provides the mother with a welcome respite from the annoying pokes and prods of curious females and juveniles obsessed with the new baby. Male baboons at Gombe Park in Tanzania and Amboseli Park in Kenya have also been seen rescuing infants from chimpanzees and lions. These several forms of male protection help to explain why females in Eburru Cliffs stuck closer to their friends in the first few months after giving birth than at any other time.

The male-infant relationship develops out of the male's friendship with the mother, but as the infant matures, this new bond takes on a life of its own. My co-worker Nancy Nicolson found that by about nine months of age, infants actively sought out their male friends when the mother was a few yards away, suggesting that the male may function as an alternative caregiver. This seemed to be especially true for infants undergoing unusually early or severe weaning. (Weaning is generally a gradual, prolonged process, but there is tremendous variation among mothers in the timing and intensity of weaning. See "Mother Baboons," *Natural History,* September 1980). After being rejected by the mother, the crying infant often

approached the male friend and sat huddled against him until its whimpers subsided. Two of the infants in Eburru Cliffs lost their mothers when they were still quite young. In each case, their bond with the mother's friend subsequently intensified, and—perhaps as a result—both infants survived.

A close bond with a male may also improve the infant's nutrition. Larger than all other troop members, adult males monopolize the best feeding sites. In general, the personal space surrounding a feeding male is inviolate, but he usually tolerates intrusions by the infants of his female friends, giving them access to choice feeding spots.

Although infants follow their male friends around rather than the reverse, the males seem genuinely attached to their tiny companions. During feeding, the male and infant express their pleasure in each other's company by sharing spirited, antiphonal grunting duets. If the infant whimpers in distress, the male friend is likely to cease feeding, look at the infant, and grunt softly, as if in sympathy, until the whimpers cease. When the male rests, the infants of his female friends may huddle behind him, one after the other, forming a "train," or, if feeling energetic, they may use his body as a trampoline.

When I returned to Eburru Cliffs four years after my initial study ended, several of the bonds formed between males and the infants of their female friends were still intact (in other cases, either the male or the infant or both had disappeared). When these bonds involved recently matured females, their long-time male associates showed no sexual interest in them, even though the females mated with other adult males. Mothers and sons, and usually maternal siblings, show similar sexual inhibitions in baboons and many other primate species.

The development of an intimate relationship between a male and the infant of his female friend raises an obvious question: Is the male the infant's father? To answer this question definitely we would need to conduct genetic analysis, which was not possible for these baboons. Instead, I estimated paternity probabilities from observations of the temporary (a few hours or days) exclusive mating relationships, or consort-ships, that estrous females form with a series of different males. These esti-mates were apt to be fairly accurate, since changes in the female's sexual swelling allow one to pinpoint the tim-ing of conception to within a few days. Most females consorted with only two or three males during this period, and these males were termed likely fathers.

In about half the friendships, the male was indeed likely to be the father of his friend's most recent infant, but in the other half he was not—in fact, he had never been seen mating with the fe-male. Interestingly, males who were friends with the mother but not likely fathers nearly always developed a rela-tionship with her infant, while males who had mated with the female but were not her friend usually did not. Thus friendship with the mother, rather than paternity, seems to mediate the develop-ment of male-infant bonds. Recently, a similar pattern was documented for South American capuchin monkeys in a laboratory study in which paternity was determined genetically.

These results fly in the face of a prominent theory that claims males will invest in infants only when they are closely related. If males are not foster-ing the survival of their own genes by caring for the infant, then why do they do so? I suspected that the key was fe-male choice. If females preferred to mate with males who had already dem-onstrated friendly behavior, then friend-ships with mothers and their infants might pay off in the future when the mothers were ready to mate again.

To find out if this was the case, I ex-amined each male's sexual behavior with females he had befriended before they resumed estrus. In most cases, males consorted considerably more often with their friends than with other females. Ba-boon females typically mate with several different males, including both friends and nonfriends, but prior friendship in-creased a male's probability of mating with a female above what it would have been otherwise.

This increased probability seemed to reflect female preferences. Females oc-casionally overtly advertised their dis-dain for certain males and their desire

for others. Zizi's behavior, described above, is a good example. Virgil was not one of her friends, but Cyclops was. Usually, however, females expressed preferences and aversions more subtly. For example, Delphi, a petite adolescent female, found herself pursued by Hec-tor, a middle-aged adult male. She did not run away or refuse to mate with him, but whenever he wasn't watching, she looked around for her friend Homer, an adolescent male. When she succeeded in catching Homer's eye, she narrowed her eyes and flattened her ears against her skull, the friendliest face one baboon can send another. This told Homer she would rather be with him. Females ex-pressed satisfaction with a current con-sort partner by staying close to him, initiating copulations, and not making advances toward other males. Baboons are very sensitive to such cues, as indi-cated by an experimental study in which rival hamadryas baboons rarely chal-lenged a male-female pair if the female strongly preferred her current partner. Similarly, in Eburru Cliffs, males were less apt to challenge consorts involving a pair that shared a long-term friendship.

Even though females usually con-sorted with their friends, they also mated with other males, so it is not sur-prising that friendships were most vul-nerable during periods of sexual activity. In a few cases, the female consorted with another male more often than with her friend, but the friendship survived nevertheless. One female, however, formed a strong sexual bond with a new male. This bond persisted after concep-tion, replacing her previous friendship. My observations suggest that adolescent and young adult females tend to have shorter, less stable friendships than do older females. Some friendships, how-ever, last a very long time. When I re-turned to Eburru Cliffs six years after my study began, five couples were still together. It is possible that friendships occasionally last for life (baboons probably live twenty to thirty years in the wild), but it will require longer studies, and some very patient scientists to find out.

By increasing both the male's chances of mating in the future and the likelihood that a female's infant will sur-vive, friendship contributes to the repro-

ductive success of both partners. This clarifies the evolutionary basis of friend-ship-forming tendencies in baboons, but what does friendship mean to a baboon? To answer this question we need to view baboons as sentient beings with feelings and goals not unlike our own in similar circumstances. Consider, for example, the friendship between Thalia and Alexander.

The affair began one evening as Alex and Thalia sat about fifteen feet apart on the sleeping cliffs. It was like watching two novices in a singles bar. Alex stared at Thalia until she turned and almost caught him looking at her. He glanced away immediately, and then she stared at him until his head began to turn to-ward her. She suddenly became en-grossed in grooming her toes. But as soon as Alex looked away, her gaze re-turned to him. They went on like this for more than fifteen minutes, always with split-second timing. Finally, Alex managed to catch Thalia looking at him. He made the friendly eyes-narrowed, ears-back face and smacked his lips to-gether rhythmically. Thalia froze, and for a second she looked into his eyes. Alex approached, and Thalia, still ner-vous, groomed him. Soon she calmed down, and I found them still together on the cliffs the next morning. Looking back on this event months later, I real-ized that it marked the beginning of their friendship. Six years later, when I returned to Eburru Cliffs, they were still friends.

If flirtation forms an integral part of baboon friendship, so does jealousy. Overt displays of jealousy, such as chas-ing a friend away from a potential rival, occur occasionally, but like humans, ba-boons often express their emotions in more subtle ways. One evening a col-league and I climbed the cliffs and settled down near Sherlock, who was friends with Cybelle, a middle-aged female still foraging on the ground below the cliffs. I observed Cybelle while my colleague watched Sherlock, and we kept up a run-ning commentary. As long as Cybelle was feeding or interacting with females, Sherlock was relaxed, but each time she approached another male, his body would stiffen, and he would stare in-tently at the scene below. When Cybelle presented politely to a male who had recently tried to befriend her, Sherlock even made threatening sounds under his breath. Cybelle was not in estrus at the time, indicating that male baboon jealousy extends beyond the sexual arena to include affiliative interactions between a female friend and other males.

Because baboon friendships are embedded in a network of friendly and antagonistic relationships, they inevitably lead to repercussions extending beyond the pair. For example, Virgil once provoked his weaker rival Cyclops into a fight by first attacking Cyclops's friend Phoebe. On another occasion, Sherlock chased Circe, Hector's best friend, just after Hector had chased Antigone, Sherlock's friend.

In another incident, the prime adult male Triton challenged Cyclops's possession of meat. Cyclops grew increasingly tense and seemed about to abandon the prey to the younger male. Then Cyclops's friend Phoebe appeared with her infant Phyllis. Phyllis wandered over to Cyclops. He immediately grabbed her, held her close, and threatened Triton away from the prey. Because any challenge to Cyclops now involved a threat to Phyllis as well, Triton risked being mobbed by Phoebe and her relatives and friends. For this reason, he backed down. Males frequently use the infants of their female friends as buffers in this way. Thus, friendship involves costs as well as benefits because it makes the participants vulnerable to social manipulation or redirected aggression by others.

Finally, as with humans, friendship seems to mean something different to each baboon. Several females in Eburru Cliffs had only one friend. They were devoted companions. Louise and Pandora, for example, groomed their friend Virgil and no other male. Then there was Leda, who, with five friends, spread herself more thinly than any other female. These contrasting patterns of friendship were associated with striking personality differences. Louise and Pandora were unobtrusive females who hung around quietly with Virgil and their close relatives. Leda seemed to be everywhere at once, playing with infants, fighting with juveniles, and making friends with males. Similar differences were apparent among the males. Some devoted a great deal of time and energy to cultivating friendships with females, while others focused more on challenging other males. Although we probably will never fully understand the basis of these individual differences, they contribute immeasurably to the richness and complexity of baboon society.

Male-female friendships may be widespread among primates. They have been reported for many other groups of savanna baboons, and they also occur in rhesus and Japanese Macaques, capuchin monkeys, and perhaps in bonobos (pygmy chimpanzees). These relationships should give us pause when considering popular scenarios for the evolution of male-female relationships in humans. Most of these scenarios assume that, except for mating, males and females had little to do with one another until the development of a sexual division of labor, when, the story goes, females began to rely on males to provide meat in exchange for gathered food. This, it has been argued, set up new selection pressures favoring the development of long-term bonds between individual males and females, female sexual fidelity, and as paternity certainty increased, greater male investment in the offspring of these unions. In other words, once women began to gather and men to hunt, presto—we had the nuclear family.

This scenario may have more to do with cultural biases about women's economic dependence on men and idealized views of the nuclear family than with the actual behavior of our hominid ancestors. The nonhuman primate evidence challenges this story in at least three ways.

First, long-term bonds between the sexes can evolve in the absence of a sexual division of labor of food sharing. In our primate relatives, such relationships rest on exchanges of social, not economic, benefits.

Second, primate research shows that highly differentiated, emotionally intense male-female relationships can occur without sexual exclusivity. Ancestral men and women may have experienced intimate friendships long before they invented marriage and norms of sexual fidelity.

Third, among our closest primate relatives, males clearly provide mothers and infants with social benefits even when they are unlikely to be the fathers of those infants. In return, females provide a variety of benefits to the friendly males, including acceptance into the group and, at least in baboons, increased mating opportunities in the future. This suggests that efforts to reconstruct the evolution of hominid societies may have overemphasized what the female must supposedly do (restrict her mating to just one male) in order to obtain male parental investment.

Maybe it is time to pay more attention to what the male must do (provide benefits to females and young) in order to obtain female cooperation. Perhaps among our ancestors, as in baboons today, sex and friendship went hand in hand. As for marriage—well, that's another story.

Dian Fossey and Digit

Sy Montgomery

Every breath was a battle to draw the ghost of her life back into her body. At age forty-two it hurt her even to breathe.

Dian Fossey had been asthmatic as a child and a heavy smoker since her teens; X-rays of her lungs taken when she graduated from college, she remembered, looked like "a road map of Los Angeles superimposed over a road map of New York." And now, after eight years of living in the oxygen-poor heights of Central Africa's Virunga Volcanoes, breathing the cold, sodden night air, her lungs were crippled. The hike to her research camp, Karisoke, at 10,000 feet, took her graduate students less than an hour; for Dian it was a gasping two-and-a-half-hour climb. She had suffered several bouts of pneumonia. Now she thought she was coming down with it again.

Earlier in the week she had broken her ankle. She heard the bone snap when she fell into a drainage ditch near her corrugated tin cabin. She had been avoiding a charging buffalo. Two days later she was bitten by a venomous spider on the other leg. Her right knee was swollen huge and red; her left ankle was black. But she would not leave the mountain for medical treatment in the small hospital down in Ruhengeri. She had been in worse shape before. Once, broken ribs punctured a lung; another time she was bitten by a dog thought to be rabid. Only when her temperature reached 105 and her symptoms clearly matched those described in her medical

book for rabies had she allowed her African staff to carry her down on a litter.

Dian was loath to leave the camp in charge of her graduate students, two of whom she had been fighting with bitterly. Kelly Steward and Sandy Harcourt, once her closest camp colleagues and confidantes, had committed the unforgivable error of falling in love with each other. Dian considered this a breach of loyalty. She yelled at them. Kelly cried and Sandy sulked.

But on this May day of 1974 Sandy felt sorry for Dian. As a gesture of conciliation, he offered to help her hobble out to visit Group 4. Splinted and steadied by a walking stick, she quickly accepted.

Group 4 was the first family of mountain gorillas Dian had contacted when she established her camp in Rwanda in September 1967. A political uprising had forced her to flee her earlier research station in Zaire. On the day she founded Karisoke—a name she coined by combining the names of the two volcanoes between which her camp nestled, Karisimbi and Visoke—poachers had led her to the group. The two Batwa tribesmen had been hunting antelope in the park—an illegal practice that had been tolerated for decades—and they offered to show her the gorillas they had encountered.

At that first contact, Dian watched the gorillas through binoculars for forty-five minutes. Across a ravine, ninety feet away, she could pick out three distinctive individuals in the fourteen-member

group. There was a majestic old male, his black form silvered from shoulder to hip. This 350-pound silverback was obviously the sultan of the harem of females, the leader of the family. One old female stood out, a glare in her eyes, her lips compressed as if she had swallowed vinegar. And one youngster was "a playful little ball of disorganized black fluff . . . full of mischief and curiosity," as Dian would later describe him in *Gorillas in the Mist*. She guessed then that he was about five years old. He tumbled about in the foliage like an animated black dustball. When the lead silverback spotted Dian behind a tree, the youngster obediently fled at his call, but Dian had the impression that the little male would rather have stayed for a longer look at the stranger. In a later contact she noticed the juvenile's swollen, extended middle finger. After many attempts at naming him, she finally called him Digit.

It was Digit, now twelve, who came over to Dian as she sat crumpled and coughing among the foliage with Sandy. Digit, a gaunt young silverback, served his family as sentry. He left the periphery of his group to knuckle over to her side. She inhaled his smell. A good smell, she noted with relief: for two years a draining wound in his neck had hunched his posture and sapped his spirit. Systemic infection had given his whole body a sour odor, not the normal, clean smell of fresh sweat. During that time Digit had become listless. Little would arouse his interest: not the sex

Excerpt from *Walking with the Great Apes* by Sy Montgomery, Chapter 3, pp. 46–66.

play between the group's lead silverback and receptive females, not even visits from Dian. Digit would sit at the edge of the group for hours, probing the wound with his fingers, his eyes fixed on some distant spot as if dwelling on a sad memory.

But today Digit looked directly into Dian's eyes. He chose to remain beside her throughout the afternoon, like a quiet visitor to a shut-in, old friends with no need to talk. He turned his great domed head to her, looking at her solemnly with a brown, cognizant gaze. Normally a prolonged stare from a gorilla is a threat. But Digit's gaze bore no aggression. He seemed to say: I know. Dian would later write that she believed Digit understood she was sick. And she returned to camp that afternoon, still limping, still sick, still troubled, but whole.

"We all felt we shared something with the gorillas," one of her students would later recall of his months at Karisoke. And it is easy to feel that way after even a brief contact with these huge, solemn beings. "The face of a gorilla," wrote nature writer David Quammen after just looking at a picture of one, "offers a shock of what feels like total recognition." To be in the presence of a mountain gorilla for even one hour simply rips your soul open with awe. They are the largest of the great apes, the most hugely majestic and powerful; but it is the gaze of a gorilla that transfixes, when its eyes meet yours. The naturalist George Schaller, whose yearlong study preceded Dian's, wrote that this is a look found in the eye of no other animal except, perhaps, a whale. It is not so much intelligence that strikes you, but understanding. You feel there has been an exchange.

The exchange between Digit and Dian that day was deep and long. By then Digit had known Dian for seven years. She had been a constant in his growing up from a juvenile to a young blackback and now to a silverback sentry. He had known her longer than he had known his own mother, who had died or left his group before he was five; he had known Dian longer than he had known his father, the old silverback who died of natural causes less than a year

after she first observed him. When Digit was nine, his three age-mates in Group 4 departed: his half sisters were "kidnaped" by rival silverbacks, as often happens with young females. Digit then adopted Dian as his playmate, and he would often leave the rest of the group to amble to her side, eager to examine her gear, sniff her gloves and jeans, tug gently at her long brown braid.

As for Dian, her relationship with Digit was stronger than her bonds with her mother, father, or stepfather. Though she longed for a husband and babies, she never married or bore children. Her relationship with Digit endured longer than that with any of her lovers and outlasted many of her human friendships.

In her slide lectures in the United States, Dian would refer to him as "my friend, Digit." "Friend," she admitted, was too weak a word, too casual; but she could find no other. Our words are something we share with other humans; but what Dian had with Digit was something she guarded as uniquely hers.

A mountain gorilla group is one of the most cohesive family units found among primates, a fact that impressed George Schaller. Adult orangutans live mostly alone, males and females meeting to mate. Chimpanzees' social groupings are so loosely organized, changing constantly in number and composition, that Jane Goodall couldn't make sense of them for nearly a decade. But gorillas live in tight-knit, clearly defined families. Typically a group contains a lead silverback, perhaps his adult brother, half brother, or nephew, and several adult females and their offspring.

A gorilla group travels, feeds, plays, and rests together. Seldom is an individual more than a hundred feet away from the others. The lead silverback slows his pace to that of the group's slowest, weakest member. All adults tolerate the babies and youngsters in the group, often with great tenderness. A wide-eyed baby, its fur still curly as black wool, may crawl over the great black bulk of any adult with impunity; a toddler may even step on the flat, leathery nose of a silverback. Usually the powerful male will gently set the baby aside

or even dangle it playfully from one of his immense fingers.

When Dian first discovered Group 4, she would watch them through binoculars from a hidden position, for if they saw her they would flee. She loved to observe the group's three infants toddle and tumble together. If one baby found the play too rough, it would make a coughing sound, and its mother would lumber over and cradle it tenderly to her breast. Dian watched Digit and his juvenile sisters play: wrestling, rolling, and chasing games often took them as far as fifty feet from the hulking adults. Sometimes a silverback led the youngsters in a sort of square dance. Loping from one palmlike *Senecio* tree to another, each gorilla would grab a trunk for a twirl, then spin off to embrace another trunk down the slope, until all the gorillas lay in a bouncing pileup of furry black bodies. And then the silverback would lead the youngsters up the slope again for another game.

Within a year, this cheerful silverback eventually took over leadership of the group, after the old leader died. Dian named him Uncle Bert, after her uncle Albert Chapin. With Dian's maternal aunt, Flossie (Dian named a Group 4 female after her as well), Uncle Bert had helped care for Dian after her father left the family when she was three. While Dian was in college Bert and Flossie gave her money to help with costs that her holiday, weekend, and summer jobs wouldn't cover. Naming the silverback after her uncle was the most tender tribute Dian could have offered Bert Chapin: his was the name given to the group's male magnet, its leader, protector—and the centerpiece of a family life whose tenderness and cohesion Dian, as a child, could not have imagined.

Dian was a lonely only child. Her father's drinking caused the divorce that took him out of her life; when her mother, Kitty, remarried when Dian was five, even the mention of George Fossey's name became taboo in the house. Richard Price never adopted Dian. Each night she ate supper in the kitchen with the housekeeper. Her stepfather did not allow her at the dinner table with him and her mother until she was ten. Though Dian's stepfather, a building contractor, seemed wealthy, she largely paid her own way through school. Once she worked as a machine operator in a factory.

Dian seldom spoke of her family to friends, and she carried a loathing for her childhood into her adult life. Long after Dian left the family home in California, she referred to her parents as "the Prices." She would spit on the ground whenever her stepfather's name was mentioned. When her Uncle Bert died, leaving Dian $50,000, Richard Price badgered her with cables to Rwanda, pressing her to contest the will for more money; after Dian's death he had her will overturned by a California court, claiming all her money for himself and his wife.

Her mother and stepfather tried desperately to thwart Dian's plans to go to Africa. They would not help her finance her lifelong dream to go on safari when she was twenty-eight. She borrowed against three years of her salary as an occupational therapist to go. And when she left the States three years later to begin her study of the mountain gorillas, her mother begged her not to go, and her stepfather threatened to stop her.

She chose to remain in the alpine rain forest, as alone as she had ever been. She chose to remain among the King Kong beasts whom the outside world still considered a symbol of savagery, watching their gentle, peaceful lives unfold.

Once Dian, watching Uncle Bert with his family, saw the gigantic male pluck a handful of white flowers with his huge black hands. As the young Digit ambled toward him, the silverback whisked the bouquet back and forth across the youngster's face. Digit chuckled and tumbled into Uncle Bert's lap, "much like a puppy wanting attention," Dian wrote. Digit rolled against the silverback, clutching himself in ecstasy as the big male tickled him with petals.

By the end of her first three months in Rwanda, Dian was following two gorilla groups regularly and observing another sporadically. She divided most of her time between Group 5's fifteen members, ranging on Visoke's southeastern slopes, and Group 4. Group 8, a family

of nine, all adult, shared Visoke's western slopes with Group 4.

Dian still could not approach them. Gorilla families guard carefully against intrusion. Each family has at least one member who serves as sentry, typically posted at the periphery of the group to watch for danger—a rival silverback or a human hunter. Gorilla groups seldom interact with other families, except when females transfer voluntarily out of their natal group to join the families of unrelated silverbacks or when rival silverbacks "raid" a neighboring family for females.

Adult gorillas will fight to the death defending their families. This is why poachers who may be seeking only one infant for the zoo trade must often kill all the adults in the family to capture the baby. Once Dian tracked one such poacher to his village; the man and his wives fled before her, leaving their small child behind.

At first Dian observed the animals from a distance, silently, hidden. Then slowly, over many months, she began to announce her presence. She imitated their contentment vocalizations, most of often the *naoom, naoom, naoom,* a sound like belching or deeply clearing the throat. She crunched wild celery stalks. She crouched, eyes averted, scratching herself loud and long, as gorillas do. Eventually she could come close enough to them to smell the scent of their bodies and see the ridges inside the roofs of their mouths when they yawned; at times she came close enough to distinguish, without binoculars, the cuticles of their black, humanlike fingernails.

She visited them daily; she learned to tell by the contour pressed into the leaves which animal had slept in a particular night nest, made from leaves woven into a bathtub shape on the ground. She knew the sound of each individual voice belching contentment when they were feeding. But it was more than two years before she knew the touch of their skin.

Peanuts, a young adult male in Group 8, was the first mountain gorilla to touch his fingers to hers. Dian was lying on her back among the foliage, her right arm outstretched, palm up. Peanuts looked at her hand intently; then he stood, extended his hand, and touched

her fingers for an instant. *National Geographic* photographer Bob Campbell snapped the shutter only a moment afterward: that the photo is blurry renders it dreamlike. The 250-pound gorilla's right hand still hangs in midair. Dian's eyes are open but unseeing, her lips parted, her left hand brought to her mouth, as if feeling for the lingering warmth of a kiss.

Peanuts pounded his chest with excitement and ran off to rejoin his group. Dian lingered after he left; she named the spot where they touched Fasi Ya Mkoni, "the Place of the Hands." With his touch, Peanuts opened his family to her; she became a part of the families she had observed so intimately for the past two years. Soon the gorillas would come forward and welcome her into their midst.

Digit was almost always the first member of Group 4 to greet her. "I received the impression that Digit really looked forward to the daily contacts," she wrote in her book. "If I was alone, he often invited play by flopping over on his back, waving stumpy legs in the air, and looking at me smilingly as if to say, 'How can you resist me?' "

At times she would be literally blanketed with gorillas, when a family would pull close around her like a black furry quilt. In one wonderful photo, Puck, a young female of Group 5, is reclining in back of Dian and, with the back of her left hand, touching Dian's cheek—the gesture of a mother caressing the cheek of a child.

Mothers let Dian hold their infants; silverbacks would groom her, parting her long dark hair with fingers thick as bananas, yet deft as a seamstress's touch. "I can't tell you how rewarding it is to be with them," Dian told a New York crowd gathered for a slide lecture in 1982. "Their trust, the cohesiveness, the tranquility . . ." Words failed her, and her hoarse, breathy voice broke. "It is really something."

Other field workers who joined Dian at Karisoke remember similar moments. Photographer Bob Campbell recalls how Digit would try to groom his sleeves and pants and, finding nothing groomable, would pluck at the hairs of his wrist; most of the people who worked there have pictures of themselves with young gorillas on their heads or in their laps.

But with Dian it was different. Ian Redmond, who first came to Karisoke in 1976, remembers one of the first times he accompanied Dian to observe Group 4. It was a reunion: Dian hadn't been out to visit the group for a while. "The animals filed past us, and each one paused and briefly looked into my face, just briefly. And then each one looked into Dian's eyes, at very close quarters, for half a minute or so. It seemed like each one was queuing up to stare into her face and remind themselves of her place with them. It was obvious they had a much deeper and stronger relationship with Dian than with any of the other workers."

In the early days Dian had the gorillas mostly to herself. It was in 1972 that Bob Campbell filmed what is arguably one of the most moving contacts between two species on record: Digit, though still a youngster, is huge. His head is more than twice the size of Dian's, his hands big enough to cover a dinner plate. He comes to her and with those enormous black hands gently takes her notebook, then her pen, and brings them to his flat, leathery nose. He gently puts them aside in the foliage and rolls over to snooze at Dian's side.

Once Dian spotted Group 4 on the opposite side of a steep ravine but knew she was not strong enough to cross it. Uncle Bert, seeing her, led the entire group across the ravine to her. This time Digit was last in line. "Then," wrote Dian, "he finally came right to me and gently touched my hair. . . . I wish I could have given them all something in return."

At times like these, Dian wept with joy. Hers was the triumph of one who has been chosen: wild gorillas would come to her.

The great intimacy of love is onlyness, of being the loved One. It is the kind of love most valued in Western culture; people choose only one "best" friend, one husband, one wife, one God. Even our God is a jealous one, demanding "Thou shalt have no other gods before Me."

This was a love Dian sought over and over again—as the only child of parents who did not place her first, as the paramour of a succession of married lovers. The love she sought most desperately was a jealous love, exclusive—not *agape,* the Godlike, spiritual love of all beings, not the uniform, brotherly love, *philia.* The love Dian sought was the love that singles out.

Digit singled Dian out. By the time he was nine he was more strongly attracted to her than were any of the other gorillas she knew. His only age-mates in his family, his half sisters, had left the group or been kidnaped. When Digit heard Dian belch-grunting a greeting, he would leave the company of his group to scamper to greet her. To Digit, Dian was the sibling playmate he lacked. And Dian recognized his longing as clearly as she knew her own image in a mirror.

Dian had had few playmates as a child. She had longed for a pet, but her stepfather wouldn't allow her to keep even a hamster a friend offered, because it was "dirty." He allowed her a single goldfish; she was devastated when it died and was never allowed another.

But Digit was no pet. "Dian's relationship with the gorillas is really the highest form of human-animal relationship," observed Ian Redmond. "With almost any other human-animal relationship, that involves feeding the animals or restraining the animals or putting them in an enclosure, or if you help an injured animal—you do something to the animal. Whereas Dian and the gorillas were on completely equal terms. It was nothing other than the desire to be together. And that's as pure as you can get."

When Digit was young, he and Dian played together like children. He would strut toward her, playfully whacking foliage; she would tickle him; he would chuckle and climb on her head. Digit was fascinated by any object Dian had with her: once she brought a chocolate bar to eat for lunch and accidentally dropped it into the hollow stump of a tree where she was sitting next to Digit. Half in jest, she asked him to get it back for her. "And according to script," she wrote her Louisville friend, Betty Schartzel, "Digit reached one long,

hairy arm into the hole and retrieved the candy bar." But the chocolate didn't appeal to him. "After one sniff he literally threw it back into the hole. The so-called 'wild gorillas' are really very discriminating in their tastes!"

Dian's thermoses, notebooks, gloves, and cameras were all worthy of investigation. Digit would handle these objects gently and with great concentration. Sometimes he handed them back to her. Once Dian brought Digit a hand mirror. He immediately approached it, propped up on his forearms, and sniffed the glass. Digit pursed his lips, cocked his head, and then uttered a long sigh. He reached behind the mirror in search of the body connected to the face. Finding nothing, he stared at his reflection for five minutes before moving away.

Dian took many photos of all the gorillas, but Digit was her favorite subject. When the Rwandan Office of Tourism asked Dian for a gorilla photo for a travel poster, the slide she selected was one of Digit. He is pictured holding a stick of wood he has been chewing, his shining eyes a mixture of innocence and inquiry. He looks directly into the camera, his lips parted and curved as if about to smile. "Come to meet him in Rwanda," exhorts the caption. When his poster began appearing in hotels, banks, and airports, "I could not help feeling that our privacy was on the verge of being invaded," Dian wrote.

Her relationship with Digit was one she did not intend to share. Hers was the loyalty and possessiveness of a silverback: what she felt for the gorillas, and especially Digit, was exclusive, passionate, and dangerous.

No animal, Dian believed, was truly safe in Africa. Africans see most animals as food, skins, money. "Dian had a compulsion to buy every animal she ever saw in Africa," remembers her friend Rosamond Carr, an American expatriate who lives in nearby Gisenyi, "to save it from torture." One day Dian, driving in her Combi van, saw some children on the roadside, swinging a rabbit by the ears. She took it from them, brought it back to camp, and built a spacious hutch for it. Another time it would be a chicken: visiting villagers sometimes brought one to camp, intending, of course, that it be eaten. Dian would keep it as a pet.

Dian felt compelled to protect the vulnerable, the innocent. Her first plan after high school had been to become a veterinarian; after failing chemistry and physics, she chose occupational therapy; with her degree, she worked for a decade with disabled children.

One day Dian came to the hotel in Gisenyi where Rosamond was working as a manager. Dian was holding a monkey. She had seen it at a market, packed in a carton. Rosamond remembers, "I look and see this rotten little face, this big ruff of hair, and I say, 'Dian. I'm sorry, you cannot have that monkey in this hotel!'" But Dian spent the night with the animal in her room anyway.

"Luckily for me, she left the next day. I have never seen anything like the mess. There were banana peels on the ceiling, sweet potatoes on the floor; it had broken the water bottle, the glasses had been smashed and had gone down the drain of the washbasin. And with that adorable animal she starts up the mountain."

Kima, as Dian named the monkey, proved no less destructive in camp. Full grown when Dian brought her, Kima bit people, urinated on Dian's typewriter, bit the head off all her matches, and terrorized students on their way to the latrine, leaping off the roof of Dian's cabin and biting them. Yet Dian loved her, built a hatchway allowing Kima free access to her cabin, bought her toys and dolls, and had her camp cook prepare special foods for her. Kima especially liked french fries, though she discarded the crunchy outsides and ate only the soft centers. "Everyone in camp absolutely hated that animal," Rosamond says. "But Dian loved her."

Another of Dian's rescue attempts occurred one day when she was driving down the main street of Gisenyi on a provisioning trip. Spotting a man walking a rack-ribbed dog on a leash, she slammed on the brakes. "I want to buy that dog," she announced. The Rwandan protested that it was not for sale. She got out of her Combi, lifted up the sickly animal, and drove off with it.

Rosamond learned of the incident from a friend named Rita who worked at the American embassy. For it was to Rita's home that the Rwandan man returned that afternoon to explain why the dog, which he had been taking to the vet for worming, had never made it to its destination. "Madame, a crazy woman stopped and stole your dog, and she went off with it in a gray van."

"Rita got her dog back," Rosamond continues. Dian had taken it to the hotel where she was staying overnight; when Rita tracked her down, she was feeding the dog steak in her room. "And that was typical of Dian. She had to save every animal she saw. And they loved her—every animal I ever saw her with simply loved her."

When Dian first came to Karisoke, elephants frequently visited her camp. Rosamond used to camp with Dian in those early days before the cabin was built. The elephants came so close that she remembers hearing their stomachs rumble at night. Once she asked Dian if she undressed at night. "Of course not, are you crazy?" Dian replied. "I go to bed in my blue jeans. I have to get up six times at night to see what's happening outside."

One night an elephant selected Dian's tent pole as a scratching post. Another time a wild elephant accepted a banana from Dian's hand. The tiny antelopes called duiker often wandered through camp; one became so tame it would follow Dian's laying hens around. A family of seven bushbucks adopted Karisoke as home, as did an ancient bull buffalo she named Mzee.

Dian's camp provided refuge from the poacher-infested, cattle-filled forest. For centuries the pygmylike Batwa had used these volcanic slopes as a hunting ground. And as Rwanda's human population exploded, the Virungas were the only source of bush meat left, and poaching pressure increased. Today you will find no elephants in these forests; they have all been killed by poachers seeking ivory.

If you look at the Parc National des Volcans from the air, the five volcanoes, their uppermost slopes puckered like the lips of an old woman, seem to be standing on tiptoe to withdraw from the flood of cultivation and people below. Rwanda is the most densely populated country in Africa, with more than 500 people per square mile. Almost every inch is cultivated, and more than 23,000 new families need new land each year. In rural Rwanda outside of the national parks, if you wander from a path you are more likely to step in human excrement than the scat of a wild animal. The Parc des Volcans is thoroughly ringed with *shambas*, little farm plots growing bananas, peanuts, beans, manioc, and with fields of pyrethrum, daisylike flowers cultivated as a natural insecticide for export. The red earth of the fields seems to bleed from all the human scraping.

The proud, tall Batutsi have few other areas to pasture their cattle, the pride of their existence; everywhere else are shambas. From the start Dian tried to evict the herders from the park, kidnaping their cows and sometimes even shooting them. On the Rwandan side of the mountain, cattle herds were so concentrated, she wrote, that "many areas were reduced to dustbowls." She felt guilty, but the cattle destroyed habitat for the gorillas and other wild animals the park was supposed to protect. Worse were the snares set by the Batwa. Many nights she stayed awake nursing a duiker or bushbuck whose leg had been mangled in a trap. Dian lived in fear that one of the gorillas would be next.

The Batwa do not eat gorillas; gorillas fall victim to their snares, set for antelope, by accident. But the Batwa have for centuries hunted gorillas, to use the fingers and genitals of silverbacks in magic rituals and potions. And now the hunters found a new reason to kill gorillas: they learned that Westerners would pay high prices for gorillas heads for trophies, gorilla hands for ashtrays, and gorilla youngsters for zoos.

In March 1969, only eighteen months into her study, a friend in Ruhengeri came to camp to tell Dian that a young gorilla had been captured from the southern slopes of Mount Karisimbi. All ten adults in the group had been killed so that the baby could be taken for display in the Cologne Zoo. The capture had been approved by the park conservator, who was paid handsomely for his cooperation. But something had gone wrong: the baby gorilla was dying.

Dian took the baby in, a three- to four-year-old female she named Coco. The gorilla's wrists and feet had been bound with wire to a pole when the hunters carried her away from the corpses of her family; she had spent two or three weeks in a coffinlike crate, fed only corn, bananas, and bread, before Dian came to her rescue. When Dian left the park conservator's office, she was sure the baby would die. She slept with Coco in her bed, awakening amid pools of the baby's watery feces.

A week later came another sick orphan, a four- to five-year-old female also intended for the zoo. Her family had shared Karisimbi's southern slope with Coco's; trying to defend the baby from capture, all eight members of her group had died. Dian named this baby Pucker for the huge sores that gave her face a puckered look.

It took Dian two months to nurse the babies back to health. She transformed half her cabin into a giant gorilla playpen filled with fresh foliage. She began to take them into the forest with her, encouraging them to climb trees and vines. She was making plans to release them into a wild group when the park conservator made the climb to camp. He and his porters descended with both gorillas in a box and shipped them to the zoo in West Germany. Coco and Pucker died there nine years later, within a month of one another, at an age when, in the wild, they would have been mothering youngsters the same age they had been when they were captured.

Thereafter Dian's antipoaching tactics became more elaborate. She learned from a friend in Ruhengeri that the trade in gorilla trophies was flourishing; he had counted twenty-three gorilla heads for sale in that town in one year. As loyal as a silverback, as wary as a sentry, Dian and her staff patrolled the forest for snares and destroyed the gear poachers left behind in their temporary shelters.

Yet each day dawned to the barking of poachers' dogs. A field report she submitted to the National Geographic Society in 1972 gave the results of the most recent gorilla census: though her study groups were still safe, the surrounding areas of the park's five volcanoes were literally under siege. On Mount Muhavura census workers saw convoys of smugglers leaving the park every forty-five minutes. Only thirteen gorillas were left on the slopes of Muhavura. On neighboring Mount Gahinga no gorillas were left. In the two previous years, census workers had found fresh remains of slain silverbacks. And even the slopes of Dian's beloved Karisimbi, she wrote, were covered with poachers' traps and scarred by heavily used cattle trails; "poachers and their dogs were heard throughout the region."

It was that same year, 1972, that a maturing Digit assumed the role of sentry of Group 4. In this role he usually stayed on the periphery of the group to watch for danger; he would be the first to defend his family if they were attacked. Once when Dian was walking behind her Rwandan tracker, the dark form of a gorilla burst from the bush. The male stood upright to his full height of five and a half feet; his jaw gaped open, exposing black gums and three-inch canines as he uttered two long, piercing screams at the terrified tracker. Dian stepped into view, shoving the tracker down behind her, and stared into the animal's face. They recognized each other immediately. Digit dropped to all fours and ran back to his group.

Dian wrote that Digit's new role made him more serious. No longer was he a youngster with the freedom to roll and wrestle with his playmate. But Dian was still special to him. Once when Dian went out to visit the group during a downpour, the young silverback emerged from the gloom and stood erect before his crouching human friend. He pulled a stalk of wild celery—a favorite gorilla food that Digit had seen Dian munch on many times—peeled it with his great hands, and dropped the stalk at her feet like an offering. Then he turned and left.

As sentry, Digit sustained the wound that sapped his strength for the next two years. Dian did not observe the fight, but she concluded from tracking clues that Digit had warded off a raid by the silverback leader of Group 8, who had previously kidnaped females from Group 4. Dian cringed each time she heard him coughing and retching. Digit sat alone, hunched and indifferent. Dian worried that his growth would be retarded. In her field notes she described his mood as one of deep dejection.

This was a time when Dian was nursing wounds of her own. She had hoped that Bob Campbell, the photographer, would marry her, as Hugo van Lawick had married Jane; but Bob left Karisoke for the last time at the end of May 1972, to return to his wife in Nairobi. Then she had a long affair with a Belgian doctor, who left her to marry the woman he had been living with. Dian's health worsened. Her trips overseas for primatology conferences and lecture tours were usually paired with hospital visits to repair broken bones and heal her fragile lungs. She feared she had tuberculosis. She noted her pain in her diary telegraphically: "Very lung-sick." "Coughing up blood." "Scum in urine."

When she was in her twenties, despite her asthma, Dian seemed as strong as an Amazon. Her large-boned but lanky six-foot frame had a coltish grace; one of her suitors, another man who nearly married her, described her as "one hell of an attractive woman," with masses of long dark hair and "eyes like a Spanish dancer." But now Dian felt old and ugly and weak. She used henna on her hair to try to cover the gray. (Dian told a friend that her mother's only comment about her first appearance on a National Geographic TV special was, "Why did you dye your hair that awful orange color?") In letters to friends Dian began to sign off as "The Fossil." She referred to her house as "the Mausoleum." In a cardboard album she made for friends from construction paper and magazine cutouts, titled "The Sage of Karisoke," she pasted a picture of a mummified corpse sitting upright on a bed. She realized that many of her students disliked her. Under the picture Dian printed a caption: "Despite their protests, she stays on."

By 1976 Dian was spending less and less time in the field. Her lungs and legs had grown too weak for daily contacts; she had hairline fractures on her feet. And she was overwhelmed with paperwork. She became increasingly testy with her staff, and her students feared to knock on her door. Her students wouldn't even see her for weeks at a time, but they would hear her pounding on her battered Olivetti, a task from which she would pause only to take another drag on an Impala *filtrée* or to munch sunflower seeds. Her students were taking the field data on the gorillas by this time: when Dian went out to see the groups, she simply visited with them.

One day, she ventured out along a trail as slippery as fresh buffalo dung to find Group 4. By the time she found them, the rain was driving. They were huddled against the downpour. She saw Digit sitting about thirty feet apart from the group. She wanted to join him but resisted; she now feared that her early contact with him had made him too human-oriented, more vulnerable to poachers. So she settled among the soaking foliage several yards from the main group. She could barely make

out the humped black forms in the heavy mist.

On sunny days there is no more beautiful place on earth than the Virungas; the sunlight makes the *Senecio* trees sparkle like fireworks in midexplosion; the gnarled old *Hagenias,* trailing lacy beards of gray-green lichen and epiphytic ferns, look like friendly wizards, and the leaves of palms seem like hands upraised in praise. But rain transforms the forest into a cold, gray hell. You stare out, tunnel-visioned, from the hood of a dripping raincape, at a wet landscape cloaked as if in evil enchantment. Each drop of rain sends a splintering chill into the flesh, and your muscles clench with cold; you can cut yourself badly on the razorlike cutty grass and not even feel it. Even the gorillas, with their thick black fur coats, look miserable and lonely in the rain.

Minutes after she arrived, Dian felt an arm around her shoulders. "I looked

up into Digit's warm, gentle brown eyes," she wrote in *Gorillas in the Mist.* He gazed at her thoughtfully and patted her head, then sat by her side. As the rain faded to mist, she laid her head down in Digit's lap.

On January 1, 1978, Dian's head tracker returned to camp late in the day. He had not been able to find Group 4. But he had found blood along their trail.

Ian Redmond found Digit's body the next day. His head and hands had been hacked off. There were five spear wounds in his body.

Ian did not see Dian cry that day. She was almost supercontrolled, he remembers. No amount of keening, no incantation or prayer could release the pain of her loss. But years later she filled a page of her diary with a single word, written over and over: "Digit Digit Digit Digit . . ."

The Mind of the Chimpanzee

Jane Goodall

Often I have gazed into a chimpanzee's eyes and wondered what was going on behind them. I used to look into Flo's, she so old, so wise. What did she remember of her young days? David Greybeard had the most beautiful eyes of them all, large and lustrous, set wide apart. They somehow expressed his whole personality, his serene self-assurance, his inherent dignity—and, from time to time, his utter determination to get his way. For a long time I never liked to look a chimpanzee straight in the eye—I assumed that, as is the case with most primates, this would be interpreted as a threat or at least as a breach of good manners. No so. As long as one looks with gentleness, without arrogance, a chimpanzee will understand, and may even return the look. And then— or such is my fantasy—it is as though the eyes are windows into the mind. Only the glass is opaque so that the mystery can never be fully revealed.

I shall never forget my meeting with Lucy, an eight-year-old home-raised chimpanzee. She came and sat beside me on the sofa and, with her face very close to mine, searched in my eyes—for what? Perhaps she was looking for signs of mistrust, dislike, or fear, since many people must have been somewhat disconcerted when, for the first time, they came face to face with a grown chimpanzee. Whatever Lucy read in my eyes clearly satisfied her for she suddenly put one arm round my neck and gave me a generous and very chimp-like kiss, her mouth wide open and laid over mine. I was accepted.

For a long time after that encounter I was profoundly disturbed. I had been

at Gombe for about fifteen years then and I was quite familiar with chimpanzees in the wild. But Lucy, having grown up as a human child, was like a changeling, her essential chimpanzeeness overlaid by the various human behaviours she had acquired over the years. No longer purely chimp yet eons away from humanity, she was man-made, some other kind of being. I watched, amazed, as she opened the refrigerator and various cupboards, found bottles and a glass, then poured herself a gin and tonic. She took the drink to the TV, turned the set on, flipped from one channel to another then, as though in disgust, turned it off again. She selected a glossy magazine from the table and, still carrying her drink, settled in a comfortable chair. Occasionally, as she leafed through the magazine she identified something she saw, using the signs of ASL, the American Sign Language used by the deaf. I, of course, did not understand, but my hostess, Jane Temerlin (who was also Lucy's 'mother'), translated: 'That dog,' Lucy commented, pausing at a photo of a small white poodle. She turned the page. 'Blue,' she declared, pointing then signing as she gazed at a picture of a lady advertising some kind of soap powder and wearing a brilliant blue dress. And finally, after some vague hand movements—perhaps signed mutterings— 'This Lucy's, this mine,' as she closed the magazine and laid it on her lap. She had just been taught, Jane told me, the use of the possessive pronouns during the thrice weekly ASL lessons she was receiving at the time.

The book written by Lucy's human 'father,' Maury Temerlin, was entitled *Lucy, Growing Up Human.* And in fact,

the chimpanzee is more like us than is any other living creature. There is close resemblance in the physiology of our two species and genetically, in the structure of the DNA, chimpanzees and humans differ by only just over one per cent. This is why medical research uses chimpanzees as experimental animals when they need substitutes for humans in the testing of some drug or vaccine. Chimpanzees can be infected with just about all known human infectious diseases including those, such as hepatitis B and AIDS, to which other non-human animals (except gorillas, orangutans and gibbons) are immune. There are equally striking similarities between humans and chimpanzees in the anatomy and wiring of the brain and nervous system, and—although many scientists have been reluctant to admit to this—in social behaviour, intellectual ability, and the emotions. The notion of an evolutionary continuity in physical structure from pre-human ape to modern man has long been morally acceptable to most scientists. That the same might hold good for mind was generally considered an absurd hypothesis—particularly by those who used, and often misused, animals in their laboratories. It is, after all, convenient to believe that the creature you are using, while it may react in disturbingly human-like ways, is, in fact, merely a mindless and, above all, unfeeling, 'dumb' animal.

When I began my study at Gombe in 1960 it was not permissible—at least not in ethological circles—to talk about an animal's mind. Only humans had minds. Nor was it quite proper to talk about animal personality. Of course everyone knew that they *did* have their own unique characters—everyone who had ever owned a dog or other pet was aware of that. But ethologists, striving to make theirs a 'hard' science, shied away from the task of trying to explain such things objectively. One respected ethologist, while acknowledging that there was 'variability between individual animals,' wrote that it was best that this fact be 'swept under the carpet.' At that time ethological carpets fairly bulged with all that was hidden beneath them.

How naive I was. As I had not had an undergraduate science education I didn't realize that animals were not supposed to have personalities, or to think, or to feel emotions or pain. I had no idea that it would have been more appropriate to assign each of the chimpanzees a number rather than a name when I got to know him or her. I didn't realize that it was not scientific to discuss behaviour in terms of motivation or purpose. And no one had told me that terms such as *childhood* and *adolescence* were uniquely human phases of the life cycle, culturally determined, not to be used when referring to young chimpanzees. Not knowing, I freely made use of all those forbidden terms and concepts in my initial attempt to describe, to the best of my ability, the amazing things I had observed at Gombe.

I shall never forget the response of a group of ethologists to some remarks I made at an erudite seminar. I described how Figan, as an adolescent, had learned to stay behind in camp after senior males had left, so that we could give him a few bananas for himself. On the first occasion he had, upon seeing the fruits, uttered loud, delighted food calls: whereupon a couple of the older males had charged back, chased after Figan, and taken his bananas. And then, coming to the point of the story, I explained how, on the next occasion, Figan had actually suppressed his calls. We could hear little sounds, in his throat, but so quiet that none of the others could have heard them. Other young chimps, to whom we tried to smuggle fruit without the knowledge of their elders, never learned such self-control. With shrieks of glee they would fall to, only to be robbed of their booty when the big

males charged back. I had expected my audience to be as fascinated and impressed as I was. I had hoped for an exchange of views about the chimpanzee's undoubted intelligence. Instead there was a chill silence, after which the chairman hastily changed the subject. Needless to say, after being thus snubbed, I was very reluctant to contribute any comments, at any scientific gatherings, for a very long time. Looking back, I suspect that everyone was interested, but it was, of course, not permissible to present a mere 'anecdote' as evidence for anything.

The editorial comments on the first paper I wrote for publication demanded that every *he* or *she* be replaced with *it,* and every *who* be replaced with *which.* Incensed, I, in my turn, crossed out the *its* and *whichs* and scrawled back the original pronouns. As I had no desire to carve a niche for myself in the world of science, but simply wanted to go on living among and learning about chimpanzees, the possible reaction of the editor of the learned journal did not trouble me. In fact I won that round: the paper when finally published did confer upon the chimpanzees the dignity of their appropriate genders and properly upgraded them from the status of mere 'things' to essential Beingness.

However, despite my somewhat truculent attitude, I did want to learn, and I was sensible of my incredible good fortune in being admitted to Cambridge. I wanted to get my PhD, if only for the sake of Louis Leakey and the other people who had written letters in support of my admission. And how lucky I was to have, as my supervisor, Robert Hinde. Not only because I thereby benefitted from his brilliant mind and clear thinking, but also because I doubt that I could have found a teacher more suited to my particular needs and personality. Gradually he was able to cloak me with at least some of the trappings of a scientist. Thus although I continued to hold to most of my convictions—that animals had personalities; that they could feel happy or sad or fearful; that they could feel pain; that they could strive towards planned goals and achieve greater success if they were highly motivated—I soon realized

that these personal convictions were, indeed, difficult to prove. It was best to be circumspect—at least until I had gained some credentials and credibility. And Robert gave me wonderful advice on how best to tie up some of my more rebellious ideas with scientific ribbon. 'You can't *know* that Fifi was jealous,' had admonished on one occasion. We argued a little. And then: 'Why don't you just say *If Fifi were a human child we would say she was jealous.*' I did.

It is not easy to study emotions even when the subjects are human. I know how I feel if I am sad or happy or angry, and if a friend tells me that he is feeling sad, happy or angry, I assume that his feelings are similar to mine. But of course I cannot know. As we try to come to grips with the emotions of beings progressively more different from ourselves the task, obviously, becomes increasingly difficult. If we ascribe human emotions to non-human animals we are accused of being anthropomorphic—a cardinal sin in ethology. But is it so terrible? If we test the effect of drugs on chimpanzees because they are biologically so similar to ourselves, if we accept that there are dramatic similarities in chimpanzee and human brain and nervous system, is it not logical to assume that there will be similarities also in at least the more basic feelings, emotions, moods of the two species?

In fact, all those who have worked long and closely with chimpanzees have no hesitation in asserting that chimps experience emotions similar to those which in ourselves we label pleasure, joy, sorrow, anger, boredom and so on. Some of the emotional states of the chimpanzee are so obviously similar to ours that even an inexperienced observer can understand what is going on. An infant who hurls himself screaming to the ground, face contorted, hitting out with his arms at any nearby object, banging his head, is clearly having a tantrum. Another youngster, who gambols around his mother, turning somersaults, pirouetting and, every so often, rushing up to her and tumbling into her lap, patting her or pulling her hand towards him in a request for tickling, is obviously filled with *joie de vivre*. There are few observers who would not unhesitatingly ascribe his behaviour to a happy, care-free state of well-being. And one cannot watch chimpanzee infants for long without realizing that they have the same emotional need for affection and reassurance as human children. An adult male, reclining in the shade after a good meal, reaching benignly to play with an infant or idly groom an adult female, is clearly in a good mood. When he sits with bristling hair, glaring at his subordinates and threatening them, with irritated gestures, if they come too close, he is clearly feeling cross and grumpy. We make these judgements because the similarity of so much of a chimpanzee's behaviour to our own permits us to empathize.

It is hard to empathize with emotions we have not experienced. I can image, to some extent, the pleasure of a female chimpanzee during the act of procreation. The feelings of her male partner are beyond my knowledge—as are those of the human male in the same context. I have spent countless hours watching mother chimpanzees interacting with their infants. But not until I had an infant of my own did I begin to understand the basic, powerful instinct of mother-love. If someone accidentally did something to frighten Grub, or threaten his well-being in any way, I felt a surge of quite irrational anger. How much more easily could I then understand the feelings of the chimpanzee mother who furiously waves her arm and barks in threat at an individual who approaches her infant too closely, or at a playmate who inadvertently hurts her child. And it was not until I knew the numbing grief that gripped me after the death of my second husband that I could even begin to appreciate the despair and sense of loss that can cause young chimps to pine away and die when they lose their mothers.

Empathy and intuition can be of tremendous value as we attempt to understand certain complex behavioral interactions, provided that the behaviour, as it occurs, is recorded precisely and objectively. Fortunately I have seldom found it difficult to record facts in an orderly manner even during times of powerful emotional involvement. And "knowing" intuitively how a chimpanzee is feeling—after an attack,

for example—may help one to understand what happens next. We should not be afraid at least to try to make use of our close evolutionary relationship with the chimpanzees in our attempts to interpret complex behaviour.

Today, as in Darwin's time, it is once again fashionable to speak of and study the animal mind. This change came about gradually, and was, at least in part, due to the information collected during careful studies of animal societies in the field. As these observations became widely known, it was impossible to brush aside the complexities of social behaviour that were revealed in species after species. The untidy clutter under the ethological carpets was brought out and examined, piece by piece. Gradually it was realized that parsimonious explanations of apparently intelligent behaviours were often misleading. This led to a succession of experiments that, taken together, clearly prove that many intellectual abilities that had been thought unique to humans were actually present, though in a less highly developed form, in other, non-human beings. Particularly, of course, in the non-human primates and especially in chimpanzees.

When first I began to read about human evolution, I learned that one of the hallmarks of our own species was that we, and only we, were capable of making tools. *Man the Toolmaker* was an oft-cited definition—and this despite the careful and exhaustive research of Wolfgang Kohler and Robert Yerkes on the tool-using and tool-making abilities of chimpanzees. Those studies, carried out independently in the early twenties, were received with scepticism. Yet both Kohler and Yerkes were respected scientists, and both had a profound understanding of chimpanzee behaviour. Indeed, Kohler's descriptions of the personalities and behaviour of the various individuals in his colony, published in his book *The Mentality of Apes,* remain some of the most vivid and colourful ever written. And his experiments, showing how chimpanzees could stack boxes, then climb the unstable constructions to reach fruit suspended from the ceiling, or join two short sticks to make a pole long enough to rake in fruit otherwise out of reach, have become classic, ap-

pearing in almost all textbooks dealing with intelligent behaviour in non-human animals.

By the time systematic observations of tool-using came from Gombe those pioneering studies had been largely forgotten. Moreover, it was one thing to know that humanized chimpanzees in the lab could use implements: it was quite another to find that this was a naturally occurring skill in the wild. I well remember writing to Louis about my first observations, describing how David Greybeard not only used bits of straw to fish for termites but actually stripped leaves from a stem and thus *made* a tool. And I remember too receiving the now oft-quoted telegram he sent in response to my letter: "Now we must redefine *tool,* redefine *Man,* or accept chimpanzees as humans."

There were initially, a few scientists who attempted to write off the termiting observations, even suggesting that I had taught the chimps! By and large, though, people were fascinated by the information and by the subsequent observations of the other contexts in which the Gombe chimpanzees used objects as tools. And there were only a few anthropologists who objected when I suggested that the chimpanzees probably passed their tool-using traditions from one generation to the next, through observations, imitation and practice, so that each population might be expected to have its own unique tool-using culture. Which, incidentally, turns out to be quite true. And when I described how one chimpanzee, Mike, spontaneously solved a new problem by using a tool (he broke off a stick to knock a banana to the ground when he was too nervous to actually take it from my hand) I don't believe there were any raised eyebrows in the scientific community. Certainly I was not attacked viciously, as were Kohler and Yerkes, for suggesting that humans were not the only beings capable of reasoning and insight.

The mid-sixties saw the start of a project that, along with other similar research, was to teach us a great deal about the chimpanzee mind. This was Project Washoe, conceived by Trixie and Allen Gardner. They purchased an infant chimpanzee and began to teach her the signs of ASL, the American Sign Language used by the deaf. Twenty years earlier another husband and wife team, Richard and Cathy Hayes, had tried, with an almost total lack of success, to teach a young chimp, Vikki, to talk. The Hayes's undertaking taught us a lot about the chimpanzee mind, but Vikki, although she did well in IQ tests, and was clearly an intelligent youngster, could not learn human speech. The Gardners, however, achieved spectacular success with their pupil, Washoe. Not only did she learn signs easily, but she quickly began to string them together in meaningful ways. It was clear that each sign evoked, in her mind, a mental image of the object it represented. If, for example, she was asked, in sign language, to fetch an apple, she would go and locate an apple that was out of sight in another room.

Other chimps entered the project, some starting their lives in deaf signing families before joining Washoe. And finally Washoe adopted an infant, Loulis. He came from a lab where no thought of teaching signs had ever penetrated. When he was with Washoe he was given no lessons in language acquisition—not by humans, anyway. Yet by the time he was eight years old he had made fifty-eight signs in their correct contexts. How did he learn them? Mostly, it seems, by imitating the behaviour of Washoe and the other three signing chimps, Dar, Moja and Tatu. Sometimes, though, he received tuition from Washoe herself. One day, for example, she began to swagger about bipedally, hair bristling, signing *food! food! food!* in great excitement. She had seen a human approaching with a bar of chocolate. Loulis, only eighteen months old, watched passively. Suddenly Washoe stopped her swaggering, went over to him, took his hand, and moulded the sign for *food* (fingers pointing towards mouth). Another time, in a similar context, she made the sign for *chewing gum*—but with *her* hand on *his* body. On a third occasion Washoe, apropos of nothing, picked up a small chair, took it over to Loulis, set it down in front of him, and very distinctly made the *chair* sign three times, watching him closely as she did so. The two food signs be-

came incorporated into Loulis's vocabulary but the sign for chair did not. Obviously the priorities of a young chimp are similar to those of a human child!

When news of Washoe's accomplishments first hit the scientific community it immediately provoked a storm of bitter protest. It implied that chimpanzees were capable of mastering a human language, and this, in turn, indicated mental powers of generalization, abstraction and concept-formation as well as an ability to understand and use abstract symbols. And these intellectual skills were surely the prerogatives of *Homo sapiens.* Although there were many who were fascinated and excited by the Gardners' findings, there were many more who denounced the whole project, holding that the data was suspect, the methodology sloppy, and the conclusions not only misleading, but quite preposterous. The controversy inspired all sorts of other language projects. And, whether the investigators were sceptical to start with and hoped to disprove the Gardners' work, or whether they were attempting to demonstrate the same thing in a new way, their research provided additional information about the chimpanzee's mind.

And so, with new incentive, psychologists began to test the mental abilities of chimpanzees in a variety of different ways; again and again the results confirmed that their minds are uncannily like our own. It had long been held that only humans were capable of what is called 'cross-modal transfer of information'—in other words, if you shut your eyes and someone allows you to feel a strangely shaped potato, you will subsequently be able to pick it out from other differently shaped potatoes simply by looking at them. And vice versa. It turned out that chimpanzees can 'know' with their eyes what they 'feel' with their fingers in just the same way. In fact, we now know that some other non-human primates can do the same thing. I expect all kinds of creatures have the same ability.

Then it was proved, experimentally and beyond doubt, that chimpanzees could recognize themselves in mirrors—that they had, therefore, some kind of self-concept. In fact, Washoe, some

years previously, had already demonstrated the ability when she spontaneously identified herself in the mirror, staring at her image and making her name sign. But that observation was merely anecdotal. The proof came when chimpanzees who had been allowed to play with mirrors were, while anaesthetized, dabbed with spots of odourless paint in places, such as the ears or the top of the head, that they could see only in the mirror. When they woke they were not only fascinated by their spotted images, but immediately investigated, with their fingers, the dabs of paint.

The fact that chimpanzees have excellent memories surprised no one. Everyone, after all, has been brought up to believe that 'an elephant never forgets' so why should a chimpanzee be any different? The fact that Washoe spontaneously gave the name-sign of Beatrice Gardner, her surrogate mother, when she saw her after a separation of eleven years was no greater an accomplishment than the amazing memory shown by dogs who recognize their owners after separations of almost as long—and the chimpanzee has a much longer life span than a dog. Chimpanzees can plan ahead, too, at least as regards the immediate future. This, in fact, is well illustrated at Gombe, during the termiting season: often an individual prepares a tool for use on a termite mound that is several hundred yards away and absolutely out of sight.

This is not the place to describe in detail the other cognitive abilities that have been studied in laboratory chimpanzees. Among other accomplishments chimpanzees possess pre-mathematical skills: they can, for example, readily differentiate between *more* and *less.* They can classify things into specific categories according to a given criterion—thus they have no difficulty in separating a pile of food into *fruits* and *vegetables* on one occasion, and, on another, dividing the same pile of food into *large* versus *small* items, even though this requires putting some vegetables with some fruits. Chimpanzees who have been taught a language can combine signs creatively in order to describe objects for which they have no symbol. Washoe, for example, puzzled her caretakers by asking, repeatedly, for a *rock berry*. Eventually it transpired that she was referring to Brazil nuts which she had encountered for the first time a while before. Another language-trained chimp described a cucumber as a *green banana,* and another referred to an Alka-Seltzer as a *listen drink*. They can even invent signs. Lucy, as she got older, had to be put on a leash for her outings. One day, eager to set off but having no sign for *leash,* she signalled her wishes by holding a crooked index finger to the ring on her collar. This sign became part of her vocabulary. Some chimpanzees love to draw, and especially to paint. Those who have learned sign language sometimes spontaneously label their works, 'This [is] apple'—or bird, or sweetcorn, or whatever. The fact that the paintings often look, to our eyes, remarkably unlike the objects depicted by the artists either means that the chimpanzees are poor draughtsmen or that we have much to learn regarding ape-style representational art!

People sometimes ask why chimpanzees have evolved such complex intellectual powers when their lives in the wild are so simple. The answer is, of course, that their lives in the wild are not so simple! They use—and need—all their mental skills during normal day-to-day life in their complex society. They are always having to make choices—where to go, or with whom to travel. They need highly developed social skills—particularly those males who are ambitious to attain high positions in the dominance hierarchy. Low-ranking chimpanzees must learn deception—to conceal their intentions or to do things in secret—if they are to get their way in the presence of their superiors. Indeed, the study of chimpanzees in the wild suggests that their intellectual abilities evolved, over the millennia, to help them cope with daily life. And now, the solid core of data concerning chimpanzee intellect collected so carefully in the lab setting provides a background against which to evaluate the many examples of intelligent, rational behaviour that we see in the wild.

It is easier to study intellectual prowess in the lab where, through carefully devised tests and judicious use of rewards, the chimpanzees can be encouraged to exert themselves, to stretch their minds to the limit. It is more meaningful to study the subject in the wild, but much harder. It is more meaningful because we can better understand the environmental pressures that led to the evolution of intellectual skills in chimpanzee societies. It is harder because, in the wild, almost all behaviours are confounded by countless variables; years of observing, recording and analysing take the place of contrived testing; sample size can often be counted on the fingers of one hand; the only experiments are nature's own, and only time— eventually—may replicate them.

In the wild a single observation may prove of utmost significance, providing a clue to some hitherto puzzling aspect of behaviour, a key to the understanding of, for example, a changed relationship. Obviously it is crucial to see as many incidents of this sort as possible. During the early years of my study at Gombe it became apparent that one person alone could never learn more than a fraction of what was going on in a chimpanzee community at any given time. And so, from 1964 onwards, I gradually built up a research team to help in the gathering of information about the behaviour of our closest living relatives.

Dim Forest, Bright Chimps

*In the rain forest of Ivory Coast, chimpanzees meet
the challenge of life by hunting cooperatively
and using crude tools*

Christophe Boesch and
Hedwige Boesch-Achermann

Taï National Park, Ivory Coast, December 3, 1985. Drumming, barking, and screaming, chimps rush through the undergrowth, little more than black shadows. Their goal is to join a group of other chimps noisily clustering around Brutus, the dominant male of this seventy-member chimpanzee community. For a few moments, Brutus, proud and self-confident, stands fairly still, holding a shocked, barely moving red colobus monkey in his hand. Then he begins to move through the group, followed closely by his favorite females and most of the adult males. He seems to savor this moment of uncontested superiority, the culmination of a hunt high up in the canopy. But the victory is not his alone. Cooperation is essential to capturing one of these monkeys, and Brutus will break apart and share this highly prized delicacy with most of the main participants of the hunt and with the females. Recipients of large portions will, in turn, share more or less generously with their offspring, relatives, and friends.

In 1979, we began a long-term study of the previously unknown chimpanzees of Taï National Park, 1,600 square miles of tropical rain forest in the Republic of the Ivory Coast (Côte d'Ivoire). Early

on, we were most interested in the chimps' use of natural hammers—branches and stones—to crack open the five species of hard-shelled nuts that are abundant here. A sea otter lying on its back, cracking an abalone shell with a rock, is a familiar picture, but no primate had ever before been observed in the wild using stones as hammers. East Africa's savanna chimps, studied for decades by Jane Goodall in Gombe, Tanzania, use twigs to extract ants and termites from their nests or honey from a bees' nest, but they have never been seen using hammerstones.

As our work progressed, we were surprised by the many ways in which the life of the Taï forest chimpanzees differs from that of their savanna counterparts, and as evidence accumulated, differences in how the two populations hunt proved the most intriguing. Jane Goodall had found that chimpanzees hunt monkeys, antelope, and wild pigs, findings confirmed by Japanese biologist Toshida Nishida, who conducted a long-term study 120 miles south of Gombe, in the Mahale Mountains. So we were not surprised to discover that the Taï chimps eat meat. What intrigued us was the degree to which they hunt cooperatively. In 1953 Raymond Dart proposed

that group hunting and cooperation were key ingredients in the evolution of *Homo sapiens*. The argument has been modified considerably since Dart first put it forward, and group hunting has also been observed in some social carnivores (lions and African wild dogs, for instance), and even some birds of prey. Nevertheless, many anthropologists still hold that hunting cooperatively and sharing food played a central role in the drama that enabled early hominids, some 1.8 million years ago, to develop the social systems that are so typically human.

We hoped that what we learned about the behavior of forest chimpanzees would shed new light on prevailing theories of human evolution. Before we could even begin, however, we had to habituate a community of chimps to our presence. Five long years passed before we were able to move with them on their daily trips through the forest, of which "our" group appeared to claim some twelve square miles. Chimpanzees are alert and shy animals, and the limited field of view in the rain forest—about sixty-five feet at best—made finding them more difficult. We had to rely on sound, mostly their vocalizations and drumming on trees. Males often drum regularly while moving through the forest: pant-hooting, they draw near a big buttress tree; then, at full speed they fly over the buttress, hitting it repeatedly with their hands and feet. Such drumming may resound more than half a mile in the forest. In the beginning, our ignorance about how they moved and who was drumming led to failure more often than not, but eventually we learned that the dominant males drummed during the day to let other group members know the direction of travel. On some days, however, intermittent drumming about dawn was the only signal for the whole day. If we were out of earshot at the time, we were often reduced to guessing.

During these difficult early days, one feature of the chimps' routine proved to be our salvation: nut cracking is a noisy business. So noisy, in fact, that in the early days of French colonial rule, one officer apparently even proposed the theory that some unknown tribe was forging iron in the impenetrable and dangerous jungle.

Guided by the sounds made by the chimps as they cracked open nuts, which they often did for hours at a time, we were gradually able to get within sixty feet of the animals. We still seldom saw the chimps themselves (they fled if we came too close), but even so, the evidence left after a session of nut cracking taught us a great deal about what types of nuts they were eating, what sorts of hammer and anvil tools they were using, and—thanks to the very distinctive noise a nut makes when it finally splits open—how many hits were needed to crack a nut and how many nuts could be opened per minute.

After some months, we began catching glimpses of the chimpanzees before they fled, and after a little more time, we were able to draw close enough to watch them at work. The chimps gather nuts from the ground. Some nuts are tougher to crack than others. Nuts of the *Panda oleosa* tree are the most demanding, harder than any of the foods processed by present-day hunter-gatherers and breaking open only when a force of 3,500 pounds is applied. The stone hammers used by the Taï chimps range from stones of ten ounces to granite blocks

of four to forty-five pounds. Stones of any size, however, are a rarity in the forest and are seldom conveniently placed near a nut-bearing tree. By observing closely, and in some cases imitating the way the chimps handle hammerstones, we learned that they have an impressive ability to find just the right tool for the job at hand. Taï chimps could remember the positions of many of the stones scattered, often out of sight, around a panda tree. Without having to run around rechecking the stones, they would select one of appropriate size that was closest to the tree. These mental abilities in spatial representation compare with some of those of nine-year-old humans.

To extract the four kernels from inside a panda nut, a chimp must use a hammer with extreme precision. Time and time again, we have been impressed to see a chimpanzee raise a twenty-pound stone above its head, strike a nut with ten or more powerful blows, and then, using the same hammer, switch to delicate little taps from a height of only four inches. To finish the job, the chimps often break off a small piece of twig and use it to extract the last tiny fragments of kernel from the shell. Intriguingly, females crack panda nuts more often than males, a gender difference in tool use that seems to be more pronounced in the forest chimps than in their savanna counterparts.

After five years of fieldwork, we were finally able to follow the chimpanzees at close range, and gradually, we gained insights into their way of hunting. One morning, for example, we followed a group of six male chimps on a three-hour patrol that had taken them into foreign territory to the north. (Our study group is one of five chimpanzee groups more or less evenly distributed in the Taï forest.) As always during these approximately monthly incursions, which seem to be for the purpose of territorial defense, the chimps were totally silent, clearly on edge and on the lookout for trouble. Once the patrol was over, however, and they were back within their own borders, the chimps shifted their attention to hunting. They were after monkeys, the most abundant mammals in the forest. Traveling in large, multi-species groups, some of the forest's ten species

of monkeys are more apt than others to wind up as a meal for the chimps. The relatively sluggish and large (almost thirty pounds) red colobus monkeys are the chimps' usual fare. (Antelope also live in the forest, but in our ten years at Taï, we have never seen a chimp catch, or even pursue, one. In contrast, Gombe chimps at times do come across fawns, and when they do, they seize the opportunity—and the fawn.)

The six males moved on silently, peering up into the vegetation and stopping from time to time to listen for the sound of monkeys. None fed or groomed; all focused on the hunt. We followed one old male, Falstaff, closely, for he tolerates us completely and is one of the keenest and most experienced hunters. Even from the rear, Falstaff set the pace; whenever he stopped, the others paused to wait for him. After thirty minutes, we heard the unmistakable noises of monkeys jumping from branch to branch. Silently, the chimps turned in the direction of the sounds, scanning the canopy. Just then, a diana monkey spotted them and gave an alarm call. Dianas are very alert and fast; they are also about half the weight of colobus monkeys. The chimps quickly gave up and continued their search for easier, meatier prey.

Shortly after, we heard the characteristic cough of a red colobus monkey. Suddenly Rousseau and Macho, two twenty-year-olds, burst into action, running toward the cough. Falstaff seemed surprised by their precipitousness, but after a moment's hesitation, he also ran. Now the hunting barks of the chimps mixed with the sharp alarm calls of the monkeys. Hurrying behind Falstaff, we saw him climb up a conveniently situated tree. His position, combined with those of Schubert and Ulysse, two mature chimps in their prime, effectively blocked off three of the monkeys' possible escape routes. But in another tree, nowhere near any escape route and thus useless, waited the last of the hunters, Kendo, eighteen years old and the least experienced of the group. The monkeys, taking advantage of Falstaff's delay and Kendo's error, escaped.

The six males moved on and within five minutes picked up the sounds of another group of red colobus. This time, the chimps approached cautiously, nobody hurrying. They screened the canopy intently to locate the monkeys, which were still unaware of the approaching danger. Macho and Schubert chose two adjacent trees, both full of monkeys, and started climbing very quietly, taking care not to move any branches. Meanwhile, the other four chimps blocked off anticipated escape routes. When Schubert was halfway up, the monkeys finally detected the two chimps. As we watched the colobus monkeys take off in literal panic, the appropriateness of the chimpanzees' scientific name—*Pan* came to mind: with a certain stretch of the imagination, the fleeing monkeys could be shepherds and shepherdesses frightened at the sudden appearance of Pan, the wild Greek god of the woods, shepherds, and their flocks.

Taking off in the expected direction, the monkeys were trailed by Macho and Schubert. The chimps let go with loud hunting barks. Trying to escape, two colobus monkeys jumped into smaller trees lower in the canopy. With this, Rousseau and Kendo, who had been watching from the ground, sped up into the trees and tried to grab them. Only a third of the weight of the chimps, however, the monkeys managed to make it to the next tree along branches too small for their pursuers. But Falstaff had anticipated this move and was waiting for them. In the following confusion, Falstaff seized a juvenile and killed it with a bite to the neck. As the chimps met in a rush on the ground, Falstaff began to eat, sharing with Schubert and Rousseau. A juvenile colobus does not provide much meat, however, and this time, not all the chimps got a share. Frustrated individuals soon started off on another hunt, and relative calm returned fairly quickly: this sort of hunt, by a small band of chimps acting on their own at the edge of their territory, does not generate the kind of high excitement that prevails when more members of the community are involved.

So far we have observed some 200 monkey hunts and have concluded that success requires a minimum of three motivated hunters acting cooperatively.

Alone or in pairs, chimps succeed less than 15 percent of the time, but when three or four act as a group, more than half the hunts result in a kill. The chimps seem well aware of the odds; 92 percent of all the hunts we observed were group affairs.

Gombe chimps also hunt red colobus monkeys, but the percentage of group hunts is much lower: only 36 percent. In addition, we learned from Jane Goodall that even when Gombe chimps do hunt in groups, their strategies are different. When Taï chimps arrive under a group of monkeys, the hunters scatter, often silently, usually out of sight of one another but each aware of the others' positions. As the hunt progresses, they gradually close in, encircling the quarry. Such movements require that each chimp coordinate his movements with those of the other hunters, as well as with those of the prey, at all times.

Coordinated hunts account for 63 percent of all those observed at Taï but only 7 percent of those at Gombe. Jane Goodall says that in a Gombe group hunt, the chimpanzees typically travel together until they arrive at a tree with monkeys. Then, as the chimps begin climbing nearby trees, they scatter as each pursues a different target. Goodall gained the impression that Gombe chimps boost their success by hunting independently but simultaneously, thereby disorganizing their prey; our impression is that the Taï chimps owe their success to being organized themselves.

Just why the Gombe and Taï chimps have developed such different hunting strategies is difficult to explain, and we plan to spend some time at Gombe in the hope of finding out. In the meantime, the mere existence of differences is interesting enough and may perhaps force changes in our understanding of human evolution. Most currently accepted theories propose that some three million years ago, a dramatic climate change in Africa east of the Rift Valley turned dense forest into open, drier habitat. Adapting to the difficulties of life under these new conditions, our ancestors supposedly evolved into cooperative hunters and began sharing food they caught. Supporters of this idea point out that plant and animal remains indicative

of dry, open environments have been found at all early hominid excavation sites in Tanzania, Kenya, South Africa, and Ethiopia. That the large majority of apes in Africa today live west of the Rift Valley appears to many anthropologists to lend further support to the idea that a change in environment caused the common ancestor of apes and humans to evolve along a different line from those remaining in the forest.

Our observations, however, suggest quite another line of thought. Life in dense, dim forest may require more sophisticated behavior than is commonly assumed: compared with their savanna relatives, Taï chimps show greater complexity in both hunting and tool use. Taï chimps use tools in nineteen different ways and have six different ways of making them, compared with sixteen uses and three methods of manufacture at Gombe.

Anthropologist colleagues of mine have told me that the discovery that some chimpanzees are accomplished users of hammerstones forces them to look with a fresh eye at stone tools turned up at excavation sites. The important role played by female Taï chimps in tool use also raises the possibility that in the course of human evolution, women may have been decisive in the development of many of the sophisticated manipulative skills characteristic of our species. Taï mothers also appear to pass on their skills by actively teaching their offspring. We have observed mothers providing their young with hammers and then stepping in to help when the inexperienced youngsters encounter difficulty. This help may include carefully showing how to position the nut or hold the hammer properly. Such behavior has never been observed at Gombe.

Similarly, food sharing, for a long time said to be unique to humans, seems more general in forest than in savanna chimpanzees. Taï chimp mothers share with their young up to 60 percent of the nuts they open, at least until the latter become sufficiently adept, generally at about six years old. They also share other foods acquired with tools, including honey, ants, and bone marrow. Gombe mothers share such foods much less often, even with their infants. Taï chimps also share meat more frequently than do their Gombe relatives, sometimes dividing a chunk up and giving portions away, sometimes simply allowing beggars to grab pieces.

Any comparison between chimpanzees and our hominid ancestors can only be suggestive, not definitive. But our studies lead us to believe that the process of hominization may have begun independently of the drying of the environment. Savanna life could even have delayed the process; many anthropologists have been struck by how slowly hominid-associated remains, such as the hand ax, changed after their first appearance in the Olduvai age.

Will we have the time to discover more about the hunting strategies or other, perhaps as yet undiscovered abilities of these forest chimpanzees? Africa's tropical rain forests, and their inhabitants, are threatened with extinction by extensive logging, largely to provide the Western world with tropical timber and such products as coffee, cocoa, and rubber. Ivory Coast has lost 90 percent of its original forest, and less than 5 percent of the remainder can be considered pristine. The climate has changed dramatically. The harmattan, a cold, dry wind from the Sahara previously unknown in the forest, has now swept through the Taï forest every year since 1986. Rainfall has diminished; all the rivulets in our study region are now dry for several months of the year.

In addition, the chimpanzee, biologically very close to humans, is in demand for research on AIDS and hepatitis vaccines. Captive-bred chimps are available, but they cost about twenty times more than wild-caught animals. Chimps taken from the wild for these purposes are generally young, their mothers having been shot during capture. For every chimp arriving at its sad destination, nine others may well have died in the forest or on the way. Such priorities—cheap coffee and cocoa and chimpanzees—do not do the economies of Third World countries any good in the long run, and they bring suffering and death to innocent victims in the forest. Our hope is that Brutus, Falstaff, and their families will survive, and that we and others will have the opportunity to learn about them well into the future. But there is no denying that modern times work against them and us.

To Catch a Colobus

*Chimpanzees in Gombe National Park band together to kill nearly
a fifth of the red colobus monkeys in their range*

Craig B. Stanford

Craig B. Stanford is an assistant professor of anthropology at the University of Southern California. His first fieldwork on primates was in Peru, where he studied tamarins. For his Ph.D. at the University of California, Berkeley, Stanford traveled to India and Bangladesh to investigate ecological influences on social behavior in capped langur monkeys. Stanford hopes to expand his research to the evolution of hunting behavior in primates, including humans.

On a sunny July morning, I am sitting on the bank of Kakombe Stream in Gombe National Park, Tanzania. Forty feet above my head, scattered through large fig trees, is a group of red colobus monkeys. This is J group, whose twenty-five members I have come to know as individuals during several seasons of fieldwork. Gombe red colobus are large, long-tailed monkeys, with males sometimes weighing more than twenty pounds. Both sexes have a crown of red hair, a gray back, and buff underparts. The highlight of this particular morning has been the sighting of a new infant, born sometime in the previous two days. As the group feeds noisily on fruit and leaves overhead, I mull over the options for possible names for the infant.

While I watch the colobus monkeys, my attention is caught by the loud and excited pant-hoots of a party of chimpanzees farther down the valley. I judge the group to be of considerable size and traveling in my direction. As the calls come closer, the colobus males begin to give high-pitched alarm calls, and mothers gather up their infants and climb higher into the tree crowns.

A moment later, a wild chorus of panthoots erupts just behind me, followed by a cacophony of colobus alarm calls, and it is obvious to both J group and to me that the chimps have arrived. The male chimps immediately climb up to the higher limbs of the tall albizia tree into which most of the colobus group have retreated. Colobus females and their offspring huddle high in the crown, while a phalanx of five adult males descends to meet the advancing ranks of four adult male chimpanzees, led by seventeen-year-old, 115-pound Frodo. Frodo is the most accomplished hunter of colobus monkeys at Gombe and the only one willing to take on several colobus males simultaneously in order to catch his prey. The other hunters keep their distance while Frodo first scans the group of monkeys, then advances upon the colobus defenders. Time and again he lunges at the colobus males, attempting to race past them and into the cluster of terrified females and infants. Each time he is driven back; at one point, the two largest males of J group leap onto Frodo's back until he retreats, screaming, a few yards away.

A brief lull in the hunt follows, during which the colobus males run to one another and embrace for reassurance, then part to renew their defense. Frodo soon charges again into the midst of the colobus males, and this time manages to scatter them long enough to pluck the newborn from its mother's abdomen. In spite of fierce opposition, Frodo has caught his quarry, and he now sits calmly and eats it while the other hunters and two female chimps—their swollen pink rumps a sign that they are in estrus, a period of sexual receptivity—sit nearby begging for meat. The surviving colobus monkeys watch nervously from a few feet away. Minutes later, the mother of the dead infant attempts to approach, perhaps to try to rescue her nearly consumed offspring. She is chased, falls from the tree to the forest floor, and is pounced upon and killed by juvenile chimpanzees that have been watching the hunt from below. Seconds later, before these would-be hunters have had a chance to begin their meal, Wilkie, the chimpanzee group's dominant male, races down the tree and steals the carcass from them. He shows off his prize by charging across the forest floor, dead colobus in hand, and then, amid a frenzy of chimps eager for a morsel, he sits down to share the meat with his ally Prof and two females from the hunting party.

Until Jane Goodall observed chimpanzees eating meat in the early 1960s, they were thought to be complete vegetarians. We now know that a small but regular portion of the diet of wild chimps consists of the meat of such mammals as bush pigs, small antelopes, and a variety of monkey species. For example, chimpanzees in the Mahale Mountains of Tanzania, the Taï forest of Ivory Coast, and in Gombe all regularly hunt red colobus monkeys. Documenting the effect of such predation on wild primate populations, however, is extremely difficult because predators—whether chimps, leopards, or eagles—are generally too shy to hunt in the presence of people. The result is that even if predation is a regular occurrence, researchers are not likely to see it, let alone study it systematically.

Gombe is one of the few primate study sites where both predators and their prey have been habituated to human observers, making it possible to witness hunts. I have spent the past four field seasons at Gombe, studying the predator-prey relationship between the 45-member Kasakela chimpanzee community and the 500 red colobus monkeys that share the same twelve square miles of Gombe National Park. Gombe's rugged terrain is composed of steep slopes of open woodland, rising above stream valleys lush with riverine forest. The chimpanzees roam across these hills in territorial communities, which divide up each day into foraging parties of from one to forty animals. So far, I have clocked in more than a thousand hours with red colobus monkeys and have regularly followed the chimps on their daily rounds, observing some 150 encounters between the monkeys and chimps and more than 75 hunts. My records, together with those of my colleagues, show that the Gombe chimps may kill more than 100 red colobus each year, or nearly one-fifth of the colobus inhabiting their range. Most of the victims are immature monkeys under two years old. Also invaluable have been the data gathered daily on the chimps for the past two decades by a team of Tanzanian research assistants.

One odd outcome of my work has been that I am in the unique position of knowing both the hunters and their victims as individuals, which makes my research intriguing but a bit heart-wrenching. In October 1992, for example, a party of thirty-three chimpanzees encountered my main study group, J, in upper Kakombe valley. The result was

devastating from the monkeys' viewpoint. During the hour-long hunt, seven were killed; three were caught and torn apart right in front of me. Nearly four hours later, the hunters were still sharing and eating the meat they had caught, while I sat staring in disbelief at the remains of many of my study subjects.

Determined to learn more about the chimp-colobus relationship, however, I continued watching, that day and many others like it. I will need several more field seasons before I can measure the full impact of chimpanzee hunting on the Gombe red colobus, but several facts about hunting and its effects on the monkeys have already emerged. One major factor that determines the outcome of a hunt in Gombe is the number of male chimps involved. (Although females also hunt, the males are responsible for more than 90 percent of all colobus kills.) Red colobus males launch a courageous counterattack in response to their chimpanzee predators, but their ability to defend their group is directly proportional to the number of attackers and does not seem to be related to the number of defenders. The outcome of a hunt is thus almost always in the hands of the chimps, and in most instances, the best the monkeys can hope to do is limit the damage to a single group member rather than several. Chimpanzees have a highly fluid social grouping pattern in which males tend to travel together while females travel alone with their infants. At times, however, twenty or more male and female chimpanzees forage together. When ten or more male chimps hunt together, they are successful nine times out of ten, and the colobus have little hope of escape.

Hunting success depends on other factors as well. Unlike the shy red-tailed and blue monkeys with which they share the forest (and which are rarely hunted by the chimps), red colobus do not flee the moment they hear or see chimps approaching. Instead, the red colobus give alarm calls and adopt a vigilant wait-and-see strategy, with males positioned nearest the potential attackers. The alarm calls increase in frequency and intensity as the chimpanzees draw closer and cease only when the chimps are sighted beneath the tree. Then, the

colobus sit quietly, watching intently, and only if the chimps decide to hunt do the colobus males launch a counterattack. The monkeys' decision to stand and fight rather than flee may seem maladaptive given their low rate of successful defense. I observed, however, that when the monkeys scatter or try to flee, the chimps nearly always pursue and catch one or more of them.

Fleeing red colobus monkeys are most likely to be caught when they have been feeding on the tasty new leaves of the tallest trees, the "emergents," which rise above the canopy. When these trees are surrounded by low plant growth, they frequently become death traps because the only way colobus can escape from attacking chimps is to leap out of the tree—often into the waiting arms of more chimpanzees on the ground below.

One of my primary goals has been to learn why a party of chimps will eagerly hunt a colobus group one day while ignoring the same group under seemingly identical circumstances on another. One determinant is the number of males in the chimp party: the more males, the more likely the group will hunt. Hunts are also undertaken mainly when a mother colobus carrying a small infant is visible, probably because of the Gombe chimps' preference for baby red colobus, which make up 75 percent of all kills. The situation is quite different in the Taï forest, where half of the chimp kills are adult colobus males (see "Dim Forest, Bright Chimps," *Natural History,* September 1991). Christophe and Hedwige Boesch have shown that the Taï chimps hunt cooperatively, perhaps because red colobus monkeys are harder to catch in the much taller canopy of the Taï rainforest. Successful Taï chimp hunters also regularly share the spoils. In contrast, each chimp in Gombe appears to have his own hunting strategy.

The single best predictor of when Gombe chimps will hunt is the presence of one or more estrous females in the party. This finding, together with the earlier observation by Geza Teleki (formerly of George Washington University) that male hunters tend to give meat preferentially to swollen females traveling with the group, indicates that Gombe chimps sometimes hunt in order to obtain meat to offer a sexually receptive female. Since hunts also occur when no estrous females are present, this trade of sex for meat cannot be the exclusive explanation, but the implications are nonetheless intriguing. Gombe chimps use meat not only for nutrition; they also share it with their allies and withhold it from their rivals. Meat is thus a social, political, and even reproductive tool. These "selfish" goals may help explain why the Gombe chimps do not cooperate during a hunt as often as do Taï chimps.

Whatever the chimps want the monkey meat for, their predation has a severe effect on the red colobus population. Part of my work involves taking repeated censuses of the red colobus groups living in the different valleys that form the hunting range of our chimpanzees. In the core area of the range, where hunting is most intense, predation by chimps is certainly the limiting factor on colobus population growth: red colobus group size in this area is half that at the periphery of the chimps' hunting range. The number of infant and juvenile red colobus monkeys is particularly low in the core area; most of the babies there are destined to become chimpanzee food.

The proportion of the red colobus population eaten by chimps appears to fluctuate greatly from year to year, and probably from decade to decade, as the number of male hunters in the chimpanzee community changes. In the early 1980s, for instance, there were five adult and adolescent males in the Kasakela chimp community, while today there are eleven; the number of colobus kills per year has risen as the number of hunters in the community has grown.

Furthermore, a single avid hunter may have a dramatic effect. I estimate that Frodo has single-handedly killed up to 10 percent of the entire red colobus population within his hunting range. I now want to learn if chimps living in forests elsewhere in Africa are also taking a heavy toll of red colobus monkeys. If they are, then they will add support to the theory that predation is an important limiting factor on wild primate populations and may also influence some aspects of behavior. Meanwhile, I will continue to watch in awe as Frodo and his fellow hunters attack my colobus monkeys and to marvel at the courageousness of the colobus males that risk their lives to protect the other members of their group.

Language Training of Apes

E. S. Savage-Rumbaugh

Can apes learn to communicate with human beings? Scientists have been attempting to answer this question since the late 1960s when it was first reported that a young chimpanzee named Washoe in Reno, Nevada had been taught to produce hand signs similar to those used by deaf humans.

Washoe was reared much like a human child. People made signs to her throughout the day and she was given freedom to move about the caravan where she lived. She could even go outdoors to play. She was taught how to make different signs by teachers who moved her hands through the motions of each sign while showing her the object she was learning to 'name'. If she began to make a portion of the hand movement on her own she was quickly rewarded, either with food or with something appropriate to the sign. For example, if she was being taught the sign for 'tickle' her reward was a tickling game.

This training method was termed 'moulding' because it involved the physical placement of Washoe's hands. Little by little, Washoe became able to produce more and more signs on her own. As she grew older, she occasionally even learned to make new signs without moulding. Once Washoe had learned several signs she quickly began to link them together to produce strings of signs such as 'you me Out'. Such sequences appeared to her teachers to be simple sentences.

Many biologists were sceptical of the claims made for Washoe. While they agreed that Washoe was able to produce different gestures, they doubted that such signs really served as names. Perhaps, to Washoe, the gestures were just tricks to be used to get the experimenter to give her things she wanted; even though Washoe knew how and when to make signs, she really did not know what words meant in the sense that people do.

If . . . Washoe used the sign 'drink' to represent any liquid beverage, then she was doing something very different—something that everyone had previously thought only humans could do.

The disagreement was more than a scholarly debate among scientists. Decades of previous work had demonstrated that many animals could learn to do complex things to obtain food, without understanding what they were doing. For example, pigeons had been taught to bat a ball back and forth in what looked liked a game of ping pong. They were also taught to peck keys with such words as 'Please', 'Thank you', 'Red' and 'Green' printed on them. They did this in a way that made it appear that they were communicating, but they were not; they had simply learned to peck each key when a special signal was given.

This type of learning is called *conditioned discrimination* learning, a term that simply means that an animal can learn to make one set of responses in one group of circumstances and another in different circumstances. Although some aspects of human language can be explained in this way, such as 'Hello', 'Goodbye', 'Please' and 'Thank you', most cannot. Human beings learn more than what to say when: they learn what words stand for.

If Washoe had simply signed 'drink' when someone held up a bottle of soda, there would be little reason to conclude that she was doing anything different from other animals. If, however, Washoe used the sign 'drink' to represent any liquid beverage, then she was doing something very different—something that everyone had previously thought only humans could do.

It was difficult to determine which of these possibilities characterised her behaviour, as the question of how to distinguish between the 'conditioned response' and a 'word' had not arisen. Before Washoe, the only organisms that used words were human beings, and to determine if a person knew what a word stood for was easy: one simply asked.

Lana uses her keyboard to type out the practice sentence 'You put piece of bread in machine?'. In response, a person would fill Lana's machine with bread. Lana could then request it by selecting the symbols for 'Please machine give piece of bread'. Lana thus had to learn to address her requests to people to fill the machine when it was empty and to the machine itself to obtain food when food was available.

This was impossible with Washoe, because her use of symbols was not advanced enough to allow her to comprehend complex questions. One- and two-year-old children are also unable to answer questions such as these. However, because children are able to answer such questions later on, the issue of determining how and when a child knows that words have meanings had not until then been seen as critical.

Teaching syntax

Several scientists attempted to solve this problem by focusing on sentences instead of words. Linguists argue that the essence of human language lies not in learning to use individual words, but rather in an ability to form a large number of word combinations that follow the same set of specific rules. These rules are seen as a genetic endowment unique to humans. If it could be shown that apes learn syntactical rules, then it must be true that they were using symbols to represent things, not just perform tricks.

Three psychologists in the 1970s each used a different method in an attempt to teach apes syntax. One group followed the method used with Washoe and began teaching another chimpanzee,

Nim, sign language. Another opted for the use of plastic symbols with the chimpanzee Sarah. Still another used geometric symbols, linked to a computer keyboard, with a chimpanzee named Lana. Both Lana and Sarah were taught a simple syntax, which required them to fill in one blank at a time in a string of words. The number of blanks was slowly increased until the chimpanzee

---

Both Sarah and Lana learned to fill in the blanks in sentences in ways that suggested they had learned the rules that govern simple sentence construction. Moreover, 6 percent of Lana's sentences were 'novel' in that they differed from the ones that she had been taught.

---

was forming a complete 'sentence'. Nim was asked to produce syntactically correct strings by making signs along with his teacher.

Without help from his teachers, Nim was unable to form sentences that displayed the kind of syntactical rules used by humans. Nim's sign usage could best be interpreted as a series of 'conditioned discriminations' similar to, albeit more complex than, behaviours seen in many less-intelligent animals. This work suggested that Nim, like circus animals but unlike human children, was using words only to obtain rewards.

However, the other attempts to teach sentences to apes arrived at a different conclusion, perhaps because a different training method was used. Both Sarah and Lana learned to fill in the blanks in sentences in ways that suggested they had learned the rules that govern simple sentence construction. Moreover, 6 per cent of Lana's sentences were 'novel' in that they differed from the ones that she had been taught. Many of these sentences, such as 'Please you move coke in cup into room', followed syntactical rules and were appropriate and meaningful communications. Other sentences followed the syntactical rules that Lana had learned, but did not make sense; for example, 'Question you give beancake shut-open'. Thus, apes appeared to be able to learn rules for sentence construction, but they did not generalise these rules in a way that suggested full comprehension of the words.

By 1980, Washoe had matured and given birth. At this time there was great interest in whether or not she would teach her offspring to sign. Unfortunately, her infant died. However, another infant was obtained and given to Washoe. This infant, Loulis, began to imitate many of the hand gestures that Washoe used, though the imitations were often quite imprecise. Washoe made few explicit attempts to mould Loulis's hands. Although Loulis began to make signs, it was not easy to determine why he was making them or what, if anything, he meant. Loulis has not yet received any tests like those that were given to Washoe to determine if he can make the correct sign when shown an object. It is clear that he learned to imitate Washoe, but it is not clear that he learned what the signs meant.

Most important, Sherman and Austin began to show an aspect of symbol usage that they had not been taught: they used symbols to say what they were going to do before they did it.

The question of whether or not apes understand words caused many developmental psychologists to study earlier and earlier aspects of language acquisition in children. Their work gave, for the first time, a detailed insight into how children use words during the 'one-word' stage of language learning and showed that children usually learn to understand words before they begin to use them. At the same time, there was a new approach to the investigation of ape language. Instead of teaching names by pairing an object with its sign or symbol and rewarding correct responses, there was a new emphasis on the communicative aspect of symbols. For example, to teach a symbol such as 'key', a desirable item was locked in a box that was given to the chimpanzee. When the chimpanzee failed to open it, he was shown how to ask for and how to use a key. On other occasions, the chimpanzee was asked to retrieve a key for the teacher, so that she might open the box.

This new approach was first used with two chimpanzees named Sherman and Austin. It resulted in a clearer symbolic use of words than that found in animals trained by other methods. In addition, because these chimpanzees were taught comprehension skills, they were able to communicate with one another and not just with the experimenters. Sherman and Austin could use their symbols to tell each other things that could not be conveyed by simple glances or by pointing. For example, they could describe foods they had seen in another room, or the types of tools they needed to solve a problem. Although other apes had been reported to sign in each other's presence, there was no evidence that they were intentionally signing to each other or that they responded to each other's signs.

Most important, Sherman and Austin began to show an aspect of symbol usage that they had not been taught; they used symbols to say what they were going to do before they did it. Symbol use by other apes had not included descriptions of intended actions; rather, communications had been begun by a teacher, or limited to simple requests.

Sherman and Austin also began to use symbols to share information about objects that were not present and they passed a particularly demanding test, which required them to look at *symbols and answer questions that could be answered* only if they knew what each symbol represented. For example, they could look at printed lexigram symbols such as 'key', 'lever', 'stick', 'wrench', 'apple', 'banana', 'pineapple' and 'juice', and state whether each lexigram belonged to the class of 'food' words or 'tool' words. They could do this without ever being told whether these lexigram symbols should be classified as foods or tools. These findings were important, because they revealed that by using symbols an ape can describe what it is about to do.

(Above) Sherman asks for a food he would like to have (cherries) while Austin watches; (centre) Austin selects some cherries for Sherman and will then have some also (Austin is smiling because he is happy that Sherman has chosen cherries); (below) they change places and Austin then uses the keyboard to tell Sherman the food he would like to eat next.

How similar is ape language to human language?

Even though it was generally agreed that apes could do something far more complex than most other animals, there still remained much disagreement as to whether ape's symbols were identical to human symbols. This uncertainty arose for two reasons: apes did not acquire

Kanzi uses his keyboard to say that he intends to drink some of the coke in the picture. Unlike other apes, Kanzi uses symbols to describe his intended actions, such as where [he] is going to go, the games he is going to play, and the food he intends to eat.

Mulika uses the joystick to chase a target on the television set and a chimpanzee version of 'Pacman'. Chimpanzees are very good at such games.

words in the same manner as children—that is, by observing others use them; and apes did not appear to use true syntactical rules to construct multiple-word utterances.

The first of these differences between ape and child has recently been challenged by a young pygmy chimpanzee or bonobo named Kanzi. Most previous studies had focused on common chimpanzees because pygmy chimpanzees are very rare (they are in great danger of having their habitat destroyed in the coming decade and have no protected parks).

In contrast to other apes, Kanzi learned symbols simply by observing human beings point to them while speaking to him. He did not need to

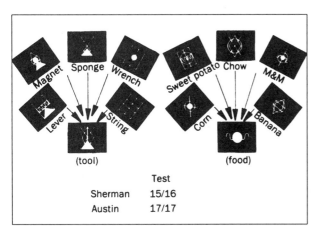

The lexigram-words that Sherman and Austin described as 'Foods' or 'Tools' the first time they were asked to classify them. They looked at each symbol and placed it In the food or the tool category. Because the symbols themselves do not look like foods or tools (nor do they resemble the food or tool words), the only way that Sherman and AustIn could have properly categorised these lexigram-words was by understanding what they symbolised.

have his arms placed in position, or to be rewarded for using a correct symbol. More important, he did not need to be taught to comprehend symbols or taught that symbols could be used for absent objects as well as those present. Kanzi spontaneously used symbols to announce his actions or intentions and, if his meaning was ambiguous, he often invented gestures to clarify it, as young children do.

Kanzi learned words by listening to speech. He first comprehended certain spoken words, then learned to read the lexigram symbols. This was possible because his caretakers pointed to these symbols as they spoke. For example, Kanzi learned 'strawberries' as he heard people mention the word when they ran across wild strawberries growing in the woods. He soon became able to lead people to strawberries whenever they asked him to do so. He similarly learned the spoken names of many other foods that grew outdoors, such as wild grapes, honeysuckle, privet berries, blackberries and mushrooms, and could take people to any of these foods upon spoken request.

Unlike previous apes reared as human children, Kanzi was reared in a semi-natural woodland. Although he could

not produce speech, he understood much of what was said to him. He could appropriately carry out novel spoken requests such as 'Will you take some hamburger to Austin?" 'Can you show your new toy to Kelly?' and 'Would you give Panzee some of your melon?'. There appeared to be no limit to the number of sentences that Kanzi could understand as long as the words in the sentences were in his vocabulary.

During the first 3 or 4 years of his life, Kanzi's comprehension of spoken sentences was limited to things that he heard often. However, when he was 5 years old, he began to respond to novel sentences upon first hearing them. For example, the first time he heard someone talk about throwing a ball in the river, he suddenly turned and threw his ball right in the water, even though he had never done this before. Similarly, when someone suggested, for fun, that he might then try to throw a potato at a turtle that was nearby, he found a potato and tossed it at the turtle. To be certain that Kanzi was not being somehow 'cued' inadvertently by people, he was tested with headphones. In this test he had to listen to a word and point to a picture of the word that he heard. Kanzi did this easily, the first time he took the test.

About this time, Kanzi also began to combine symbols. Unlike other apes, he did not combine symbols ungrammatically to get the experimenter to give something that was purposefully being held back. Kanzi's combinations had a primitive English word order and

Kanzi understands spoken English words, so the ability that is reflected in language comprehension is probably an older evolutionary adaptation than is the ability to talk.

conveyed novel information. For example, he formed utterances such as 'Ball go group room' to say that he wanted to play with a specific ball—the one he had seen in the group room on the previous day. Because the experimenter was not attempting to get Kanzi to say this, and was indeed far from the group room, such a sentence conveyed something that only Kanzi— not the experimenter—knew before Kanzi spoke.

Thus Kanzi's combinations differed from those of other apes in that they often referred to things or events that were absent and were known only to Kanzi; they contained a primitive grammar and were not imitations of the experimenter. Nor did the experimenter ask rhetorical questions such as 'What is this?' to elicit them. Kanzi's combinations include sentences such as 'Tickle bite', 'Keep-way balloon' and 'Coke chase'. As almost nothing is yet known of how pygmy chimpanzees communicate, they could use a form of simple language in the wild. Kanzi understands spoken English words, so the ability that is reflected in language comprehension is probably an older evolutionary adaptation than is the ability to talk.

Studying ape language presents a serious challenge to the long-held view that only humans can talk and think. Certainly, there is now no doubt that apes communicate in much more complex and abstract ways than dogs, cats and other familiar animals. Similarly, apes that have learned some language skills are also able to do some remarkable non-linguistic tasks. For example, they can recognise themselves on television and even determine whether an image is taped or live. They can also play video games, using a joy-stick to catch and trap a video villain.

Scientists have only just begun to discover ways of tapping the hidden talents for language and communication of our closest relatives. Sharing 98 per cent of their DNA with human beings, it has long been wondered why African apes seem so much like us at a biological level, but so different when it comes to behaviour. Ape-language studies continue to reveal that apes are more like us than we ever imagined.

COMMENTARY

Are We in Anthropodenial?

BY FRANS DE WAAL

WHEN GUESTS ARRIVE AT THE Yerkes Regional Primate Research Center in Georgia, where I work, they usually pay a visit to the chimpanzees. And often, when she sees them approaching the compound, an adult female chimpanzee named Georgia will hurry to the spigot to collect a mouthful of water. She'll then casually mingle with the rest of the colony behind the mesh fence, and not even the sharpest observer will notice anything unusual. If necessary, Georgia will wait minutes, with her lips closed, until the visitors

wish to claim that she understands Dutch, but she must have sensed that I knew what she was up to, and that I was not going to be an easy target.

Now, no doubt even a casual reader will have noticed that in describing Georgia's actions, I've implied human qualities such as intentions, the ability to interpret my own awareness, and a tendency toward mischief. Yet scientific tradition says I should avoid such language—I am committing the sin of anthropomorphism, of turning nonhumans into humans. The word comes from the

sponses shaped by rewards and punishments rather than the result of internal decision making, emotions, or intentions. They would say that Georgia was not "up to" anything when she sprayed water on her victims. Far from planning and executing a naughty plot, Georgia merely fell for the irresistible reward of human surprise and annoyance. Whereas any person acting like her would be scolded, arrested, or held accountable, Georgia is somehow innocent.

Behaviorists are not the only scientists who have avoided thinking about

To endow animals with human emotions has long been a scientific taboo. But if we do not, we risk missing something fundamental, about both animals and us.

come near. Then there will be shrieks, laughs, jumps—and sometimes falls—when she suddenly sprays them.

I have known quite a few apes that are good at surprising people, naive and otherwise. Heini Hediger, the great Swiss zoo biologist, recounts how he—being prepared to meet the challenge and paying attention to the ape's every move—got drenched by an experienced chimpanzee. I once found myself in a similar situation with Georgia; she had taken a drink from the spigot and was sneaking up to me. I looked her straight in the eye and pointed my finger at her, warning in Dutch, "I have seen you!" She immediately stepped back, let some of the water dribble from her mouth, and swallowed the rest. I certainly do not

Greek, meaning "human form," and it was the ancient Greeks who first gave the practice a bad reputation. They did not have chimpanzees in mind: the philosopher Xenophanes objected to Homer's poetry because it treated Zeus and the other gods as if they were people. How could we be so arrogant, Xenophanes asked, as to think that the gods should look like us? If horses could draw pictures, he suggested mockingly, they would no doubt make their gods look like horses.

Nowadays the intellectual descendants of Xenophanes warn against perceiving animals to be like ourselves. There are, for example, the behaviorists, who follow psychologist B. F. Skinner in viewing the actions of animals as re-

the inner life of animals. Some sociobiologists—researchers who look for the roots of behavior in evolution—depict animals as "survival machines" and "pre-programmed robots" put on Earth to serve their "selfish" genes. There is a certain metaphorical value to these concepts, but is has been negated by the misunderstanding they've created. Such language can give the impression that only genes are entitled to an inner life. No more delusively anthropomorphizing idea has been put forward since the pet-rock craze of the 1970s. In fact, during evolution, genes—a mere batch of molecules—simply multiply at different rates, depending on the traits they produce in an individual. To say that genes are selfish is like saying a snowball growing in size as it rolls down a hill is greedy for snow.

Logically, these agnostic attitudes toward a mental life in animals can be valid only if they're applied to our own species as well. Yet it's uncommon to find researchers who try to study human behavior as purely a matter of reward and punishment. Describe a person as having intentions, feelings, and thoughts and you most likely won't encounter much resistance. Our own familiarity with our inner lives overrules whatever some school of thought might claim

lemma when Charles Darwin came along: If we descended from such automatons, were we not automatons ourselves? If not, how did we get to be so different?

Each time we must ask such a question, another brick is pulled out of the dividing wall, and to me this wall is beginning to look like a slice of Swiss cheese. I work on a daily basis with animals from which it is about as hard to distance yourself as from "Lucy," the famed 3.2-million-year-old fossil australopithecine. If we owe Lucy the respect of an ancestor, does this not force a dif-

animals. When my book *Chimpanzee Politics* came out in France, in 1987, my publisher decided (unbeknownst to me) to put François Mitterrand and Jacques Chirac on the cover with a chimpanzee between them. I can only assume he wanted to imply that these politicians acted like "mere" apes. Yet by doing so he went completely against the whole point of my book, which was not to ridicule people but to show that chimpanzees live in complex societies full of alliances and power plays that in some ways mirror our own.

Bonobos have been known to assist companions new to their quarters in zoos, taking them by the hand to guide them through the maze of corridors connecting parts of their building.

about us. Yet despite this double standard toward behavior in humans and animals, modern biology leaves us no choice other than to conclude that we *are* animals. In terms of anatomy, physiology, and neurology we are really no more exceptional than, say, an elephant or a platypus is in its own way. Even such presumed hallmarks of humanity as warfare, politics, culture, morality, and language may not be completely unprecedented. For example, different groups of wild chimpanzees employ different technologies—some fish for termites with sticks, others crack nuts with stones—that are transmitted from one generation to the next through a process reminiscent of human culture.

Given these discoveries, we must be very careful not to exaggerate the uniqueness of our species. The ancients apparently never gave much thought to this practice, the opposite of anthropomorphism, and so we lack a word for it. I will call it anthropodenial: a blindness to the humanlike characteristics of other animals, or the animal-like characteristics of ourselves.

Those who are in anthropodenial try to build a brick wall to separate humans from the rest of the animal kingdom. They carry on the tradition of René Descartes, who declared that while humans possessed souls, animals were mere automatons. This produced a serious di-

ferent look at the apes? After all, as far as we can tell, the most significant difference between Lucy and modern chimpanzees is found in their hips, not their craniums.

AS SOON AS WE ADMIT THAT ANIMALS are far more like our relatives than like machines, then anthropodenial becomes impossible and anthropomorphism becomes inevitable—and scientifically acceptable. But not *all* forms of anthropomorphism, of course. Popular culture bombards us with examples of animals being humanized for all sorts of purposes, ranging from education to entertainment to satire to propaganda. Walt Disney, for example, made us forget that Mickey is a mouse, and Donald a duck. George Orwell laid a cover of human societal ills over a population of livestock. I was once struck by an advertisement for an oil company that claimed its propane saved the environment, in which a grizzly bear enjoying a pristine landscape had his arm around his mate's shoulders. In fact, bears are nearsighted and do not form pair-bonds, so the image says more about our own behavior than theirs.

Perhaps that was the intent. The problem is, we do not always remember that, when used in this way, anthropomorphism can provide insight only into human affairs and not into the affairs of

You can often hear similar attempts at anthropomorphic humor in the crowds that form around the monkey exhibit at a typical zoo. Isn't it interesting that antelopes, lions, and giraffes rarely elicit hilarity? But people who watch primates end up hooting and yelling, scratching themselves in exaggeration, and pointing at the animals while shouting, "I had to look twice, Larry. I thought it was you!" In my mind, the laughter reflects anthropodenial: it is a nervous reaction caused by an uncomfortable resemblance.

That very resemblance, however, can allow us to make better use of anthropomorphism, but for this we must view it as a means rather than an end. It should not be our goal to find some quality in an animal that is precisely equivalent to an aspect of our own inner lives. Rather, we should use the fact that we are similar to animals to develop ideas we can test. For example, after observing a group of chimpanzees at length, we begin to suspect that some individuals are attempting to "deceive" others—by giving false alarms to distract unwanted attention from the theft of food or from forbidden sexual activity. Once we frame the observation in such terms, we can devise testable predictions. We can figure out just what it would take to demonstrate deception on the part of chimpanzees. In this way, a speculation is turned into a challenge.

Naturally, we must always be on guard. To avoid making silly interpretations based on anthropomorphism, one must always interpret animal behavior in the wider context of a species' habits and natural history. Without experience with primates, one could imagine that a grinning rhesus monkey must be delighted, or that a chimpanzee running toward another with loud grunts must be in an aggressive mood. But primatologists know from many hours of observation that rhesus monkeys bare their teeth when intimidated, and that chimpanzees often grunt when they meet and embrace. In other words, a grinning rhesus monkey signals submission, and a chimpanzee's grunting often serves as a greeting. A careful observer may thus arrive at an informed anthropomorphism that is at odds with extrapolations from human behavior.

One must also always be aware that some animals are more like ourselves than others. The problem of sharing the experiences of organisms that rely on different senses is a profound one. It was expressed most famously by the philosopher Thomas Nagel when he asked, "What is it like to be a bat?" A bat perceives its world in pulses of reflected sound, something we creatures of vision would have a hard time imagining. Perhaps even more alien would be the experience of an animal such as the star-nosed mole. With 22 pink, writhing tentacles around its nostrils, it is able to feel microscopic textures on small objects in the mud with the keenest sense of touch of any animal on Earth.

Humans can barely imagine a star-nosed mole's *Umwelt*—a German term for the environment as perceived by the animal. Obviously, the closer a species is to us, the easier it is to enter its *Umwelt*. This is why anthropomorphism is not only tempting in the case of apes but also hard to reject on the grounds that we cannot know how they perceive the world. Their sensory systems are essentially the same as ours.

LAST SUMMER, AN APE SAVED A three-year-old boy. The child, who had fallen 20 feet into the primate exhibit at Chicago's Brookfield Zoo, was scooped up and carried to safety by Binti Jua, an eight-year-old western lowland female gorilla. The go-rilla sat down on a log in a stream, cradling the boy in her lap and patting his back, and then carried him to one of the exhibit doorways before laying him down and continuing on her way.

Binti became a celebrity overnight, figuring in the speeches of leading politicians who held her up as an example of much-needed compassion. Some scientists were less lyrical, however. They cautioned that Binti's motives might have been less noble than they appeared, pointing out that this gorilla had been raised by people and had been taught parental skills with a stuffed animal. The whole affair might have been one of a confused maternal instinct, they claimed.

The intriguing thing about this flurry of alternative explanations was that nobody would think of raising similar doubts when a person saves a dog hit by a car. The rescuer might have grown up around a kennel, have been praised for being kind to animals, have a nurturing personality, yet we would still see his behavior as an act of caring. Whey then, in Binti's case, was her background held against her? I am not saying that I know what went through Binti's head, but I do know that no one had prepared her for this kind of emergency and that it is unlikely that, with her own 17-month-old infant on her back, she was "maternally confused." How in the world could such a highly intelligent animal mistake a blond boy in sneakers and a red T-shirt for a juvenile gorilla? Actually, the biggest surprise was how surprised most people were. Students of ape behavior did not feel that Binti had done anything unusual. Jörg Hess, a Swiss gorilla expert, put it most bluntly, "The incident can be sensational only for people who don't know a thing about gorillas."

Binti's action made a deep impression mainly because it benefited a member of our own species, but in my work on the evolution of morality and empathy, I have encountered numerous instances of animals caring for one another. For example, a chimpanzee consoles a victim after a violent attack, placing an arm around him and patting his back. And bonobos (or pygmy chimpanzees) have been known to as-sist companions new to their quarters in zoos, taking them by the hand to guide them through the maze of corridors connecting parts of their building. These kinds of cases don't reach the newspapers but are consistent with Binti's assistance to the unfortunate boy and the idea that apes have a capacity for sympathy.

The traditional bulwark against this sort of cognitive interpretation is the principle of parsimony—that we must make as few assumptions as possible when trying to construct a scientific explanation, and that assuming an ape is capable of something like sympathy is too great a leap. But doesn't that same principle of parsimony argue against assuming a huge cognitive gap when the evolutionary distance between humans and apes is so small? If two closely related species act in the same manner, their underlying mental processes are probably the same, too. The incident at the Brookfield Zoo shows how hard it is to avoid anthropodenial and anthropomorphism at the same time: in trying to avoid thinking of Binti as a human being, we run straight into the realization that Binti's actions make little sense if we refuse to assume intentions and feelings.

In the end we must ask: What kind of risk we are willing to take—the risk of underestimating animal mental life or the risk of overestimating it? There is no simple answer. But from an evolutionary perspective, Binti's kindness, like Georgia's mischief, is most parsimoniously explained in the same way we explain our own behavior—as the result of a complex, and familiar, inner life.

FRANS DE WAAL is a professor of psychology at Emory University and research professor at the Yerkes Regional Primate Research Center in Atlanta. He is the author of several books, including *Chimpanzee Politics* and *Good Natured: The Origins of Right and Wrong in Humans and Other Animals.* His latest book, in collaboration with acclaimed wildlife photographer Frans Lanting, is *Bonobo: The Forgotten Ape,* published by the University of California Press (1997).

Unit 3

Unit Selections

13. **These Are Real Swinging Primates,** Shannon Brownlee
14. **The Myth of the Coy Female,** Carol Tavris
15. **First, Kill the Babies,** Carl Zimmer
16. **A Woman's Curse?** Meredith F. Small
17. **What's Love Got to Do with It?** Meredith F. Small
18. **Apes of Wrath,** Barbara Smuts

Key Points to Consider

❖ How can the muriqui monkeys be sexually competitive and yet gregarious and cooperative?

❖ How does human sexuality differ from that of other creatures?

❖ What implications does bonobo sexual behavior have for understanding human evolution?

❖ Why do cultures the world over treat menstruating women as taboo?

❖ How do social bonds provide females with protection against abusive males?

 **Links**　　www.dushkin.com/online/

These sites are annotated on pages 4 and 5.

Any account of hominid evolution would be remiss if it did not at least attempt to explain that most mystifying of all human experiences—our sexuality.

No other aspect of our humanity—whether it be upright posture, tool-making ability, or intelligence in general—seems to elude our intellectual grasp at least as much as it dominates our subjective consciousness. While we are a long way from reaching a consensus as to why it arose and what it is all about, there is widespread agreement that our very preoccupation with sex is in itself one of the hallmarks of being human. Even as we experience it and analyze it, we exalt it and condemn it. Beyond seemingly irrational fixations, however, there is the further tendency to project our own values upon the observations we make and the data we collect.

There are many who argue quite reasonably that the human bias has been more male- than female-oriented and that the recent "feminization" of anthropology has resulted in new kinds of research and refreshingly new theoretical perspectives. (See "The Myth of the Coy Female" by Carol Tavris and "First, Kill the Babies" by Carl Zimmer). Not only should we consider the source when evaluating the old theories, so goes the reasoning, but we should

also welcome the source when considering the new. To take one example, traditional theory would have predicted that the reproductive competitiveness of muriqui monkeys, as described in "These Are Real Swinging Primates" by Shannon Brownlee, would be associated with greater size and aggression among males. That this is not so, that making love can be more important than making war, and that females do not necessarily have to live in fear of competitive males, just goes to show that, even among monkeys, nothing can be taken for granted. The very idea that females are helpless in the face of male aggression is called into question by Barbara Smuts in "Apes of Wrath."

Finally, there is the question of the social significance of sexuality in humans. In "What's Love Got to Do with It?" Meredith Small shows that the chimp-like bonobos of Zaire use sex to reduce tensions and cement social relations and, in so doing, have achieved a high degree of equality between the sexes. Whether we see parallels in the human species, says Small, depends on our willingness to interpret bonobo behavior as a "modern version of our own ancestors' sex play," and this, in turn, may depend on our prior theoretical commitments.

Sex and Society

These Are Real Swinging Primates

There's a good evolutionary reason why the rare muriqui of Brazil should heed the dictum 'Make love, not war'

Shannon Brownlee

When I first heard of the muriqui four years ago, I knew right away that I had to see one. This is an unusual monkey, to say the least. To begin with, it's the largest primate in South America; beyond that, the males have very large testicles. We're talking gigantic, the size of billiard balls, which means that the 30-pound muriqui has *cojones* that would look more fitting on a 400-pound gorilla.

But it wasn't prurience that lured me to Brazil. My interest in the muriqui was intellectual, because more than this monkey's anatomy is extraordinary. Muriqui society is untroubled by conflict: troops have no obvious pecking order; males don't compete overtly for females; and, most un-monkeylike, these monkeys almost never fight.

The muriqui is also one of the rarest monkeys in the world. It lives in a single habitat, the Atlantic forest of southeastern Brazil. This mountainous region was once blanketed with forest from São Paulo to Salvador (*see map*), but several centuries of slash-and-burn agriculture have reduced it to fragments.

In 1969 Brazilian conservationist Alvaro Coutinho Aguirre surveyed the remaining pockets of forest and estimated that 2,000 to 3,000 muriquis survived. His data were all but ignored until Russell Mittermeier, a biologist, trained his sights on the muriquis ten years later. Known as Russel of the Apes to his colleagues, Mittermeier, an American, directs the primate program for the World

Wildlife Fund. He hopscotches from forest to forest around the world looking for monkeys in trouble and setting up conservation plans for them. In 1979 he and Brazilian zoologist Celio Valle retraced Aguirre's steps and found even fewer muriquis. Today only 350 to 500 are left, scattered among four state and national parks and six other privately held plots.

In 1981 Karen Strier, then a graduate student at Harvard, approached Mittermeier for help in getting permission to observe the muriqui. He took her to a coffee plantation called Montes Claros, near the town of Caratinga, 250 miles north of Rio de Janeiro. Over the next four years she studied the social behavior of the muriqui there—and came up with a provocative theory about how the monkey's unconventional behavior, as well as its colossal testicles, evolved. She reasoned that the evolution of both could be explained, at least in part, by the muriquis' need to avoid falling out of trees.

Last June I joined Strier, now a professor at Beloit (Wis.) College, on one of her periodic journeys to Montes Claros—clear mountains, in Portuguese. We arrived there after a disagreeable overnight bus trip over bad roads. As we neared the plantation, I found it difficult to believe there was a forest—much less a monkey—within miles. Through the grimy windows of the bus I saw hillsides stripped down to russet dirt and

dotted with spindly coffee plants and stucco farmhouses. There wasn't anything taller than a banana tree in sight. As the bus lurched around the last curve before our stop the forest finally appeared, an island of green amid thousands of acres of coffee trees and brown pastures.

Strier was eager to start looking for the muriquis—"There's a chance we won't see them the whole four days you're here," she said—so no sooner had we dropped our bags off at a cottage on the plantation than we set out along a dirt road into the forest. The trees closed around us—and above us, where they gracefully arched to form a vault of green filigree. Parrots screeched; leaves rustled; a large butterfly flew erratically by on transparent wings. By this time Strier had guided me onto a steep trail, along which she stopped from time to time to listen for the monkeys.

They appeared soon enough, but our first meeting was less than felicitous. After we had climbed half a mile, Strier motioned for me to stop. A muffled sound, like that of a small pig grunting contentedly, came from up ahead. We moved forward a hundred yards. Putting a finger to her lips, Strier sank to her haunches and looked up.

I did the same; twelve round black eyes stared back at me. A group of six muriquis squatted, silent, 15 feet above in the branches, watching us intently. They began to grunt again. A sharp smell with undertones of cinnamon permeated the air. A light rain began to fall. I held out my palm to catch a drop. It was warm.

"Hey, this isn't rain!" I said.

Strier grinned and pointed to her head. "That's why I wear a hat," she said.

My enthusiasm for the muriquis waned slightly after that. We left them at dusk and retired to the cottage, where Strier described her arrival at Montes Claros four years earlier. Mittermeier acted as guide and interpreter during the first few days of her pilot study. He introduced her to the owner of the 5,000-acre plantation, Feliciano Miguel Abdala, then 73, who had preserved the 2,000-acre forest for more than 40 years. His is one of the only remaining tracts of Atlantic forest, and he agreed to let Strier use it as the site of her study. Then Mittermeier introduced her to the muriquis, assuring her they would be easy to see.

They weren't, and observing them closely is a little like stargazing on a rainy night: not only do you run the risk of getting wet, but you can also spend a lot of time looking up and never see a thing. Mittermeier was adept at spotting the monkeys in the forest, and helped Strier acquire this skill.

But brief glimpses of the monkeys weren't enough. "My strategy was to treat them like baboons, the only other species I'd ever studied," she says. "I thought I couldn't let them out of my sight." She tried to follow on the ground as they swung along in the trees. "They went berserk," she says. They threw branches, shrieked, urinated on her—or worse—and fled.

Even after the muriquis grew accustomed to her, keeping up with them wasn't easy. They travel as much as two miles a day, which is tough for someone picking her way through thick growth on the forest floor. As Strier and a Brazilian assistant learned the muriquis' habitual routes and daily patterns, they cleared trails. These helped, but the muriquis could still travel much faster than she could. "I've often thought the thing to have would be a jet pack," Strier says. "It would revolutionize primatology. Your National Science Foundation grant would include binoculars, pencils, and a jet pack."

The monkeys move by brachiating, swinging hand over hand from branch to branch, much like a child on a jungle gym. Only one other group of monkeys brachiates; the rest clamber along branches on all fours. The muriquis' closest relatives are two other Latin American genera, the woolly monkeys and the spider monkeys—hence woolly spider monkey, its English name. But the muriqui is so unlike them that it has its own genus, *Brachyteles,* which refers to its diminutive thumb, an adaptation for swinging through the trees. Its species name is *arachnoides,* from the Greek for spider, which the muriqui resembles when its long arms, legs, and tail are outstretched.

Brachiating is a specialization that's thought to have evolved because it enables primates to range widely to feed on fruit. Curiously, though, muriquis have a stomach designed for digesting leaves. Strier found that their diet consists of a combination of the two foods. They eat mostly foliage, low-quality food for a monkey, but prefer flowers and fruits, like figs and the *caja manga,* which is similar to the mango. Year after year they return to certain trees when they bloom and bear fruit. The rest of the time the muriquis survive on leaves by passing huge quantities of them through their elongated guts, which contain special bacteria to help them digest the foliage. By the end of the day their bellies are so distended with greenery that even the males look pregnant.

We returned to the trail the next morning just after dawn. Condensation trickled from leaves; howler monkeys roared and capuchins cooed and squeaked; a bird sang with the sweet, piercing voice of a piccolo. Then Strier had to mention snakes. "Watch out for snakes," she said blithely, scrambling on all fours up a steep bank. I followed her, treading cautiously.

The muriquis weren't where we had left them the day before. Strier led me along a ridge through a stand of bamboo, where a whisper of movement drifted up from the slope below. Maybe it was just the wind, but she thought it was the muriquis, so we sat down to wait. After a couple of hours, she confessed, "This part of research can get kind of boring."

By noon the faint noise became a distinct crashing. "That's definitely them," she said. "It's a good thing they're so noisy, or I'd never be able to find them." The monkeys, perhaps a dozen of them, swarmed uphill, breaking branches, chattering, uttering their porcine grunts as they swung along. At the crest of the ridge they paused, teetering in indecision while they peered back and forth before settling in some legume trees on the ridgetop. We crept down out of the bamboo to within a few feet of them, so close I noticed the cinnamon scent again—only this time I kept out of range.

Each monkey had its own feeding style. One hung upside down by its tail and drew the tip of a branch to its mouth; it delicately plucked the tenderest shoots with its rubbery lips. Another sat upright, grabbing leaves by the handful and stuffing its face. A female with twins—"Twins have never been seen in this species," Strier whispered as she excitedly scribbled notes—ate with one hand while hanging by the other and her tail. Her babies clung to the fur on her belly.

I had no trouble spotting the males. Their nether parts bulged unmistakably—blue-black or pink-freckled, absurd-looking monuments to monkey virility. I asked Strier what sort of obscene joke evolution was playing on the muriquis when it endowed them thus.

We were about to consider this question when a high-pitched whinnying began a few hundred yards away. Immediately a monkey just overhead pulled itself erect and let out an ear-splitting shriek, which set the entire troop to neighing like a herd of nervous horses. Then they took off down into the valley.

Strier and I had to plunge pell-mell into the underbrush or risk losing them for the rest of the day. "They're chasing the other troop," she said as we galloped downhill. A group of muriquis living on the opposite side of the forest had made a rare foray across the valley.

The monkeys we were observing swung effortlessly from tree to tree; we wrestled with thorny vines, and fell farther and farther behind. An impenetrable thicket forced us to backtrack in search of another route. By the time we caught up to the muriquis, they were lounging in a tree, chewing on unripe fruit and chuckling in a self-satisfied sort of way. The intruding troop was nowhere to be seen. "They must have scared the hell out of those other guys," said Strier, laughing.

Such confrontations occur infrequently; muriquis ordinarily tolerate another troop's incursions. Strier thinks they challenge intruders only when there's a valuable resource to defend—like the fruit tree they were sitting in.

Tolerance of another troop is odd behavior for monkeys, but not as odd as the fact that members of a muriqui troop never fight among themselves. "They're remarkably placid," said Strier. "They wait in line to dip their hands into water collected in the bole of a tree. They have no apparent pecking order or dominance hierarchy. Males and females are equal in status, and males don't squabble over females." No other primate society is known to be so free of competition, not even that of gorillas, which have lately gained a reputation for being the gentle giants of the primate world.

Strier's portrayal of the muriqui brought to mind a bizarre episode that Katharine Milton, an anthropologist at the University of California at Berkeley, once described. While studying a troop of muriquis in another patch of the Atlantic forest, she observed a female mating with a half a dozen males in succession; that a female monkey would entertain so many suitors came as no surprise, but Milton was astonished at the sight of the males lining up behind the female "like a choo-choo train" and politely taking turns copulating. They continued in this manner for two days, stopping only to rest and eat, and never even so much as bared their teeth.

Primates aren't known for their graciousness in such matters, and I found Milton's report almost unbelievable. But Strier confirms it. She says that female muriquis come into heat about every two and a half years, after weaning their latest offspring, and repeatedly copulate during that five- to seven-day period with a number of males. Copulations, "cops" in animal-behavior lingo, last as long as 18 minutes, and average six, which for most primates (including the genus *Homo,* if Masters and Johnson are correct) would be a marathon. Yet no mat-

ter how long a male muriqui takes, he's never harassed by suitors-in-waiting.

Strier has a theory to explain the muriqui's benignity, based on a paper published in 1980 by Richard Wrangham, a primatologist at the University of Michigan. He proposed that the social behavior of primates could in large part be predicted by what the females eat.

This isn't a completely new idea. For years primatologists sought correlations between ecological conditions and social structure, but few patterns emerged—until Wrangham's ingenious insight that environment constrains the behavior of each sex differently. Specifically, food affects the sociability of females more than males.

Wrangham started with the generally accepted premise that both sexes in every species have a common aim: to leave as many offspring as possible. But each sex pursues this goal in its own way. The best strategy for a male primate is to impregnate as many females as he can. All he needs, as Wrangham points out, is plenty of sperm and plenty of females. As for the female, no matter how promiscuous she is, she can't match a male's fecundity. On average, she's able to give birth to only one offspring every two years, and her success in bearing and rearing it depends in part upon the quality of food she eats. Therefore, all other things being equal, male primates will spend their time cruising for babes, while females will look for something good to eat.

Wrangham perceived that the distribution of food—that is, whether it's plentiful or scarce, clumped or evenly dispersed—will determine how gregarious the females of a particular species are. He looked at the behavior of 28 species and found that, in general, females forage together when food is plentiful and found in large clumps—conditions under which there's enough for all the members of the group and the clumps can be defended against outsiders. When clumps become temporarily depleted, the females supplement their diet with what Wrangham calls subsistence foods. He suggests that female savanna baboons, for example, live in groups because their favorite foods, fruits and flowers, grow in large clumps that are easy to defend. When these are exhausted

they switch to seeds, insects, and grasses. The females form long-lasting relationships within their groups, and establish stable dominance hierarchies.

Chimpanzees provide an illustration of how females behave when their food isn't in clumps big enough to feed everybody. Female chimps eat flowers, shoots, leaves, and insects, but their diet is composed largely of fruits that are widely scattered and often not very plentiful. They may occasionally gather at a particularly abundant fruit tree, but when the fruit is gone they disperse to forage individually for other foods. Members of the troop are constantly meeting at fruit trees, splitting up, and gathering again.

These two types of female groups, the "bonded" savanna baboons and "fissioning" chimps, as Wrangham calls them, pose very different mating opportunities for the males of their species. As a consequence, the social behavior of the two species is different. For a male baboon, groups of females represent the perfect opportunity for him to get cops. All he has to do is exclude other males. A baboon troop includes a clan of females accompanied by a number of males, which compete fiercely for access to them. For baboons there are few advantages to fraternal cooperation, and many to competition.

Male chimpanzees fight far less over females than male baboons do, principally because there's little point—the females don't stick together. Instead, the males form strong alliances with their fellows. They roam in gangs looking for females in heat, and patrol their troop's borders against male interlopers.

Wrangham's theory made so much sense, Strier says, that it inspired researchers to go back into the field with a new perspective. She saw the muriqui as an excellent species for evaluating the model, since Wrangham had constructed it before anyone knew the first thing about this monkey. His idea would seem all the more reasonable if it could predict the muriqui's behavior.

It couldn't, at least not entirely. Strier has found that the females fit Wrangham's predictions: they stick together and eat a combination of preferred and subsistence foods, defending the preferred from other troops. But the males don't conform to the theory. "Considering that the females are foraging together, there should be relatively low pressure on the males to cooperate," she says. "It's odd: the males should compete, but they don't."

She thinks that limitations on male competition may explain muriqui behavior. First, the muriquis are too big to fight in trees. "I think these monkeys are at about the limit of size for rapid brachiation," she says. "If they were bigger, they couldn't travel rapidly through the trees. They fall a lot as it is, and it really shakes them up. I've seen an adult fall about sixty feet, nearly to the ground, before catching hold of a branch. That means that whatever they fight about has got to be worth the risk of falling out of a tree."

Moreover, fighting may require more energy than the muriquis can afford. Milton has estimated the caloric value of the food eaten by a muriqui each day and compared it to the amount of energy she would expect a monkey of that size to need. She concluded that the muriqui had little excess energy to burn on combat.

The restriction that rapid brachiation sets on the muriqui's size discourages competition in more subtle ways, as well.

Given that muriquis are polygynous, the male should be bigger than the female, as is almost invariably the case among other polygynous species—but he's not. The link between larger males and polygyny is created by sexual selection, an evolutionary force that Darwin first recognized, and which he distinguished from natural selection by the fact that it acts exclusively on one sex. Sexual selection is responsible for the manes of male lions, for instance, and for the large canines of male baboons.

In a polygynous society, the advantages to being a large male are obvious: he who's biggest is most likely to win the battles over females—and pass on his genes for size. But sexual selections' push toward large males has been thwarted in the muriqui, says Strier. Any competitive benefits greater size might bring a male would be offset in part by the excessive demands on his energy and the costs of falling out of trees.

She believes that the constraints on the males' size have had a profound effect on the muriquis' social behavior. Most important, says Strier, with males and females being the same size, the females can't be dominated, which means they can pick their mates. Most female

primates aren't so fortunate: if they copulate with subordinate males, they risk being attacked by dominant ones. But a female muriqui in heat can easily refuse a suitor, simply by sitting down or by moving away. Fighting not only doesn't help the male muriqui in his quest for cops; it may even harm his chances, since females can shun an aggressive male. Strier believes that females may also be responsible for the male muriquis' canine teeth not being oversized. As a rule, the male's canines are the same size as the female's only in monogamous primate species, but over the generations female muriquis may have mated more readily with males whose teeth were no bigger than their own. In sum, Strier thinks, for a male muriqui the costs of competing are far outweighted by the benefits of avoiding it.

But he has the means to vie for reproductive success and still come across as Mr. Nice Guy: his sperm. Sperm competition, as it's called, is a hot new idea in sociobiology, originally proposed to explain male bonding in chimpanzees, and, as Milton was the first to suggest, it may explain why the muriqui has such enormous testicles.

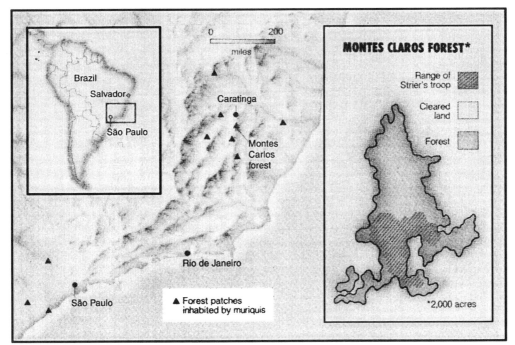

JOE LEMONNIER

The 350 to 500 surviving muriquis live in ten patches of the Atlantic forest of southeastern Brazil.

The competition is something like a game of chance. Imagine a bucket with a hole in the bottom just big enough for a marble to pass through. People gather round, each with a handful of marbles. They drop their marbles in the bucket, mix them up, and one comes out the bottom. Whoever owns that marble is the winner.

In the sperm competition among male muriquis, the bucket is a female, the marbles are sperm, and winning means becoming a father. No male can be sure it will be his sperm that impregnates a female, since she mates with a number of his fellows. His chances are further complicated by the fact that the female muriqui, like all New World monkeys, gives no visible indication of ovulation; there may be nothing that signals the male (or the female) when during her heat that occurs. So it's to the male's advantage to continue mating as often as the female will have him.

This may sound like monkey heaven, but it puts the male on the horns of a dilemma. If he copulates as often as possible, he could run low on sperm just when the female is ovulating. On the other hand, if he refrains from copulating to save sperm, he may miss his chance at procreating altogether. Selection may have come to his aid, Strier reasons, by acting on his testicles.

Here's a plausible scenario. Suppose a male came along that could produce more sperm than the average muriqui because his testicles were bigger than average. That male would clean up in the reproductive arena. The ratio of testicle size to body weight has been correlated with high sperm count and repeated copulation over a short period in other mammals, and bigger testicles probably also increase the percentage of viable and motile sperm.

If the muriqui's testicles are anything like those of other species, then a male with extra big ones has a slight reproductive advantage. Like a player with more marbles to put in the bucket, a male that can produce more and better sperm has a better than average chance of impregnating females and passing on this advantageous trait to his sons. Just as important, the outsized organs probably don't cost him much in metabolic energy. Thus, over generations, the muriqui's testicles have grown larger and larger.

Strier's theory has five years of data behind it, and it's the kind of theory that will stimulate researchers to reexamine their ideas about other species. Yet it isn't her only concern; she concentrates equally on the muriqui's uncertain future. On our last day in the forest we watched the monkeys cross a six-foot gap in the canopy 60 feet above us. One by one they stood poised for a moment on the end of a branch before launching themselves. Strier counted them as they appeared in silhouette against a grey sky. The total was 33, including the twins. "They're up from twenty-two in 1982," she said. "That's a very fast increase."

The muriquis at Montes Claros make up almost one-tenth of the total population of the species, and they're critical to its survival—as are all the other isolated and widely separated troops. Each groups's genetic pool is limited, and eventually the troops could suffer inbreeding depression, a decline in fecundity that often appears in populations with little genetic variability.

Strier and Mittermeier predict that one day muriquis will have to be managed, the way game species are in the U.S. They may be transported between patches of forest to provide some gene flow. But that's a dangerous proposition now. There are too few muriquis to risk it, and none has ever bred or survived for long in captivity. "Before my study, conservationists would probably have moved males between forests," Strier says. "That would've been a mistake. I have tentative evidence that in a natural situation the females may be the ones that do the transferring between groups."

For now, though, she thinks the biggest concern isn't managing the monkeys but preventing their habitat from disappearing. Preserving what remains of the Atlantic forest won't be easy, and no one knows this better than Feliciano Miguel Abdala, the man responsible for there being any forest at all at Montes Claros.

Abdala has little formal education, but he's rich; he owns nine plantations beside Montes Claros. His family lives in relative splendor in Caratinga, but he likes to spend the weekdays here. His house is just beyond the edge of the forest, and sunlight filters through the bougainvillea vine entwining the front porch. Chickens can be seen through the cracks in the floorboards, scratching in the dirt under the house. Electric cords are strung crazily from the rafters, and a bare bulb dangles in the center of his office. Abdala removes his straw hat decorously and places it on a chair before sitting at his desk.

Abdala bought the 5,000 acres of Montes Claros in 1944. The region was barely settled then, and smoke still rose from the great burning heaps of slash left from clearing the forest. Abdala's land included one of the last stands of trees. I ask him why he saved it. "I am a conservationist," he says. "For a long time the local people thought I was crazy because I wouldn't cut the forest. I told them not to shoot the monkeys, and they stopped. Now all my workers are crazy, too."

I ask Abdala about his plans for his forest. He rubs his head distractedly and says, vaguely. "I hope it will continue."

Abdala believes the government should buy Montes Claros—plantation and rain forest—to create a nature reserve. He'll probably maintain the forest as long as he lives, but the land is quite valuable, and his heirs might not share his lofty sentiments.

As important as the muriquis have become to understanding social systems, and as much as U.S. conservationists may wish to see these monkeys preserved, Strier thinks that in the end it's up to the Brazilians to save them. She's expecting a three-year grant from the National Science Foundation; part of the money will go toward allowing her to observe the monkeys in other forest patches, watching for variation in their behavior as a test of her ideas. Studies like hers will be critical not only for proving theories but also for ensuring that plans for managing the muriquis will work. The rest of the money will permit her to train seven Brazilian graduate students, because she says, "the future of the muriqui lies with the Brazilians."

The Myth of the Coy Female

Carol Tavris

[Thus] we arrived at the important conclusion that polygamy is the natural order among human beings, just as it is in most species of the animal kingdom . . . monogamy is responsible for the high incidence of divorce and female grievances in modern society, as well as the genetic deevolution and behavioral degeneration of civilization as a whole. . . . Culture is to blame, and fortunately *culture can be changed.* Mating is the key. [Emphasis in original.]

> —Sam Kash Kachigan,
> *The Sexual Matrix*

Sam Kash Kachigan is not a social scientist; he's just a regular fellow who thinks that the theories of sociobiology offer the best hope of improving relations between women and men. "Mating is the key," he argues. The mating he has in mind, it turns out, would (if we were truly to follow our evolutionary heritage) occur between rich old men and beautiful young girls. Among the annoying contemporary practices that Kachigan laments is the habit of beautiful young girls marrying boys their own age. To Kachigan, in any truly civilized society—that is, one in which our practices fit our sociobiological natures—girls would marry men who were old enough to demonstrate their "true potential":

In every respect, then, it makes much more sense for young women to mate with *older* men, who will have *proven* their genetic endowment as well as their financial and emotional capacity for raising children. [Emphasis in original.]

Why do I suspect that Kachigan is such a man?

The basic ideas behind sociobiology date back to Charles Darwin, who in 1871 described what he considered to be a basic dichotomy in the sexual natures of males and females of all species. Males actively pursue females; they are promiscuous; and those who are strongest, most fit in evolutionary terms, succeed in their sexual conquest. Females, said Darwin, are "comparatively passive"; they may choose their preferred suitor, but then remain monogamous and faithful. That this dichotomy conveniently fit Victorian dating and mating patterns was, naturally, pure coincidence.

For a century after Darwin, research on sexual selection and sexual behavior was based on the belief that males are passionate and undiscriminating (any female in a storm will do), whereas females are restrained, cautious, and highly discriminating in their choice of partner (only a male who meets her shopping list of qualifications will do). According to primatologist Sarah Blaffer Hrdy, this stereotype of "the coy female" has persisted in the public mind—and she adds a phrase that by now should be familiar to us—*"despite the accumulation of abundant openly available evidence contradicting it"* [my emphasis].

The stereotype of the coy female got a major boost in an important paper published in 1948 by Angus John Bateman. Bateman was a distinguished plant geneticist who did dozens of experiments with Drosophila, the tiny fruit fly

that many people remember from science experiments in junior high school. Bateman found that successful male fruit flies could, with multiple matings, produce nearly three times as many offspring as the most reproductively successful female. As Hrdy explains, "whereas a male could always gain by mating just one more time, and hence benefit from a nature that made him undiscriminatingly eager to mate, a female, already breeding near capacity after just one copulation, could gain little from multiple mating and should be quite uninterested in mating more than once or twice."

What, you may ask, does a human man have in common with a fruit fly? When it comes to sexual strategies, said Bateman, the answer is everything. Generalizing from his sixty-four experiments with Drosophila to all species, Bateman concluded that there is a universally lopsided division in the sexual natures of all creatures, apart from "a few very primitive organisms." Quite simply, males profit, evolutionarily speaking, from frequent mating, and females do not. This is why, said Bateman, "there is nearly always a combination of an undiscriminating eagerness in the males and a discriminating passivity in the females."

The modern field of sociobiology took this idea still further, attempting to account for complex human social arrangements and customs—warfare and corporate raiding, feeding infants and giving children karate lessons—in terms of the individual's basic need to repro-

Excerpt from *The Mismeasure of Woman* by Carol Tavris, pp. 212–221. © 1992 by Carol Tavris. Reprinted by permission of Simon & Schuster, Inc.

duce his or her genes. Women and men, sociobiologists believe, adopt highly different strategies in order to do this. Males compete with other males for access to desirable females, and their goal is to inseminate as many females as possible. Females, in contrast, are motivated to attach themselves to genetically "superior" males because of the female's greater "investment" in terms of time and energy in her offspring; this, according to sociobiologists, is why females are more faithful and nurturant than males. As biologist Ruth Hubbard observes, "Thus, from the seemingly innocent asymmetries between eggs and sperm [say the sociobiologists] flow such major social consequences as female fidelity, male promiscuity, women's disproportional contribution to the care of children, and the unequal distribution of labor by sex."

Sociobiological explanations of competitive, promiscuous men and choosy, inhibited but flirtatious women fit right in with many elements within the popular culture. "And so it was," Hrdy says, "that 'coyness' came to be the single most commonly mentioned attribute of females in the literature on sociobiology."

It all seems a cruel joke of nature. Certainly many people are convinced, as the King of Siam sings in *The King and I,* that the male is like the honeybee, flitting from flower to flower, "gathering all he can," whereas the female has "honey for just one man." But notice that it is the King who sings that song; until relatively recently, no one was asking Queens for their view of things. Nor were male observers asking why, if human females were so naturally chaste, coy, and monogamous, social taboos from ostracism to death had to be placed on females who indulged in forbidden sexual relationships. For that matter, why did non-marital affairs need to be forbidden anyway, if females have "honey for just one man"?

Sociobiologists attempt to explain human social customs by drawing on research on nonhuman animals, from the fields of primatology, evolutionary biology, anthropology, and related disciplines. In the last two decades, however, there has been an explosion of new research that casts doubt on many socio-

biological assumptions, a change that is largely a result of the growing numbers of women who have entered these fields. Most of the women saw animal behavior in a different light from most of the male observers who had preceded them. Male primatologists, for example, had tended to observe and emphasize male-male competition and the number of times the male animals "got lucky"; the female animals, to the human men observing them, seemed mysterious and unpredictable. This is not unlike the ways in which human females have seemed mysterious and unpredictable to the human males who have observed *them.*

At first, women who went into these research fields saw the world as they had been taught to see it, through the academic perspective of their mentors. But after a while, they began to ask different questions and to bring different expectations to their observations. Hrdy recalls her own first glimpse of a female langur

> . . . moving away from her natal group to approach and solicit males in an all-male band. At the time, I had no context for interpreting behavior that merely seemed strange and incomprehensible to my Harvard-trained eyes. Only in time, did I come to realize that such wandering and such seemingly "wanton" behavior were recurring events in the lives of langurs.

Eventually, Hrdy learned that female langurs often leave their troops to join up with bands of males; and she also found that often a female, for reasons unknown, "simply takes a shine to the resident male of a neighboring troop." In fact, female langurs (and many other primate species) are able to shift from being in heat once a month to being continuously receptive for weeks at a time, a state not unlike the first phase of (human) love. In many primates, female receptivity is often *situation specific,* rather than being dependent exclusively on cyclical periods of being in heat.

As a result of the efforts of many pioneers like Hrdy, we now know that the females of many animal species do not behave like the patient, coy fruit fly. On the contrary, the females are sexually ar-

dent and can even be called polyandrous (having many male partners). Further, their sexual behavior does not depend simply on the goal of being fertilized by the male, because in many cases females actively solicit males when they are not ovulating, and even when they are already pregnant. Here are a few illustrations from hundreds of research studies:

• Many species of female birds are promiscuous. In one study, researchers vasectomized the "master" of a blackbird harem . . . but the females nevertheless conceived.
• Many species of female fish are promiscuous. A female shiner perch who is not ovulating will nevertheless mate with many males, collecting sperm and storing them internally until she is ready to ovulate.
• Many species of female cats, notably leopards, lions, and pumas, are promiscuous. A lioness may mate dozens of times with many different partners during the week she is in estrus.
• Many species of female primates are promiscuous. Among savanna baboons and Barbary macaques, females initiate many different brief sexual encounters. Among chimpanzees, Hrdy reports, some females form partnerships with one male, but others engage in communal mating with all males in the vicinity. And among wild tamarin monkeys, a species long thought to be monogamous (at least in captivity), supposedly faithful females will mate with several males. So do female Hanuman langurs, blue monkeys, and redtail monkeys, all primates that were formerly believed to be one-man women. The old notion that primate females typically form "one-male breeding units," as primatologists would say, is now seriously called into question.

In spite of rapidly accumulating evidence that females of many different and varied species do mate "promiscuously" (a word that itself has evaluative overtones), it was not until 1980 or so that researchers realized that this fact threw, well, a monkey wrench into traditional evolutionary theories. Why would females have more copulations than are necessary for conception? Why would they go off with some guy from a neighboring town, whom none of her friends approves of? Why risk losing the genetic father's support by joining the baboon equivalent of Hell's Angels? And the brooding question over all of them, why did female primates develop continuous sexual receptivity?

These questions stimulated a flurry of new theories to explain why female philandering would make as much survival sense as its male counterpart. Most of these new explanations directly resulted from considering the world from the female's point of view. Traditional theories of sexual selection, after all, were based exclusively on the perspective of the male: Males compete for *access* to the female, who apparently is just hanging around waiting to go out and party with the winner. And it's only from a male point of view that multiple female matings can be considered "excessive," or that female sexual interest is even described as her time of "receptivity." Is she passively "receptive" to the active intentions of the male? The word implies that she's just putting up with his annoying lustfulness yet again.

New hypotheses argue that there are genetic benefits for the offspring of sexually adventurous mothers. According to Hrdy's review of these explanations, the "fertility backup" hypothesis assumes that females need sperm from a number of males in order to assure conception by the healthiest sperm. The "inferior cuckold" hypothesis suggests that a female who has a genetically inferior mate will sneak off with a genetically superior male when she is likely to conceive. (I suppose she knows this by the size of his income.) And the "diverse paternity" hypothesis argues that when the environment is unpredictable, females diversify. Over a reproductive lifetime, females who have numerous partners, and thus different fathers for their offspring, improve their offspring's chances for survival.

Other theories look for the social and environmental benefits of female promiscuity to the mother and her infants. The "therapeutic hypothesis" suggests that having lots of partners and multiple orgasms (in some species) makes intercourse and conception more pleasurable, and therefore more likely to occur. The

"keep 'em around" hypothesis maintains that females actively solicit lower-status males (with the tacit approval of dominant males), a behavior that prevents weaker males from leaving the group. Hrdy's own favored theory is what she calls the "manipulation hypothesis," the idea that females mate with numerous males precisely because paternity becomes uncertain. The result is that male partners will be more invested in, and tolerant of, the female's infants. This idea, Hrdy explains,

grew out of a dawning awareness that, first of all, individual females could do a great deal that would affect the survival of their offspring, and second, that males, far from mere dispensers of sperm, were critical features on the landscape where infants died or survived. That is, females were more political, males more nurturing (or at least not neutral), than some earlier versions of sexual selection theory would lead us to suppose.

Both of these points are essential: Not only are females more than passive receptacles of sperm, but also males are more than "mere dispensers of sperm." They don't just mate and run. They have a key role in determining whether infants survive or die. Among primates, there is enormous variation in the extent to which males nurture and protect offspring:

- Among the ruffed lemur, the male tends the nest while the female forages for food.
- Among New World monkeys, males directly care for offspring in half of all species; often, the male is the primary caretaker, carrying the infant on his back, sharing food with it.
- In a rare study of a monogamous species of night money, an observer found that during one infant's first week of life, the mother carried it 33 percent of the time, the father 51 percent of the time, and a juvenile member of the troop the remaining time.
- Among baboons, males do not have much direct contact with infants, but they hover nearby protectively and offer what Hrdy calls "quality" time in a very real sense: They increase the infant's chances of survival. They discourage attacks on the infant from

males who are unknown, in both the literal and the Biblical sense, to the mother.

Hrdy's "manipulation hypothesis" assumes that primate males respond more benevolently to the offspring of females with whom they have mated, so the females derive obvious benefits from mating with more than one male. In numerous primate species, the mother's multiple sexual partners act like godfathers to the infant, as primatologist Jeanne Altmann calls them. Each of these males will help care for the female's offspring. Baboon males, many of whom could have served as the model for *Three Men and a Baby*, develop special relationships with the infant, carrying it on their backs on times of danger and protecting it from strangers and hazards. These affectionate bonds are possible because of the mother's closeness to the males, says Hrdy, and because the infant comes to trust these males and seek them out.

The manipulation hypothesis may or may not hold up with further research, as Hrdy acknowledges. It certainly does not apply to most human societies, where husbands do not look too kindly on their wives' "special relationships" with other men, let alone their previous lovers, husbands, and wooers. Hrdy's work, nonetheless, shows that theories depend, first and foremost, on what an observer *observes*, and then on how those observations can be blurred by unconscious expectations. Hrdy initially regarded those "wanton" female langurs as aberrations because their behavior did not fit the established theory. Not until researchers began to speculate on the potential benefits of female promiscuity did they come up with different questions and answers about female sexual behavior than had sociobiologists.

In evolutionary biology, if not in the popular press, the myth of the coy female (and, for that matter, the myth of the absent father) is dead. Hrdy is encouraged by the speed with which primatologists, once aware of the male bias that permeated their discipline, have produced "a small stampede by members of both sexes to study female reproductive strategies." This she takes to be a healthy sign, as I do. But Hrdy cautions against "substituting a new set of biases for the old ones":

> That is, among feminist scholars it is now permissible to say that males and females are different, provided one also stipulates that females are more cooperative, more nurturing, more supportive—not to mention equipped with unique moral sensibilities. . . .

Perhaps it is impossible, as biologist Donna Haraway suggests, for any of us to observe the behavior of other species, let alone our own, in a way that does not mirror the assumptions of our own way of life. It is disconcerting, says Hrdy wryly, that primatologists were finding "politically motivated females and nurturing males at roughly the same time that a woman runs for vice president of the United States and [Garry] Trudeau starts to poke fun at 'caring males' in his cartoons." Informally, scientists admit that their prejudices—such as the tendency to identify with the same sex of the species they are studying—affect their research. One woman primatologist told Hrdy, "I sometimes identify with female baboons more than I do with males of my own species."

The recognition of a male-centered bias in primatology and biology proved to be an enormous step forward, allowing scientists of both sexes to revise their theories of animal behavior. Sociobiologists (and their fans like Sam Kash Kachigan) can no longer justify traditional sex roles, particularly male dominance and female nurturance and chastity, by appealing to the universality of such behavior in other species. Other species aren't cooperating.

But that is not the only moral of the Parable of the Primates. The female perspective is invaluable, but, as Hrdy warns, a female-centered bias will provide its own set of distortions. Cultural feminists who look to evolutionary biology to explain women's allegedly sweeter, more cooperative ways are on as shaky ground as the antifeminists they would replace.

If the sociobiological heroine is the coy female who is so different from males, the heroine of modern sexology is the lusty female who is just like them. I like her better, but I'm afraid that she, too, is (as a student of mine once inadvertently said) a fig leaf of the imagination. . . .

First, Kill the Babies

In the fierce evolutionary battle to pass on one's genes, says one controversial hypothesis, everyone else is a potential competitor—even the infants.

BY CARL ZIMMER

TWENTY-FIVE YEARS AGO THIS SUMMER a Harvard graduate student named Sarah Hrdy went to northwestern India and met the monkeys that would make her famous. The immediate impetus for the trip was a series of lectures by Stanford ecologist Paul Ehrlich on the dangers of overpopulation. Though Ehrlich was speaking about humans, what Hrdy thought of were the Indian monkeys known as Hanuman langurs. The langurs are considered sacred by many Indians and so are regularly fed by the people with whom they come into contact. Consequently, near town, Hanuman langurs live in extremely dense populations, and apparently this unnatural density had led to unnatural, pathologically violent behavior. There had been several reports of adult males killing infants. "So there

---

CARL ZIMMER is a senior editor of DISCOVER. While traveling in Indonesia last year, Zimmer visited a group of primatologists who study Thomas langurs, monkeys that have been reported to commit infanticide. "I can see why the reports have been so rare," says Zimmer. "You have to get up before dawn and find the tree where the monkeys are sleeping and then follow them as they leap around the canopy. They jump from branch to branch while we humans stumble around in the brush. It's a wonder any science gets done at all."

I was, listening to Ehrlich," says Hrdy, "with this adolescent desire to go do something relevant with my life, and I thought, 'I am going to go study the effects of crowding on behavior.'"

Hrdy traveled to dry, deforested Mount Abu and began to get acquainted with the sandy-bodied, dark-faced Hanumans. Before long she decided the assumption that had propelled her to India had been wrong. "It happened pretty fast. I was watching these very crowded animals, and here were these infants playing around, bouncing on these males like trampolines, pulling on their tails, and so forth. These guys were aloof but totally tolerant. They might show some annoyance occasionally, but there was nothing approaching pathological hostility toward offspring. The trouble seemed to be when males came into the troop from outside it."

The langurs of Abu are arranged into two kinds of groups. In the first, a single male—or, rarely, two—lives with a group of females and their infants. The infant females, when they grow up, stay put; the males leave to join the other kind of group, a small all-male band.

Eventually a grown male lucky enough to be in a troop of females will come under attack, either from the all-male bands or from the male of another mixed troop. Odds are that sooner or later the resident male will be chased out by a new one. Hrdy witnessed many such takeovers, and she noticed that af-

terward the new male would often chase after the babies in the troop, presumably all of which were offspring of the old male's. Before long some of these infants would disappear. She didn't actually see what happened to the infants, but townspeople around Abu told her that they had seen a male killing baby langurs. Soon after these takeovers, the new resident male would mate with the females.

"I realized I needed a new explanatory model," says Hrdy. Her new model would become one of modern biology's most famous—and in some circles, notorious—hypotheses about animal behavior. There was nothing pathological about langur infanticide, she suggested. On the contrary, it actually made a chilling kind of sense: While a langur mother nurses, she cannot conceive; when she stops nursing, she can. Thus if a male langur kills her infant—one that is not related to him—she can bear the infanticidal male's own offspring. In Hanuman langur society, in which a male's sojourn with a harem averages a little over two years, the time saved can be critical. After all, for any offspring to survive, they should ideally be weaned before a new, potentially infanticidal male shows up. Seen in this light, infanticide could actually be an "adaptive" evolutionary strategy for fathering as many offspring as possible.

In the quarter century since Hrdy first conceived this idea, naturalists have reported cases of infanticide among a wide range of animals. Some now argue

---

that the threat of infanticide is such a pervasive and powerful influence that it can shape animal societies. A few theorists even claim that infanticide was an important factor in human evolution.

Yet when Hrdy first published her hypothesis, she was immediately attacked, most of all by other researchers studying langurs. They contended that Hanuman langurs living in natural conditions, in remote forests, had never been observed killing infants. At Abu, they said, human feeding had crowded the langurs into conditions evolution had never prepared them for, and as a result the transfer of males into new groups became drenched with aggression. In other words, the langurs of Abu were simply not normal.

Underlying this species-specific dustup, though, was a deeper conflict. Before the 1970s most researchers viewed animal societies as smooth-running systems in which each member knew its proper role and played it for the good of the many. Animal societies—and primate societies in particular—were often portrayed as utopias that we humans would do well to emulate. But then biologists such as E. O. Wilson and Robert Trivers (both mentors of Hrdy's at Harvard) argued that such a view didn't make sense in a world shaped by evolution. Just as a bird's wing is the product of natural selection, so are the ways the individual bird interacts with other birds. Its social behavior, like its body, is ultimately designed for one purpose: to get its own genes duplicated as much as possible. Rather than being a peaceful group of community-minded role players, an animal society was made up of individuals trying to maximize their reproductive gain, with cooperation always a compromise between competing genetic interests. As a biological explanation for society, this school of thought came to be known as sociobiology.

When Hrdy hypothesized about langur infanticide, then, she wasn't just explaining the odd behavior of a few monkeys. She was pushing sociobiology to its logical extreme, in which a male's drive to reproduce was so strong that it would resort to the decidedly antisocial act of killing the young of its own species.

OVER TIME, HRDY ADDED SOME NUances to her stark hypothesis. For example, she noted that female langurs did not passively sit by as invading males tried to rob them of their genetic legacy. Rather, females banded together to help fend off males bent on infanticide. Once an infanticidal male was in charge, however, they might choose a different tack. Female langurs can continue to have sex even after they conceive, and by mating with an invading male, they might trick him into thinking the infant was his own.

Hrdy also began noting reports of infanticide among other animals. Her "outsider male" infanticide, she realized was clearly not the only kind of adaptive strategy practiced in the animal world. A mother might resort to infanticide if she didn't have the resources to raise all her children. Adults might also kill the infants of strangers simply for food or to eliminate the competition for limited resources.

In the decade and a half since Hrdy's work first appeared, infanticide has been reported among mice and ground squirrels, bears and deer, prairie dogs and foxes, fish and dwarf mongooses and wasps and bumblebees and dung beetles. Although the evidence from the wild has often been sketchy, most of the strategies appear to fall into one of those that Hrdy sketched out. In a few cases researchers have been able to test the hypothesis by performing natural experiments. In 1987, for example, Cornell ornithologist Stephen Emlen was studying the jacana, a Panamanian bird in which the common sex roles are switched: males sit on the eggs and raise the young alone while the females rove around their territories, mating with many males and fighting off intruding females. Essentially, if you turn Hanuman langurs into birds and switch the sex roles, you get jacanas—and theoretically, under the right conditions, you should also see infanticide. Emlen needed to shoot some birds for DNA testing, and he chose two females with male partners caring for nests of babies.

"I shot a female one night, and the next morning was just awesome. By first light a new female was already on the turf. I saw terrible things—pecking and picking up and throwing down chicks until they were dead. Within hours she was soliciting the male, and he was mounting her the same day. The next night I shot the other female, then came out the next morning and saw the whole thing again."

Among mammals, one of the best documented killers of infants has proved to be the lion. Though actual killings have only rarely been witnessed (the total is about a dozen), massive indirect evidence for the phenomenon has been gathered by the husband-and-wife team of Craig Packer and Anne Pusey, both behavioral ecologists at the University of Minnesota. From their observations of lions in the Serengeti, they've found that whenever a new male comes into the pride, the death rate of nursing cubs—and nursing cubs only—shoots up. Within six months none of the cubs are left alive.

Other primates have also joined the ranks of the infanticidal. The first reports were of only a few species such as the red howlers of Venezuela, the gorillas of Rwanda, and the blue monkeys of Uganda. But in the past few years there have been more reported instances of primate infanticide, some of which demand some expansion of Hrdy's ideas. Lemurs in Madagascar, for example, breed once a year. If a male kills another male's nursing infant, he doesn't hasten his own fatherhood, since he still has to wait until the breeding season to mate. Nonetheless, researchers have seen male lemurs sinking their fangs into babies. One observer, Michael Pereira of Bucknell University, offers an idea as to why it happens. Madagascar has a harsh climate, with a long dry season that keeps female lemurs on a knife-edge of survival. Pregnancy and the first few months of nursing take place in the harshest time of the year, and, says Pereira, successfully raising an infant one year may reduce the chances a mother will be able to raise the next year's baby. "Females who lose their infants are much fatter than females whose infants survive," Pereira explains. "If by your killing the infant she's more likely to be successful during your reign, then it's to your advantage."

In Sumatra, Dutch researchers have been studying a relative of the Hanumans known as the Thomas langur. Thomas langurs were essentially a mystery when the Utrecht University primatologists began to observe them in 1988. Now, after eight years of relentlessly tracking the animals through the forest and painstakingly recording their daily habits, the researchers are finding the langurs to be all too revealing. "Infanticide does occur. I've seen the attacks," explains Romy Steenbeek, who ran the program for four years. "We saw the body of a baby with canine slices in its belly. I saw a male attack a baby, and the baby disappeared. One baby received big wounds in an attack by a neighboring male, and she died after two very bad weeks. The males run a few hundred meters to the troop, silently attack, and when they leave they loud-call."

Like lemurs, these langurs require yet one more variation on Hrdy's grisly theme. Thomas langurs have the same all-male bands and one-male/many-female troops of other langurs, but what distinguishes this species is that the males don't make hostile takeovers of groups of females—at least not directly. Instead outside males make harassing raids on a troop, chasing the infants and sometimes killing them. One by one the females abandon their male, the childless females first, the others as soon as their babies are weaned or die. Steenbeek suspects that the infanticidal male langurs are trying to discredit the harem male, demonstrating how incompetent he is by killing the troop's infants. The females are continually judging the contest, and if they sense that their male is getting weak, they abandon him. "Sometimes at the end of the tenure, a male stops protecting the troop," says Steenbeek. "It's like he's just given up."

If infanticide has long been a natural part of animal behavior, then so too, one would expect, has been the fight against it. Over evolutionary time, both currents would shape new behaviors and social organizations. Lionesses live in prides, according to Packer and Pusey, in large part to protect their young against murderous males. One result is that lionesses must tolerate cubs not their own

stealing milk—something rarely seen in other carnivores. Female mice can somehow tell if a male approaching their litter is infanticidal or not; if he is, they leap into battle. And apparently even the babies have evolved a protection against infanticide: they call much more frequently in the presence of an infanticidal male.

Evidence is emerging that primates may face similar pressure. Female red howler monkeys in Venezuela, for example, tend to travel in small groups—generally under five members—and are hostile to new female howlers who want to join them. What determines their group size? In some animals the availability of food is the key: if the group is too big, the competition among individuals grows too intense. Yet evidently food competition isn't a problem for red howlers. What is a problem, it seems, is that the bigger a female group, the more likely it is to suffer an infanticidal attack by a male. The benefits for a male of taking over a big group make those groups good targets, and as a result females keep the groups small.

We can only speculate on the role of infanticide in human evolution. Our complex lives hide its effects as the forest hides the secrets of langurs.

Carel van Schaik, a primatologist with Duke University and the Wildlife Conservation Society, thinks infanticide's effects may reach even further into the core of primate life. He first began thinking seriously about Hrdy's ideas in the late 1980s, while studying gibbons with Robin Dunbar, who was then at University College, London. These Asian monkeys are for the most part monogamous, although it's hard to see why. It's not that gibbon babies need the extra parental care, because the fathers don't give any. And calculations suggested that males might do better,

from a genetic viewpoint, if they tried to impregnate as many females as possible rather than just one. Van Schaik and Dunbar concluded that male gibbons were staying close to home to guard their infants from other males, and the females were choosing good protectors as mates. That would explain why on the one hand gibbon couples make calls together—advertising to neighboring males that the infant is well guarded—but why on the other hand a nursing mother who becomes widowed falls silent. It doesn't matter to Van Schaik that gibbon infanticide has never been reported—it's not seen, he thinks, because the animals do such a good job of preventing it.

Van Schaik now suspects that the ever-present threat of infanticide has a similar effect on all primates. Among mammals, primate males and females are far and away the most likely to form a long-term bond. "That raises the issue: Why primates?" says Van Schaik. The answer, he thinks, is that primate babies are particularly vulnerable to infanticide. They take a long time to mature, and compared with other young mammals, they are defenseless and exposed, more often than not clinging to their mother. Female primates also tend to stay in a given territory, thereby giving males an added incentive for disposing of unrelated infants. "If you do commit infanticide, there is a good probability that you will have a future opportunity for mating," says Van Schaik. Thus the incentive and opportunity for infanticide have driven primates more than other mammals into long-term bonds, in order that males can defend their young.

If Van Schaik is right, he will add considerable weight to speculations Hrdy made in the 1980s, that protection against infanticide may have had a profound impact on primates, including early humans. Nursing is a contraceptive among humans, as it is in langurs, and it can lower a woman's fertility for up to two or three years. That could make the incentive for infanticide on the part of a new mate enormous, as would the incentives to guard against it. Such a scenario would fly in the face of the conventional view that long-term bonds between men and women evolved so that extra parental care can help their babies survive. Instead, Van Schaik suggests that infanticide may have been the prime mover behind these bonds, and only later did the added advantage of help from a father come into play.

Not surprisingly, perhaps, such ideas are not easily accepted. Most anthropologists and psychologists still view humans much as biologists once viewed animals. "Anthropology has a long history of believing that everything is for the good of the social group," says anthropologist Kim Hill of the University of New Mexico, and in that context Hrdy's ideas about infanticide don't make sense; such instances as do occur among humans can only be explained as a result of a particular cultural bias (favoring male babies over females, say) or of individual pathology. But there are a few disturbing data points. Over the past 16 years Martin Daly and Margo Wilson, both psychologists at McMaster University in Hamilton, Ontario, have collected child abuse data from governments and humane associations. One of the most startling statistics they've uncovered is this: a preschool American stepchild is 60 times more likely than a biological child to be the victim of infanticide.

Hill himself, with his studies of the Ache people of Paraguay, has gathered some of the best infanticide data available on non-Western cultures. The Ache still go on long hunting-and-foraging expeditions, as their ancestors did for 10,000 years. When a man kills an animal, he gives it to another man, who then distributes the meat to the entire band. Congenial as this may seem, natural selection creates inevitable tension: by giving most of his food away, a man allows his efforts to be diverted from his own family. This cost is outweighed by the benefits of cooperation, but when a child's father dies, the tension reveals itself. If a child loses a father, his chances of becoming the victim of infanticide at the hands of another man increase fourfold. It's not uncommon for orphaned children to be thrown into their father's grave.

But Hill does not think that the pattern is purely a cultural tradition. "If you ask them why they're killing all these babies, their first answer is 'That's our custom,' " says Hill. "And then if you push them on that, they say, 'They don't have parents, and we have to take care of them, and that makes us mad.'"

THE RESISTANCE TO INFANTICIDE AS a reproductive strategy is still shared by many researchers. In some cases they've tested some of the predictions and found them wanting. Agustin Fuentes of the University of California at Berkeley, for example, studies the Mentawai langur, which lives on the islands of the same name, off the west coast of Sumatra. Like the gibbon, the Mentawai langur is monogamous, but it doesn't behave as Van Schaik said it should. For example, when a bonded couple are close to a solitary (and supposedly infanticidal) male, they do not become hostile or even make calls to show they are together.

Deborah Overdorff of the University of Texas at Austin studies rufous lemurs, and among these primates, at least, doubts the reality of infanticide. While rufous lemurs travel in large groups, male and female pairs will often be seen staying close together. That might seem to fit the notion of males protecting their young. Not to Overdorff. "I've found that the male is not necessarily the one the females mated with. Sometimes they turn out to be brothers. Infanticide is probably not a good explanation for pair-bonding."

Others criticize the quality of the data. They complain that most reports are inferred from indirect evidence, such as the disappearance of a baby. And except for Hanuman langurs, the few witnessed infanticides have not been followed up to see how much reproductive success the killing males have had. Given the difficulty of observing monkeys in the wild, the scrappiness of the data shouldn't be surprising, and some primatologists—including some who think that infanticide is real—worry that the theory is getting too far ahead of the data.

The biggest opposition results from the application of Hrdy's ideas to humans. Popular accounts of the theory, Hrdy complains, are "very quick to jump from the langur case to cases of strange-male-in-the-household infanticide, but the underpinnings, the groundwork for that extrapolation, aren't there." After all, a stepfather can't speed up his own reproduction through infanticide. Hrdy and Daly agree that this kind of abuse has more to do with resource competition—the resource being the mother. Moreover, they don't envision a stepfather consciously trying to eliminate that competition—rather, he may simply have a lower threshold of irritation toward the child. Such a threshold is suggested by a recent study by Daly and Wilson, in which they compared the ways in which biological fathers and stepfathers killed their children. In most cases, biological fathers shoot or suffocate their offspring (and then often kill themselves), while stepfathers kill by striking—hinting that a "lashing out" reflex is at work.

Another point of criticism is the matter of how infanticide can be carried down through the generations, and again confusion abounds. One magazine article Hrdy mentions, for example, contains a reference to an "infanticide gene." She scoffs. "I don't talk about genes." While it's true that Hrdy doesn't dabble in oversimplified genetic determinism, some of sociobiology's early pioneers did—and sometimes with great abandon. These days, however, a much suppler view exists. In any species, each individual keeps a Darwinian account book, and whenever it has to "choose" an action, it weighs the immediate and long-term costs and benefits. "Selection hasn't molded an animal that's altruistic or infanticidal," says Emlen. "It has molded an animal capable of showing a whole range of subtly different behaviors under different circumstances, but they're all predictable."

"Under one set of circumstances, a female might behave by abandoning a baby, but under another set of circumstances she would care for a baby," says Hrdy. "These are both maternal behaviors. In the first case, presumably selection has operated on her to postpone raising her young because there is the option that she might have a better chance of pulling a baby through at a future date. So it's not nonreproductive, it's not nonadaptive; it's simply a question of an animal over the course of a lifetime gauging herself." Infanticide is thus at one extreme in a spectrum of parental care. Hrdy herself has recently been exploring the ways in which European parents have historically lowered their investment in children, such as hiring a wet nurse or leaving a child at a monastery.

For animals, and to some degree ourselves, this "gauging" happens unconsciously. And while it might seem hard to believe that animals can make careful decisions, many experiments have revealed that they can. Few of these accounting experiments have been done on infanticide, though, and there isn't anywhere near enough data to test in primates, let alone humans. While almost 20 percent of Ache children fall victim to infanticide before age ten, the rate is zero among many other foraging cultures. Until researchers can explain the variation,

all speculation on the role of infanticide in early human evolution must be put on hold. Our long, complex social lives and our dizzying array of cultures hide the effects of evolution as the high, obscuring leaves of the Sumatran forest hide the secrets of Thomas langurs.

Yet those who believe that infanticide is a Darwinian reality think that we need to keep looking through the foliage. "Sure, you can deny all these results—at your own peril," says Van Schaik. "What is it that makes males infanticidal, and what is it that stops males from being infanticidal? If you know these things better, you know what to do, take certain measures, counsel people. It arms us."

A look back at the infanticide hypothesis on its silver anniversary makes clear how long it takes to test and flesh out the shocking ideas of sociobiology.

Hrdy herself sees this as the necessary pace of any science. She often describes her job as creating "imaginary worlds" that other scientists can then explore to see if they can help us understand the real one. "I see scientists working in different phases. Some people are better at one phase than another. Theoreticians think of other people as technicians; technicians think of theoreticians as people in outer space, not connected to the real world. But for the whole process, you need these phases, and in the initial phase, you're selecting a project, you're coming up with assumptions, you're trying to model what might be true and to generate the hypotheses that you want to look at. Then you have the actual collection of data and all the methodologies that go into that. Imaginary worlds have a place in science."

A Woman's Curse?

Why do cultures the world over treat menstruating women as taboo?
An anthropologist offers a new answer—and a challenge to Western
ideas about contraception

By Meredith F. Small

THE PASSAGE FROM GIRLHOOD TO WOM-anhood is marked by a flow of blood from the uterus. Without elaborate ceremony, often without discussion, girls know that when they begin to menstruate, their world is changed forever. For the next thirty years or so, they will spend much energy having babies, or trying not to, reminded at each menstruation that either way, the biology of reproduction has a major impact on their lives.

Anthropologists have underscored the universal importance of menstruation by documenting how the event is interwoven into the ideology as well as the daily activities of cultures around the world. The customs attached to menstruation take peculiarly negative forms: the so-called menstrual taboos. Those taboos may prohibit a woman from having sex with her husband or from cooking for him. They may bar her from visiting sacred places or taking part in sacred activities. They may forbid her to touch certain items used by men, such as hunting gear or weapons, or to eat certain foods or to wash at certain times. They may also require that a woman paint her face red or wear a red hip cord, or that she segregate herself in a special hut while she is menstruating. In short,

the taboos set menstruating women apart from the rest of their society, marking them as impure and polluting.

Anthropologists have studied menstrual taboos for decades, focusing on the negative symbolism of the rituals as a cultural phenomenon. Perhaps, suggested one investigator, taking a Freudian perspective, such taboos reflect the anxiety that men feel about castration, an anxiety that would be prompted by women's genital bleeding. Others have suggested that the taboos serve to prevent menstrual odor from interfering with hunting, or that they protect men from microorganisms that might otherwise be transferred during sexual intercourse with a menstruating woman. Until recently, few investigators had considered the possibility that the taboos—and the very fact of menstruation—might instead exist because they conferred an evolutionary advantage.

In the mid-1980s the anthropologist Beverly I. Strassmann of the University of Michigan in Ann Arbor began to study the ways men and women have evolved to accomplish (and regulate) reproduction. Unlike traditional anthropologists, who focus on how culture affects human behavior, Strassmann was convinced that the important role played

by biology was being neglected. Menstruation, she suspected, would be a key for observing and understanding the interplay of biology and culture in human reproductive behavior.

To address the issue, Strassmann decided to seek a culture in which making babies was an ongoing part of adult life. For that she had to get away from industrialized countries, with their bias toward contraception and low birthrates. In a "natural-fertility population," she reasoned, she could more clearly see the connection between the physiology of women and the strategies men and women use to exploit that physiology for their own reproductive ends.

Strassmann ended up in a remote corner of West Africa, living in close quarters with the Dogon, a traditional society whose indigenous religion of ancestor worship requires that menstruating women spend their nights at a small hut. For more than two years Strassmann kept track of the women staying at the hut, and she confirmed the menstruations by testing urine samples for the appropriate hormonal changes. In so doing, she amassed the first long-term data describing how a traditional society appropriates a physiological event—menstruation—and re-

Reprinted by permission of *The Sciences*, January/February 1999, pp. 24-29. © 1999 by the New York Academy of Science. Individual subscriptions are $28 per year. Write to: The Sciences, 2 East 63rd Street, New York, NY 10021.

fracts that event through a prism of behaviors and beliefs.

What she found explicitly challenges the conclusions of earlier investigators about the cultural function of menstrual taboos. For the Dogon men, she discovered, enforcing visits to the menstrual hut serves to channel parental resources into the upbringing of their own children. But more, Strassmann, who also had training as a reproductive physiologist, proposed a new theory of why menstruation itself evolved as it did—and again, the answer is essentially a story of conserving resources. Finally, her observations pose provocative questions about women's health in industrialized societies, raising serious doubts about the tactics favored by Western medicine for developing contraceptive technology.

MENSTRUATION IS THE VISIBLE stage of the ovarian cycle, orchestrated primarily by hormones secreted by the ovaries: progesterone and a family of hormones called estrogens. At the beginning of each cycle (by convention, the first day of a woman's period) the levels of the estrogens begin to rise. After about five days, as their concentrations increase, they cause the blood- and nutrient-rich inner lining of the uterus, called the endometrium, to thicken and acquire a densely branching network of blood vessels. At about the middle of the cycle, ovulation takes place, and an egg makes its way from one of the two ovaries down one of the paired fallopian tubes to the uterus. The follicle from which the egg was released in the ovary now begins to secrete progesterone as well as estrogens, and the progesterone causes the endometrium to swell and become even richer with blood vessels—in short, fully ready for a pregnancy, should conception take place and the fertilized egg become implanted.

If conception does take place, the levels of estrogens and progesterone continue to rise throughout the pregnancy. That keeps the endometrium thick enough to support the quickening life inside the uterus. When the baby is born and the new mother begins nursing, the estrogens and progesterone fall

to their initial levels, and lactation hormones keep them suppressed. The uterus thus lies quiescent until frequent lactation ends, which triggers the return to ovulation.

If conception does not take place after ovulation, all the ovarian hormones also drop to their initial levels, and menstruation—the shedding of part of the uterine lining—begins. The lining is divided into three layers: a basal layer that is constantly maintained, and two superficial layers, which shed and regrow with each menstrual cycle. All mammals undergo cyclical changes in the state of the endometrium. In most mammals the sloughed-off layers are resorbed into the body if fertilization does not take place. But in some higher primates, including humans, some of the shed endometrium is not resorbed. The shed lining, along with some blood, flows from the body through the vaginal opening, a process that in humans typically lasts from three to five days.

OF COURSE, PHYSIOLOGICAL FACTS alone do not explain why so many human groups have infused a bodily function with symbolic meaning. And so in 1986 Strassmann found herself driving through the Sahel region of West Africa at the peak of the hot season, heading for a sandstone cliff called the Bandiagara Escarpment, in Mali. There, permanent Dogon villages of mud or stone houses dotted the rocky plateau. The menstrual huts were obvious: round, low-roofed buildings set apart from the rectangular dwellings of the rest of the village.

The Dogon are a society of millet and onion farmers who endorse polygyny, and they maintain their traditional culture despite the occasional visits of outsiders. In a few Dogon villages, in fact, tourists are fairly common, and ethnographers had frequently studied the Dogon language, religion and social structure before Strassmann's arrival. But her visit was the first time someone from the outside wanted to delve into an intimate issue in such detail.

It took Strassmann a series of hikes among villages, and long talks with male elders under the thatched-roof shelters where they typically gather, to

find the appropriate sites for her research. She gained permission for her study in fourteen villages, eventually choosing two. That exceptional welcome, she thinks, emphasized the universality of her interests. "I'm working on all the things that really matter to [the Dogon]—fertility, economics—so they never questioned my motives or wondered why I would be interested in these things," she says. "It seemed obvious to them." She set up shop for the next two and a half years in a stone house in the village, with no running water or electricity. Eating the daily fare of the Dogon, millet porridge, she and a research assistant began to integrate themselves into village life, learning the language, getting to know people and tracking visits to the menstrual huts.

Following the movements of menstruating women was surprisingly easy. The menstrual huts are situated outside the walled compounds of the village, but in full view of the men's thatched-roof shelters. As the men relax under their shelters, they can readily see who leaves the huts in the morning and returns to them in the evening. And as nonmenstruating women pass the huts on their way to and from the fields or to other compounds, they too can see who is spending the night there. Strassmann found that when she left her house in the evening to take data, any of the villagers could accurately predict whom she would find in the menstrual huts.

THE HUTS THEMSELVES ARE CRAMPED, dark buildings—hardly places where a woman might go to escape the drudgery of work or to avoid an argument with her husband or a co-wife. The huts sometimes become so crowded that some occupants are forced outside—making the women even more conspicuous. Although babies and toddlers can go with their mothers to the huts, the women consigned there are not allowed to spend time with the rest of their families. They must cook with special pots, not their usual household possessions. Yet they are still expected to do their usual jobs, such as working in the fields.

Why, Strassmann wondered, would anyone put up with such conditions?

The answer, for the Dogon, is that a menstruating woman is a threat to the sanctity of religious altars, where men pray and make sacrifices for the protection of their fields, their families and their village. If menstruating women come near the altars, which are situated both indoors and outdoors, the Dogon believe that their aura of pollution will ruin the altars and bring calamities upon the village. The belief is so ingrained that the women themselves have internalized it, feeling its burden of responsibility and potential guilt. Thus violations of the taboo are rare, because a menstruating woman who breaks the rules knows that she is personally responsible if calamities occur.

NEVERTHELESS, STRASSMANN STILL thought a more functional explanation for menstrual taboos might also exist, one closely related to reproduction. As she was well aware, even before her studies among the Dogon, people around the world have a fairly sophisticated view of how reproduction works. In general, people everywhere know full well that menstruation signals the absence of a pregnancy and the possibility of another one. More precisely, Strassmann could frame her hypothesis by reasoning as follows: Across cultures, men and women recognize that a lack of menstrual cycling in a woman implies she is either pregnant, lactating or menopausal. Moreover, at least among natural-fertility cultures that do not practice birth control, continual cycles during peak reproductive years imply to people in those cultures that a woman is sterile. Thus, even though people might not be able to pinpoint ovulation, they can easily identify whether a woman will soon be ready to conceive on the basis of whether she is menstruating. And that leads straight to Strassmann's insightful hypothesis about the role of menstrual taboos: information about menstruation can be a means of tracking paternity.

"There are two important pieces of information for assessing paternity," Strassmann notes: timing of intercourse and timing of menstruation. "By forcing women to signal menstruation, men are trying to gain equal access to one part of that critical information." Such information, she explains, is crucial to Dogon men, because they invest so many resources in their own offspring. Descent is marked through the male line; land and the food that comes from the land is passed down from fathers to sons. Information about paternity is thus crucial to a man's entire lineage. And because each man has as many as four wives, he cannot possibly track them all. So forcing women to signal their menstrual periods, or lack thereof, helps men avoid cuckoldry.

TO TEST HER HYPOTHESIS, STRASSmann tracked residence in the menstrual huts for 736 consecutive days, collecting data on 477 complete cycles. She noted who was at each hut and how long each woman stayed. She also collected urine from ninety-three women over a ten-week period, to check the correlation between residence in the menstrual hut and the fact of menstruation.

Requiring Dogon women to signal their menstrual periods, or lack thereof, helps Dogon men avoid cuckoldry.

The combination of ethnographic records and urinalyses showed that the Dogon women mostly play by the rules. In 86 percent of the hormonally detected menstruations, women went to the hut. Moreover, none of the tested women went to the hut when they were not menstruating. In the remaining 14 percent of the tested menstruations, women stayed home from the hut, in violation of the taboo, but some were near menopause and so not at high risk for pregnancy. More important, none of the women who violated the taboo did it twice in a row. Even they were largely willing to comply.

Thus, Strassmann concluded, the huts do indeed convey a fairly reliable signal, to men and to everyone else, about the status of a woman's fertility. When she leaves the hut, she is considered ready to conceive. When she stops going to the hut, she is evidently pregnant or menopausal. And women of prime reproductive age who visit the hut on a regular basis are clearly infertile.

It also became clear to Strassmann that the Dogon do indeed use that information to make paternity decisions. In several cases a man was forced to marry a pregnant woman, simply because everyone knew that the man had been the woman's first sexual partner after her last visit to the menstrual hut. Strassmann followed one case in which a child was being brought up by a man because he was the mother's first sexual partner after a hut visit, even though the woman soon married a different man. (The woman already knew she was pregnant by the first man at the time of her marriage, and she did not visit the menstrual hut before she married. Thus the truth was obvious to everyone, and the real father took the child.)

In general, women are cooperative players in the game because without a man, a woman has no way to support herself or her children. But women follow the taboo reluctantly. They complain about going to the hut. And if their husbands convert from the traditional religion of the Dogon to a religion that does not impose menstrual taboos, such as Islam or Christianity, the women quickly cease visiting the hut. Not that such a religious conversion quells a man's interest in his wife's fidelity: far from it. But the rules change. Perhaps the sanctions of the new religion against infidelity help keep women faithful, so the men can relax their guard. Or perhaps the men are willing to trade the reproductive advantages of the menstrual taboo for the economic benefits gained by converting to the new religion. Whatever the case, Strassmann found an almost perfect correlation between a husband's religion and his wives' attendance at the hut. In sum, the taboo is established by men, backed by supernatural forces, and internalized and accepted by women until the men release them from the belief.

BUT BEYOND THE CULTURAL MACHINA-tions of men and women that Strassmann expected to find, her data show something even more fundamental—and surprising—about female biology. On average, she calculates, a woman in a natural-fertility population such as the Dogon has only about 110 menstrual periods in her lifetime. The rest of the time she will be prepubescent, pregnant, lactating or menopausal. Women in industrialized cultures, by contrast, have more than three times as many cycles: 350 to 400, on average, in a lifetime. They reach menarche (their first menstruation) earlier—at age twelve and a half, compared with the onset age of sixteen in natural-fertility cultures. They have fewer babies, and they lactate hardly at all. All those factors lead women in the industrialized world to a lifetime of nearly continuous menstrual cycling.

The big contrast in cycling profiles during the reproductive years can be traced specifically to lactation. Women in more traditional societies spend most of their reproductive years in lactation amenorrhea, the state in which the hormonal changes required for nursing suppress ovulation and inhibit menstruation. And it is not just that the Dogon bear more children (eight to nine on average); they also nurse each child on demand rather than in scheduled bouts, all through the night as well as the day, and intensely enough that ovulation simply stops for about twenty months per child. Women in industrialized societies typically do not breast-feed as intensely (or at all), and rarely breast-feed each child for as long as the Dogon women do. (The average for American women is four months.)

The Dogon experience with menstruation may be far more typical of the human condition over most of evolutionary history than is the standard menstrual experience in industrialized nations. If so, Strassmann's findings alter some of the most closely held beliefs about female biology. Contrary to what the Western medical establishment might think, it is not particularly "normal" to menstruate each month. The female body, according to Strassmann, is biologically designed to spend much more time in lactation amenorrhea than in menstrual cycling. That in itself suggests that oral contraceptives, which alter hormone levels to suppress ovulation and produce a bleeding, could be forcing a continual state of cycling for which the body is ill-prepared. Women might be better protected against reproductive cancers if their contraceptives mimicked lactation amenorrhea and depressed the female reproductive hormones, rather than forcing the continual ebb and flow of menstrual cycles.

Oral contraceptives could be forcing a continual state of menstrual cycling for which a woman's body is ill-prepared.

Strassmann's data also call into question a recently popularized idea about menstruation: that regular menstrual cycles might be immunologically beneficial for women. In 1993 the controversial writer Margie Profet, whose ideas about evolutionary and reproductive biology have received vast media attention, proposed in *The Quarterly Review of Biology* that menstruation could have such an adaptive value. She noted that viruses and bacteria regularly enter the female body on the backs of sperm, and she hypothesized that the best way to get them out is to flush them out. Here, then, was a positive, adaptive role for something unpleasant, an evolutionary reason for suffering cramps each month. Menstruation, according to Profet, had evolved to rid the body of pathogens. The "anti-pathogen" theory was an exciting hypothesis, and it helped win Profet a MacArthur Foundation award. But Strassmann's work soon showed that Profet's ideas could not be supported because of one simple fact: under less-industrialized conditions, women menstruate relatively rarely.

Instead, Strassmann notes, if there is an adaptive value to menstruation, it is ultimately a strategy to conserve the body's resources. She estimates that maintaining the endometrial lining during the second half of the ovarian cycle takes substantial metabolic energy. Once the endometrium is built up and ready to receive a fertilized egg, the tissue requires a sevenfold metabolic increase to remain rich in blood and ready to support a pregnancy. Hence, if no pregnancy is forthcoming, it makes a lot of sense for the body to let part of the endometrium slough off and then regenerate itself, instead of maintaining that rather costly but unneeded tissue. Such energy conservation is common among vertebrates: male rhesus monkeys have shrunken testes during their nonbreeding season, Burmese pythons shrink their guts when they are not digesting, and hibernating animals put their metabolisms on hold.

Strassmann also suggests that periodically ridding oneself of the endometrium could make a difference to a woman's long-term survival. Because female reproductive hormones affect the brain and other tissues, the metabolism of the entire body is involved during cycling. Strassmann estimates that by keeping hormonal low through half the cycle, a woman can save about six days' worth of energy for every four nonconceptive cycles. Such caloric conservation might have proved useful to early hominids who lived by hunting and gathering, and even today it might be helpful for women living in less affluent circumstances than the ones common in the industrialized West.

BUT PERHAPS THE MOST PROVOCATIVE implications of Strassmann's work have to do with women's health. In 1994 a group of physicians and anthropologists published a paper, also in *The Quarterly Review of Biology,* suggesting that the reproductive histories and lifestyles of women in industrialized cultures are at odds with women's naturally evolved biology, and that the differences lead to greater risks of reproductive cancers. For example, the investigators estimated that women in affluent cultures may have a hundredfold greater risk of breast cancer than do women who subsist by hunting and gathering. The increased risk is probably caused not only by low levels of exercise and a high-fat diet, but also by a relatively high number of menstrual cycles over a lifetime. Repeated exposure to the hormones of the ovarian cycle—because of early menarche, late menopause, lack of pregnancy and little or no breast-feeding—is implicated in other reproductive cancers as well.

Those of us in industrialized cultures have been running an experiment on ourselves. The body evolved over millions of years to move across the landscape looking for food, to live in small kin-based groups, to make babies at intervals of four years or so and to invest heavily in each child by nursing intensely for years. How many women now follow those traditional patterns? We move little, we rely on others to get our food, and we rarely reproduce or lactate. Those culturally initiated shifts in lifestyles may pose biological risks.

Our task is not to overcome that biology, but to work with it. Now that we have a better idea of how the female body was designed, it may be time to rework our lifestyles and change some of our expectations. It may be time to borrow from our distant past or from our contemporaries in distant cultures, and treat our bodies more as nature intended.

MERIDITH F. SMALL is a professor of anthropology at Cornell University in Ithaca, New York. Her latest book, OUR BABIES, OURSELVES: HOW BIOLOGY AND CULTURE SHAPE THE WAY WE PARENT, *was published in May 1998 [see Laurence A. Marschall's review in Books in Brief, November/December 1998].*

What's Love Got to Do With It?

Sex Among Our Closest Relatives Is a Rather Open Affair

Meredith F. Small

Maiko and Lana are having sex. Maiko is on top, and Lana's arms and legs are wrapped tightly around his waist. Lina, a friend of Lana's, approaches from the right and taps Maiko on the back, nudging him to finish. As he moves away, Lina enfolds Lana in her arms, and they roll over so that Lana is now on top. The two females rub their genitals together, grinning and screaming in pleasure.

This is no orgy staged for an X-rated movie. It doesn't even involve people—or rather, it involves them only as observers. Lana, Maiko, and Lina are bonobos, a rare species of chimplike ape in which frequent couplings and casual sex play characterize every social relationship—between males and females, members of the same sex, closely related animals, and total strangers. Primatologists are beginning to study the bonobos' unrestrained sexual behavior for tantalizing clues to the origins of our own sexuality.

In reconstructing how early man and woman behaved, researchers have generally looked not to bonobos but to common chimpanzees. Only about 5 million years ago human beings and chimps shared a common ancestor, and we still have much behavior in common: namely, a long period of infant dependency, a reliance on learning what to eat and how to obtain food, social bonds that persist over generations, and the need to deal as a

group with many everyday conflicts. The assumption has been that chimp behavior today may be similar to the behavior of human ancestors.

Bonobo behavior, however, offers another window on the past because they, too, shared our 5-million-year-old ancestor, diverging from chimps just 2 million years ago. Bonobos have been less studied than chimps for the simple reason that they are difficult to find. They live only on a small patch of land in Zaire, in central Africa. They were first identified, on the basis of skeletal material, in the 1920s, but it wasn't until the 1970s that their behavior in the wild was studied, and then only sporadically.

Bonobos, also known as pygmy chimpanzees, are not really pygmies but welterweights. The largest males are as big as chimps, and the females of the two species are the same size. But bonobos are more delicate in build, and their arms and legs are long and slender.

On the ground, moving from fruit tree to fruit tree, bonobos often stand and walk on two legs—behavior that makes them seem more like humans than chimps. In some ways their sexual behavior seems more human as well, suggesting that in the sexual arena, at least, bonobos are the more appropriate ancestral model. Males and females frequently copulate face-to-face,

which is an uncommon position in animals other than humans. Males usually mount females from behind, but females seem to prefer sex face-to-face. "Sometimes the female will let a male start to mount from behind," says Amy Parish, a graduate student at the University of California at Davis who's been watching female bonobo sexual behavior in several zoo colonies around the world. "And then she'll stop, and of course he's really excited, and then she continues face-to-face." Primatologists assume the female preference is dictated by her anatomy: her enlarged clitoris and sexual swellings are oriented far forward. Females presumably prefer face-to-face contact because it feels better.

"Sex is fun. Sex makes them feel good and keeps the group together."

Like humans but unlike chimps and most other animals, bonobos separate sex from reproduction. They seem to treat sex as a pleasurable activity, and they rely on it as a sort of social glue, to make or break all sorts of relationships. "Ancestral humans behaved like this," proposes Frans de Waal, an ethologist at the Yerkes Regional Primate Research Center at Emory University. "Later, when we developed the family system, the use of sex for this sort of purpose became more limited, mainly occurring within families. A lot of the things we see, like pedophilia and homosexuality, may be leftovers that some now consider unacceptable in our particular society."

Depending on your morals, watching bonobo sex play may be like watching humans at their most extreme and perverse. Bonobos seem to have sex more often and in more combinations than the average person in any culture, and most of the time bonobo sex has nothing to do with making babies. Males mount females and females sometimes mount them back; females rub against other females just for fun; males stand rump to rump and press their scrotal areas together. Even juveniles participate by rubbing their genital areas against adults, although ethologists don't think that males actually insert their penises into juvenile females. Very young animals also have sex with each other: little males suck on each other's penises or French-kiss. When two animals initiate sex, others freely join in by poking their fingers and toes into the moving parts.

One thing sex does for bonobos is decrease tensions caused by potential competition, often competition for food. Japanese primatologists observing bonobos in Zaire were the first to notice that when bonobos come across a large fruiting tree or encounter piles of provisioned sugarcane, the sight of food triggers a binge of sex. The atmosphere of this sexual free-for-all is decidedly friendly, and it eventually calms the group down. "What's striking is how rapidly the sex drops off," says Nancy Thompson-Handler of the State University of New York at Stony Brook, who has observed bonobos at a site in Zaire called Lomako. "After ten minutes, sexual behavior decreases by fifty percent." Soon the group turns from sex to feeding.

But it's tension rather than food that causes the sexual excitement. "I'm sure the more food you give them, the more sex you'll get," says De Waal. "But it's not really the food, it's competition that triggers this. You can throw in a cardboard box and you'll get sexual behavior." Sex is just the way bonobos deal with competition over limited resources and with the normal tensions caused by living in a group. Anthropologist Frances White of Duke University, a bonobo observer at Lomako since 1983, puts it simply: "Sex is fun. Sex makes them feel good and therefore keeps the group together."

"Females rule the business. It's a good species for feminists, I think."

Sexual behavior also occurs after aggressive encounters, especially among males. After two males fight, one may reconcile with his opponent by presenting his rump and backing up against the other's testicles. He might grab the penis of the other male and stroke it. It's the male bonobo's way of shaking hands and letting everyone know that the conflict has ended amicably.

Researchers also note that female bonobo sexuality, like the sexuality of female humans, isn't locked into a monthly cycle. In most other animals, including chimps, the female's interest in sex is tied to her ovulation cycle. Chimp females sport pink swellings on their hind ends for about two weeks, signaling their fertility, and they're only approachable for sex during that time. That's not the case with humans, who show no outward signs that they are ovulating, and can mate at all phases of the cycle. Female bonobos take the reverse tack, but with similar results. Their large swellings are visible for weeks before and after their fertile periods, and there is never any discernibly wrong time to mate. Like humans, they have sex whether or not they are ovulating.

What's fascinating is that female bonobos use this boundless sexuality in all their relationships. "Females rule the business—sex and food," says De Waal. "It's a good species for feminists, I think." For instance, females regularly use sex to cement relationships with other females. A genital-genital rub, bet-ter known as GG-rubbing by observers, is the most frequent behavior used by bonobo females to reinforce social ties or relieve tension. GG-rubbing takes a variety of forms. Often one female rolls on her back and extends her arms and legs. The other female mounts her and they rub their swellings right and left for several seconds, massaging their clitorises against each other. GG-rubbing occurs in the presence of food because food causes tension and excitement, but the intimate contact has the effect of making close friends.

Sometimes females would rather GG-rub with each other than copulate with a male. Parish filmed a 15-minute scene at a bonobo colony at the San Diego Wild Animal Park in which a male, Vernon, repeatedly solicited two females, Lisa and Loretta. Again and again he arched his back and displayed his erect penis—the bonobo request for sex. The females moved away from him, tactfully turning him down until they crept behind a tree and GG-rubbed with each other.

Unlike most primate species, in which males usually take on the dangerous task of leaving home, among bonobos females are the ones who leave the group when they reach sexual maturity, around the age of eight, and work their way into unfamiliar groups. To aid in their assimilation into a new community, the female bonobos make good use of their endless sexual favors. While watching a bonobo group at a feeding tree, White saw a young female systematically have sex with each member before feeding. "An adolescent female, presumably a recent transfer female, came up to the tree, mated with all five males, went into the tree, and solicited GG-rubbing from all the females present," says White.

Once inside the new group, a female bonobo must build a sisterhood from scratch. In groups of humans or chimps, unrelated females construct friendships through the rituals of shopping together or grooming. Bonobos do it sexually. Although pleasure may be the motivation behind a female-female assignation, the function is to form an alliance.

These alliances are serious business, because they determine the pecking order at food sites. Females with powerful friends eat first, and subordinate females may not get any food at all if the resource is small. When times are rough, then, it pays to have close female friends. White describes a scene at Lomako in which an adolescent female, Blanche, benefited from her established friendship with Freda. "I was following Freda and her boyfriend, and they found a tree that they didn't expect to be there. It was a small tree, heavily in fruit with one of their favorites. Freda went straight up the tree and made a food call to Blanche. Blanche came tearing over—she was quite far away—and went tearing up the tree to join Freda, and they GG-rubbed like crazy."

Alliances also give females leverage over larger, stronger males who otherwise would push them around. Females have discovered there is strength in numbers. Unlike other species of primates, such as chimpanzees or baboons (or, all too often, humans), where tensions run high between males and females, bonobo females are not afraid of males, and the sexes mingle peacefully. "What is consistently different from chimps," says Thompson-Handler, "is the composition of parties. The vast majority are mixed, so there are males and females of all different ages."

Female bonobos cannot be coerced into anything, including sex. Parish recounts an interaction between Lana and

a male called Akili at the San Diego Wild Animal Park. "Lana had just been introduced into the group. For a long time she lay on the grass with a huge swelling. Akili would approach her with a big erection and hover over her. It would have been easy for him to do a mount. But he wouldn't. He just kept trying to catch her eye, hovering around her, and she would scoot around the ground, avoiding him. And then he'd try again. She went around full circle." Akili was big enough to force himself on her. Yet he refrained.

In another encounter, a male bonobo was carrying a large clump of branches. He moved up to a female and presented his erect penis by spreading his legs and

HIDDEN HEAT

Standing upright is not a position usually—or easily—associated with sex. Among people, at least, anatomy and gravity prove to be forbidding obstacles. Yet our two-legged stance may be the key to a distinctive aspect of human sexuality: the independence of women's sexual desires from a monthly calendar.

Males in the two species most closely related to us, chimpanzees and bonobos, don't spend a lot of time worrying, "Is she interested or not?" The answer is obvious. When ovulatory hormones reach a monthly peak in female chimps and bonobos, and their eggs are primed for fertilization, their genital area swells up, and both sexes appear to have just one thing on their mind. "These animals really turn on when this happens. Everything else is dropped," says primatologist Frederick Szalay of Hunter College in New York.

Women, however, don't go into heat. And this departure from our relatives' sexual behavior has long puzzled researchers. Clear signals of fertility and the willingness to do something about it bring major evolutionary advantages: ripe eggs lead to healthier pregnancies, which leads to more of your genes in succeeding generations, which is what evolution is all about. In addition, male chimps give females that are waving these red flags of fertility first chance at high-protein food such as meat.

So why would our ancestors give this up? Szalay and graduate student Robert Costello have a simple explanation. Women gave heat up, they say, because our ancestors stood up. Fossil footprints

indicate that somewhere around 3.5 million years ago hominids—non-ape primates—began walking on two legs. "In hominids, something dictated getting up. We don't know what it was," Szalay says. "But once it did, there was a problem with the signaling system." The problem was that it didn't work. Swollen genital areas that were visible when their owners were down on all fours became hidden between the legs. The mating signal was lost.

"Uprightness meant very tough times for females working with the old ovarian cycle," Szalay says. Males wouldn't notice them, and the swellings themselves, which get quite large, must have made it hard for two-legged creatures to walk around.

Those who found a way out of this quandary, Szalay suggests, were females with small swellings but with a little less hair on their rears and a little extra fat. It would have looked a bit like the time-honored mating signal. They got more attention, and produced more offspring. "You don't start a completely new trend in signaling," Szalay says. "You have a little extra fat, a little nakedness to mimic the ancestors. If there was an ever-so-little advantage because, quite simply, you look good, it would be selected for."

And if a little nakedness and a little fat worked well, Szalay speculates, then a lot of both would work even better. "Once you start a trend in sexual signaling, crazy things happen," he notes. "It's almost like: let's escalate, let's add more. That's what happens in horns with sheep. It's a particular part of the body that brings an ad-

vantage." In a few million years human ancestors were more naked than ever, with fleshy rears not found in any other primate. Since these features were permanent, unlike the monthly ups and downs of swellings, sex was free to become a part of daily life.

It's a provocative notion, say Szalay's colleagues, but like any attempt to conjure up the past from the present, there's no real proof of cause and effect. Anthropologist Helen Fisher of the American Museum of Natural History notes that Szalay is merely assuming that fleshy buttocks evolved because they were sex signals. Yet their mass really comes from muscles, which chimps don't have, that are associated with walking. And anthropologist Sarah Blaffer Hrdy of the University of California at Davis points to a more fundamental problem: our ancestors may not have had chimplike swellings that they needed to dispense with. Chimps and bonobos are only two of about 200 primate species, and the vast majority of those species don't have big swellings. Though they are our closest relatives, chimps and bonobos have been evolving during the last 5 million years just as we have, and swollen genitals may be a recent development. The current unswollen human pattern may be the ancestral one.

"Nobody really knows what happened," says Fisher. "Everybody has an idea. You pays your money and you takes your choice."

—Joshua Fischman

arching his back. She rolled onto her back and they copulated. In the midst of their joint ecstasy, she reached out and grabbed a branch from the male. When he pulled back, finished and satisfied, she moved away, clutching the branch to her chest. There was no tension between them, and she essentially traded copulation for food. But the key here is that the male allowed her to move away with the branch—it didn't occur to him to threaten her, because their status was virtually equal.

Although the results of sexual liberation are clear among bonobos, no one is sure why sex has been elevated to such a high position in this species and why it is restricted merely to reproduction among chimpanzees. "The puzzle for me," says De Waal, "is that chimps do all this bonding with kissing and embracing, with body contact. Why do bonobos do it in a sexual manner?" He speculates that the use of sex as a standard way to underscore relationships began between adult males and adult females as an extension of the mating process and later spread to all members of the group. But no one is sure exactly how this happened.

It is also unclear whether bonobo sexually became exaggerated only after their split from the human lineage or whether the behavior they exhibit today is the modern version of our common ancestor's sex play. Anthropologist Adrienne Zihlman of the University of California at Santa Cruz, who has used the evidence of fossil bones to argue that our earliest known non-ape ancestors, the australopithecines, had body proportions similar to those of bonobos, says, "The path of evolution is not a straight line from either species, but what I think is important is that the bonobo information gives us more possibilities for looking at human origins."

Some anthropologists, however, are reluctant to include the details of bonobo life, such as wide-ranging sexuality and a strong sisterhood, into scenarios of human evolution. "The researchers have all these commitments to male dominance [as in chimpanzees], and yet bonobos have egalitarian relationships," says De Waal. "They also want to see humans as unique, yet bonobos fit very nicely into many of the scenarios, making humans appear less unique."

Our divergent, non-ape path has led us away from sex and toward a culture that denies the connection between sex and social cohesion. But bonobos, with their versatile sexuality, are here to remind us that our heritage may very well include a primordial urge to make love, not war.

Apes of Wrath

Barbara Smuts

Barbara Smuts is a professor of psychology and anthropology at the University of Michigan. She has been doing fieldwork in animal behavior since the early 1970s, studying baboons, chimps, and dolphins. "In my work I combine research in animal behavior with an abiding interest in feminist perspectives on science," says Smuts. She is the author of Sex and Friendship in Baboons.

Nearly 20 years ago I spent a morning dashing up and down the hills of Gombe National Park in Tanzania, trying to keep up with an energetic young female chimpanzee, the focus of my observations for the day. On her rear end she sported the small, bright pink swelling characteristic of the early stages of estrus, the period when female mammals are fertile and sexually receptive. For some hours our run through the park

Some female primates use social bonds to escape male aggression. Can women?

was conducted in quiet, but then, suddenly, a chorus of male chimpanzee pant hoots shattered the tranquility of the forest. My female rushed forward to join the males. She greeted each of them, bowing and then turning to present her swelling for inspection. The males examined her perfunctorily and resumed grooming one another, showing no further interest.

At first I was surprised by their indifference to a potential mate. Then I re-

alized that it would be many days before the female's swelling blossomed into the large, shiny sphere that signals ovulation. In a week or two, I thought, these same males will be vying intensely for a chance to mate with her.

The attack came without warning. One of the males charged toward us, hair on end, looking twice as large as my small female and enraged. As he rushed by he picked her up, hurled her to the ground, and pummeled her. She cringed and screamed. He ran off, rejoining the other males seconds later as if nothing had happened. It was not so easy for the female to return to normal. She whimpered and darted nervous glances at her attacker, as if worried that he might renew his assault.

In the years that followed I witnessed many similar attacks by males against females, among a variety of Old World primates, and eventually I found this sort of aggression against females so puzzling that I began to study it systematically—something that has rarely been done. My long-term research on olive baboons in Kenya showed that, on average, each pregnant or lactating female was attacked by an adult male about once a week and seriously injured about once a year. Estrous females were the target of even more aggression. The obvious question was, Why?

In the late 1970s, while I was in Africa among the baboons, feminists back in the United States were turning their attention to male violence against women. Their concern stimulated a wave of research documenting disturbingly high levels of battering, rape, sexual harassment, and murder. But although scientists investigated this kind of behavior from many perspec-

tives, they mostly ignored the existence of similar behavior in other animals. My observations over the years have convinced me that a deeper understanding of male aggression against females in other species can help us understand its counterpart in our own.

Researchers have observed various male animals—including insects, birds, and mammals—chasing, threatening, and attacking females. Unfortunately, because scientists have rarely studied such aggression in detail, we do not know exactly how common it is. But the males of many of these species are most aggressive toward potential mates, which suggests that they sometimes use violence to gain sexual access.

Jane Goodall provides us with a compelling example of how males use violence to get sex. In her 1986 book, *The Chimpanzees of Gombe,* Goodall describes the chimpanzee dating game. In one of several scenarios, males gather around attractive estrous females and try to lure them away from other males for a one-on-one sexual expedition that may last for days or weeks. But females find some suitors more appealing than others and often resist the advances of less desirable males. Males often rely on aggression to counter female resistance. For example, Goodall describes how Evered, in "persuading" a reluctant Winkle to accompany him into the forest, attacked her six times over the course of five hours, twice severely.

Sometimes, as I saw in Gombe, a male chimpanzee even attacks an estrous female days before he tries to mate with her. Goodall thinks that a male uses such aggression to train a female to fear him so that she will be more likely to surrender to his subsequent sexual advances. Similarly, male hamadryas baboons, who form small harems by kidnapping child brides, maintain a tight rein over their females through threats and intimidation. If, when another male is nearby, a hamadryas female strays even a few feet from her mate, he shoots her a threatening stare and raises his brows. She usually responds by rushing to his side; if not, he bites the back of her neck. The neck bite is ritualized—the male does not actually sink his razor-sharp canines into her flesh—but the threat of injury is clear. By repeating this behavior hundreds of times, the male lays claim to particular females months or even years before mating with them. When a female comes into estrus, she solicits sex only from her harem master, and other males rarely challenge his sexual rights to her.

In some species, females remain in their birth communities their whole lives, joining forces with related females to defend vital food resources against other females.

These chimpanzee and hamadryas males are practicing sexual coercion: male use of force to increase the chances that a female victim will mate with him, or to decrease the chances that she will mate with someone else. But sexual coercion is much more common in some primate species than in others. Orangutans and chimpanzees are the only nonhuman primates whose males in the wild force females to copulate, while males of several other species, such as vervet monkeys and bonobos (pygmy chimpanzees), rarely if ever try to coerce females sexually. Between the two extremes lie many species, like hamadryas baboons, in which males do not force

copulation but nonetheless use threats and intimidation to get sex.

These dramatic differences between species provide an opportunity to investigate which factors promote or inhibit sexual coercion. For example, we might expect to find more of it in species in which males are much larger than females—and we do. However, size differences between the sexes are far from the whole story. Chimpanzee and bonobo males both have only a slight size advantage, yet while male chimps frequently resort to force, male bonobos treat the fair sex with more respect. Clearly, then, although size matters, so do other factors. In particular, the social relationships females form with other females and with males appear to be as important.

In some species, females remain in their birth communities their whole lives, joining forces with related females to defend vital food resources against other females. In such "female bonded" species, females also form alliances against aggressive males. Vervet monkeys are one such species, and among these small and exceptionally feisty African monkeys, related females gang up against males. High-ranking females use their dense network of female alliances to rule the troop; although smaller than males, they slap persistent suitors away like annoying flies. Researchers have observed similar alliances in many other female-bonded species, including other Old World monkeys such as macaques, olive baboons, patas and rhesus monkeys, and gray langurs; New World monkeys such as the capuchin; and prosimians such as the ring-tailed lemur.

Females in other species leave their birth communities at adolescence and spend the rest of their lives cut off from their female kin. In most such species, females do not form strong bonds with other females and rarely support one another against males. Both chimpanzees and hamadryas baboons exhibit this pattern, and, as we saw earlier, in both species females submit to sexual control by males.

This contrast between female-bonded species, in which related females gang together to thwart males, and non-female-bonded species, in which they don't, breaks down when we come to the bonobo. Female bonobos, like their close

relatives the chimpanzees, leave their kin and live as adults with unrelated females. Recent field studies show that these unrelated females hang out together and engage in frequent homoerotic behavior, in which they embrace face-to-face and rapidly rub their genitals together; sex seems to cement their bonds. Examining these studies in the context of my own research has convinced me that one way females use these bonds is to form alliances against males, and that, as a consequence, male bonobos do not dominate females or attempt to coerce them sexually. How and why female bonobos, but not chimpanzees, came up with this solution to male violence remains a mystery.

Female primates also use relationships with males to help protect themselves against sexual coercion. Among olive baboons, each adult female typically forms long-lasting "friendships" with a few of the many males in her troop. When a male baboon assaults a female, another male often comes to her rescue; in my troop, nine times out of ten the protector was a friend of the female's. In return for his protection, the defender may enjoy her sexual favors the next time she comes into

Some of the factors that influence female vulnerability to male sexual coercion in different species may also help explain such variation among different groups in the same species.

estrus. There is a dark side to this picture, however. Male baboons frequently threaten or attack their female friends—when, for example, one tries to form a friendship with a new male. Other males apparently recognize friendships and rarely intervene. The female, then, becomes less vulnerable to aggression from males in general, but more vulnerable to aggression from her male friends.

As a final example, consider orangutans. Because their food grows so sparsely adult females rarely travel with anyone but their dependent offspring. But orangutan females routinely fall victim to forced copulation. Female orangutans, it seems, pay a high price for their solitude.

Some of the factors that influence female vulnerability to male sexual coercion in different species may also help explain such variation among different groups in the same species. For example, in a group of chimpanzees in the Taï Forest in the Ivory Coast, females form closer bonds with one another than do females at Gombe. Taï females may consequently have more egalitarian relationships with males than their Gombe counterparts do.

Such differences between groups especially characterize humans. Among the South American Yanomamö, for instance, men frequently abduct and rape women from neighboring villages and severely beat their wives for suspected adultery. However, among the Aka people of the Central African Republic, male aggression against women has never been observed. Most human societies, of course, fall between these two extremes.

How are we to account for such variation? The same social factors that help explain how sexual coercion differs among nonhuman primates may deepen our understanding of how it varies across different groups of people. In most traditional human societies, a woman leaves her birth community when she marries and goes to live with her husband and his relatives. Without strong bonds to close female kin, she will probably be in danger of sexual coercion. The presence of close female kin, though, may protect her. For example, in a community in Belize, women live near their female relatives. A man will sometimes beat his wife if he becomes jealous or suspects her of infidelity, but when this happens, onlookers run to tell her female kin. Their arrival on the scene, combined with the presence of other glaring women, usually shames the man enough to stop his aggression.

Even in societies in which women live away from their families, kin may provide protection against abusive husbands, though how much protection varies dramatically from one society to the next. In some societies a woman's kin, including her father and brothers, consistently support her against an abusive husband, while in others they rarely help her. Why?

The key may lie in patterns of male-male relationships. Alliances between males are much more highly developed in humans than in other primates, and men frequently rely on such alliances to compete successfully against other men. They often gain more by supporting their male allies than they do by supporting female kin. In addition, men often use their alliances to defeat rivals and abduct or rape their women, as painfully illustrated by recent events in Bosnia. When women live far from close kin, among men who value their alliances with other men more than their bonds with women, they may be even more vulnerable to sexual coercion than many nonhuman primate females.

Even in societies in which women live away from their families, kin may provide protection against abusive husbands.

Like nonhuman primate females, many women form bonds with unrelated males who may protect them from other males. However, reliance on men exacts a cost—women and other primate females often must submit to control by their protectors. Such control is more elaborate in humans because allied men agree to honor one another's proprietary rights over women. In most of the world's cultures, marriage involves not only the exclusion of other men from sexual access to a man's wife—which protects the woman against rape by other men—but also entails the husband's right to complete control over his wife's sexual life, including the right to punish her for real or suspected adultery, to have sex with her whenever he wants, and even to restrict her contact with other people, especially men.

In modern industrial society, many men—perhaps most—maintain such traditional notions of marriage. At the same time, many of the traditional sources of support for women, including censure of abusive husbands by the woman's kinfolk or other community members, are eroding as more and more people end up without nearby kin or long-term neighbors. The increased vulnerability of women isolated from their birth communities, however, is not just a by-product of modern living. Historically, in highly patriarchal societies like those found in China and northern India, married women lived in households ruled by their husband's mother and male kin, and their ties with their own kin were virtually severed. In these societies, today as in the past, the husband's female kin often view the wife as a competitor for resources. Not only do they fail to support her against male coercive control, but they sometimes actively encourage it. This scenario illustrates an important point: women do not invariably support other women against men, in part because women may perceive their interests as best served through alliances with men, not with other women. When men have most of the power and control most of the resources, this looks like a realistic assessment.

Decreasing women's vulnerability to sexual coercion, then, may require fundamental changes in social alliances. Women gave voice to this essential truth with the slogan SISTERHOOD IS POWERFUL—a reference to the importance of women's ability to cooperate with unrelated women as if they were indeed sisters. However, among humans, the male-dominant social system derives support from political, economic, legal, and ideological institutions that other primates can't even dream of. Freedom from male control—including male sexual coercion—therefore requires women to form alliances with one another (and with like-minded men) on a scale beyond that shown by nonhuman primates and humans in the past. Although knowledge of other primates can provide inspiration for this task, its achievement depends on the uniquely human ability to envision a future different from anything that has gone before.

Unit 4

Unit Selections

19. **Sunset on the Savanna,** James Shreeve
20. **Early Hominid Fossils from Africa,** Meave Leakey and Alan Walker
21. **A New Human Ancestor?** Elizabeth Culotta
22. **Asian Hominids Grow Older,** Elizabeth Culotta
23. **Scavenger Hunt,** Pat Shipman
24. **New Clues to the History of Male and Female Size Differences,** John Noble Wilford

Key Points to Consider

❖ Under what circumstances did bipedalism evolve?

❖ How would you draw the early hominid family tree?

❖ Under what circumstances did size difference between the sexes diminish in the human line of evolution, and why?

 **Links** **www.dushkin.com/online/**

22. **The African Emergence and Early Asian Dispersals of the Genus *Homo***
    http://www.sigmaxi.org/amsci/subject/EvoBio.html

23. **Anthropology, Archaeology, and American Indian Sites on the Internet**
    http://dizzy.library.arizona.edu/users/jlcox/first.html

24. **Long Foreground: Human Prehistory**
    http://www.wsu.edu:8001/vwsu/gened/learn-modules/top_longfor/lfopen-index.html

These sites are annotated on pages 4 and 5.

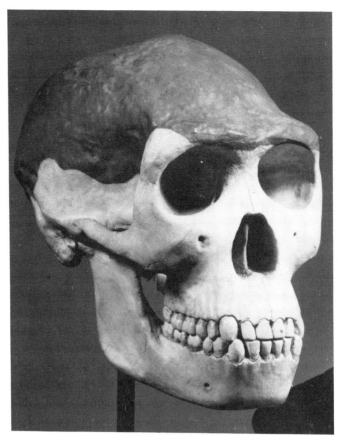

The Fossil Evidence

A primary focal point of this book, as well as of the whole of biological anthropology, is the search for and interpretation of fossil evidence for hominid (meaning human or humanlike) evolution. Paleoanthropologists are those who carry out this task by conducting the painstaking excavations and detailed analyses that serve as a basis for understanding our past. Every fragment found is cherished like a ray of light that may help to illuminate the path taken by our ancestors in the process of becoming "us." At least, that is what we would like to believe. In reality, each discovery leads to further mystery, and for every fossil-hunting paleoanthropologist who thinks his or her find supports a particular theory, there are many others anxious to express their disagreement. (See "Early Hominid Fossils from Africa" by Meave Leakey and Alan Walker and "A New Human Ancestor?" by Elizabeth Culotta.)

How wonderful it would be, we sometimes think in moments of frustration over inconclusive data, if the fossils would just speak for themselves, and every primordial piece of humanity were to carry with it a self-evident explanation for its place in the evolutionary story. Paleoanthropology would then be more a quantitative problem of amassing enough material to reconstruct our ancestral development than a qualitative problem of interpreting what it all means. It would certainly be a simpler process, but would it be as interesting?

Most scientists tolerate, welcome, or even (dare it be said?) thrive on controversy, recognizing that diversity of opinion refreshes the mind, rouses students, and captures the imagination of the general public. After all, where would paleo-anthropology be without the gadflies, the near-mythic heroes, and, lest we forget, the research funds they generate? Consider, for example, the issue of the differing roles played by males and females in the transition to humanity and all that it implies with regard to bipedalism, tool making, and the origin of the family. Should the primary theme of human evolution be summed up as "man the hunter" or "woman the gatherer?" How the fossil evidence bears upon this issue is dealt with in "New Clues to the History of Male and Female Size Differences" by John Noble Wilford.

Virginia Morell cites new evidence that challenges the long-held view that australo-pithecines were restricted to East Africa (see box "African Origins: West Side Story" in the article "Asian Hominids Grow Older"), and James Shreeve questions the notion that bipedalism evolved in the grasslands ("Sunset on the Savanna").

Not all the research and theoretical speculation taking place in the field of paleoanthropology is so controversial. Most scientists, in fact, go about their work quietly and methodically, generating hypotheses that are much less explosive and yet have the cumulative effect of enriching our understanding of the details of human evolution. In "Scavenger Hunt," for instance, Pat Shipman tells how modern technology, in the form of the scanning electron microscope, combined with meticulous detailed analysis of cut marks on fossil animal bones, can help us better understand the locomotor and food-getting adaptations of our early hominid ancestors. In one stroke, she is able to challenge the traditional "man the hunter" theme that has pervaded most early hominid research and writing and to simultaneously set forth an alternative hypothesis that will, in turn, inspire further research.

As we mull over the controversies outlined in this unit, we should not take them as reflecting an inherent weakness of the field of paleoanthropology, but rather as symbolic of its strength: the ability and willingness to scrutinize, question, and reflect (seemingly endlessly) on every bit of evidence.

Contrary to the way that the creationists would have it, an admission of doubt is not an expression of ignorance but simply a frank recognition of the imperfect state of our knowledge. If we are to increase our understanding of ourselves, we must maintain an atmosphere of free inquiry without preconceived notions and an unquestioning commitment to a particular point of view. To paraphrase anthropologist Ashley Montagu, whereas creationism seeks certainty without proof, science seeks proof without certainty.

Sunset on the Savanna

Why do we walk? For decades anthropologists said that we became bipedal to survive on the African savanna. But a slew of new fossils have destroyed that appealing notion and left researchers groping for a new paradigm.

James Shreeve

James Shreeve is a contributing editor of DISCOVER. *He is co-author, with anthropologist Donald Johanson, of* Lucy's Child: The Discovery of a Human Ancestor. *His latest book is* The Neandertal Enigma. *Shreeve is currently at work on his first novel.*

IT'S A WONDERFUL STORY. *Once upon a time, there was an ape who lived in the middle of a dark forest. It spent most of its days in the trees, munching languidly on fruits and berries. But then one day the ape decided to leave the forest for the savanna nearby. Or perhaps it was the savanna that moved, licking away at the edge of the forest one tree at a time until the fruits and berries all the apes had found so easily weren't so easy to find anymore.*

In either case, the venturesome ape found itself out in the open, where the air felt dry and crisp in its lungs. Life was harder on the savanna: there might be miles between one meal and another, there were seasons of drought to contend with, and large, fierce animals who didn't mind a little ape for lunch. But the ape did not run back into the forest. Instead it learned to adapt, walking from one place to another on two legs. And it learned to live by its wits. As the years passed, the ape grew smarter and smarter until it was too smart to be called an ape anymore. It lived anywhere it wanted and gradually made the whole world turn to its own purposes. Meanwhile, back in the forest, the other apes

went on doing the same old thing, lazily munching on leaves and fruit. Which is why they are still just apes, even to this day.

The tale of the ape who stood up on two legs has been told many times over the past century, not in storybooks or nursery rhymes but in anthropology texts and learned scientific journals. The retellings have differed from one another in many respects: the name of the protagonist, for instance, the location in the world where his transformation took place, and the immediate cause of his metamorphosis. One part of the story, however, has remained remarkably constant: the belief that it was the shift from life in the forest to life in a more open habitat that set the ape apart by forcing it onto two legs. Bipedalism allowed hominids to see over tall savanna grass, perhaps, or escape predators, or walk more efficiently over long distances. In other scenarios, it freed the hands to make tools for hunting or gathering plants. A more recent hypothesis suggests that an erect posture exposes less skin to the sun, keeping body temperature lower in open terrain. Like the painted backdrop to a puppet theater, the savanna can accommodate any number of dramatic scenarios and possible plots.

But now that familiar stage set has come crashing down under the weight of a spectacular crop of new hominid fossils from Africa, combined with revelations about the environment of our earliest ancestors. The classic savanna hypothesis is clearly wrong, and while

some still argue that the open grasslands played some role in the origins of bipedalism, a growing number of researchers are beginning to think the once unthinkable: the savanna may have had little or nothing to do with the origins of bipedalism.

"The savanna paradigm has been overthrown," says Phillip Tobias, a distinguished paleoanthropologist at the University of Witwatersrand in Johannesburg and formerly a supporter of the hypothesis. "We have to look now for some other explanation for bipedalism."

The roots of the savanna hypothesis run deep. More than 100 years ago, Charles Darwin thought that mankind's early ancestor moved from "some warm, forest-clad land" owing to "a change in its manner of procuring subsistence, or to a change in the surrounding conditions." In his view, the progenitor assumed a two-legged posture to free the hands for fashioning tools and performing other activities that in turn nourished the development of an increasingly refined intelligence. Darwin believed that this seminal event happened in Africa, where mankind's closest relatives, the African apes, still lived.

By the turn of this century, however, most anthropologists believed that the critical move to the grasslands had occurred in Asia. Though the bones of a primitive hominid—which later came to be called *Homo erectus*—had been discovered on the Southeast Asian island of Java, the change of venue had more to do with cultural values and racist reasoning than with hard evidence. Africa

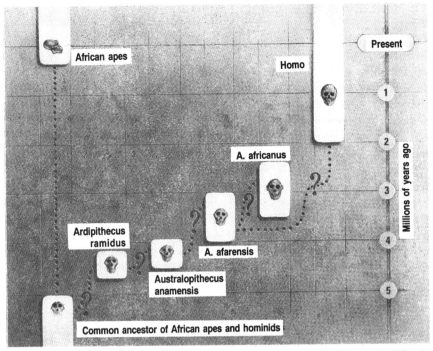

Nenad Jakesevic

New hominid fossils have filled out our evolutionary tree, but the question of exactly how they are related to one another (and ourselves) remains unanswered.

was the "dark continent," where progress was slowed by heat, disease, and biotic excess. In such a place, it was thought, the mind would vegetate—witness the "regressive" races that inhabited the place in modern times. The plains of central Asia, on the other hand, seemed just the sort of daunting habitat that would call out the best in an enterprising ape. In that environment, wrote the American paleontologist Henry Fairfield Osborn, "the struggle for existence was severe and evoked all the inventive and resourceful faculties of man . . . ; while the anthropoid apes were luxuriating in the forested lowlands of Asia and Europe, the Dawn Men were evolving in the invigorating atmosphere of the relatively dry uplands."

In hindsight, the contrasts made between dark and light, forest and plain, slovenly ape and resourceful man seem crudely moralistic. Higher evolution, one would think, was something reserved for the primate who had the guts and wits to go out there and grab it, as if the entrepreneurial spirit of the early twentieth century could be located in our species' very origins. But at the time the idea was highly influential. Among those impressed was a young anatomist in South

Africa named Raymond Dart. in 1925 Dart announced the discovery, near the town of Taung, of what he believed to be the skull of a juvenile "man-ape," which he called *Australopithecus africanus*. While Darwin had been correct in supposing Africa to be the home continent of our ancestors, it seemed that Osborn had been right about the creature's habitat: there are no forests around Taung, and scientists assumed there hadn't been any for millions of years. Forests might provide apes with "an easy and sluggish solution" to the problems of existence, wrote Dart, but "for the production of man a different apprenticeship was needed to sharpen the wits and quicken the higher manifestations of intellect—a more open veld country where competition was keener between swiftness and stealth, and where adroitness of thinking and movement played a preponderating role in the preservation of the species."

In the decades that followed Dart's discovery, more early hominids emerged from eastern and southern Africa, and most researchers concluded that they made their homes in the savanna as well. The question was less *whether* the savanna played a part in the origin of bipedalism—

that was obvious—than how. Dart originally proposed that *Australopithecus* had taken to two legs to avoid predators ("for sudden and swift bipedal movement, to elude capture"), but later he reversed this scenario and imagined his "killer ape" the eater rather than the eaten, forsaking the trees for "the more attractive fleshy foods that lay in the vast savannas of the southern plains."

Later studies suggested that the environments in eastern and southern Africa where early hominids lived were not the vast, unchanging plains Dart imagined. Instead they appeared to be variable, often characterized by seasonally semiarid terrain, a plain studded with scraggly trees and patches of denser woodland. But no matter: This "savanna mosaic" was still drier and more open than the thick forest that harbors the African apes today. It just made good sense, moreover, that our ancestors would have come down to the ground and assumed their bipedal stance in a habitat where there were not as many trees to climb around in. The satisfying darkness-into-light theme of early hominid development held up, albeit with a little less wattage.

For decades the popularity of the savanna hypothesis rested on the twin supports of its moral resonance and general plausibility: our origin should have happened this way, and it would make awfully good sense if it did. In East Africa, geography seemed to reinforce the sheer rightness of the hypothesis. Most of the earliest hominid fossils have come from the eastern branch of the great African Rift Valley; researchers believed that when these hominids were alive, the region was much like the dry open grasslands that dominate it today. The lusher, more forested western branch, meanwhile, is home to those lazy chimps and gorillas—but to no hominid fossils.

Still, despite such suggestive correspondence, until 20 years ago something was missing from the hypothesis: some hard data to link environment to human evolution. Then paleontologist Elisabeth Vrba of Yale began offering what was the strongest evidence that the drying up of African environments helped shape early human evolution. In studying the bones of antelope and other bovids from hominid sites in South Africa, Vrba noticed a dramatic change occurring between 2.5 and 2 million years ago. Many species that were adapted to wooded environments, she saw, suddenly disappeared from the fossil record, while those suited to grassy regions appeared and multiplied. This "turnover pulse" of extinctions and origins coincides with a sudden global cooling, which may have triggered the spread of savannas and the fragmentation of forests.

Other investigators, meanwhile, were documenting an earlier turnover pulse around 5 million years ago. For humans, both dates are full of significance. This earlier pulse corresponds to the date when our lineage is thought to have diverged from that of the apes and become bipedal. Vrba's second, later pulse marks the appearance of stone tools and the arrival on the scene of new hominid species, some with brains big enough

to merit inclusion in the genus *Homo*. The inference was clear: our early ancestors were savanna born and savanna bred.

"All the evidence, as I see it," Vrba wrote in 1993, "indicates that the lineage of upright primates known as australopithecines, the first hominids, was one of the founding groups of the great African savanna biota."

With empirical evidence drawn from two different sources, the turnover pulse is a great improvement on the traditional savanna hypothesis (which in retrospect looks not so much like a hypothesis as a really keen idea). Best of all, Vrba's hypothesis is testable. Let's say that global climate changes did indeed create open country in East Africa, which in turn triggered a turnover of species and pushed ahead the evolution of hominids. If so, then similar turnovers

in animal species should have appeared in the fossil record whenever global change occurred.

Over the last 15 years, Andrew Hill and John Kingston, both also at Yale, have been looking for signs of those dramatic shifts at some 400 sites in the Tugen Hills of Kenya. In the heart of the fossil-rich eastern branch of the Rift Valley, the Tugen hills offer a look at a succession of geologic layers from 16 million years ago to a mere 200,000 years ago—studded with fragmentary remains of ancient apes and hominids. To gauge the past climate of the Tugen Hills, the researchers have looked at the signatures of the ancient soils preserved in rock. Different plants incorporate different ratios of isotopes of carbon in their tissues, and when those plants die and decompose, that distinctive ratio remains in the soil. Thus grasslands and

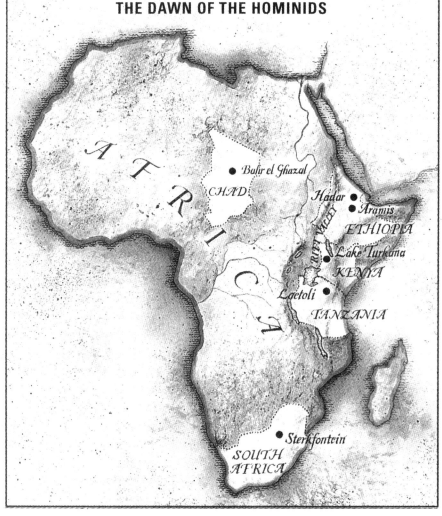

ARAMIS AGE: 4.4 million years old SPECIES: *Ardipithecus ramidus* BIPEDAL? Unknown ENVIRONMENT: Dense woodlands

LAKE TURKANA AGE: 4.2 million years old SPECIES: *Australopithecus anamensis* BIPEDAL? Yes ENVIRONMENT: Lakeside forests in an arid region

LAETOLI AGE: 3.5 million years old SPECIES: *Australopithecus afarensis* BIPEDAL? Yes ENVIRONMENT: Primarily grasslands, with some forests

THE DAWN OF THE HOMINIDS

STERKFONTEIN AGE: 3 to 3.5 million years old SPECIES: Probably *Australopithecus africanus* BIPEDAL? Yes, but also still a tree climber ENVIRONMENT: Forests

BAHR EL GHAZAL AGE: 3 to 3.5 million years old SPECIES: Similar to *Australopithecus afarensis* BIPEDAL? Probably ENVIRONMENT: Gallery forests with grassy patches

HADAR AGE: 3.2 million years old SPECIES: *Australopithecus afarensis* ("Lucy") BIPEDAL? Yes ENVIRONMENT: Forests and bushlands

Nenad Jakesevic

forests leave distinguishing isotopic marks. When Hill and Kingston looked at soils formed during Vrba's turnover pulses, however, they found nothing like the radical shifts to grasslands that she predicted. Instead of signs that the environment was opening up, they found that there was a little bit of grass all the time, with no dramatic changes, and no evidence that early hominids there ever encountered an open grassland.

"Elisabeth's turnover pulse hypothesis is very attractive," says Hill. "It would have been lovely if it had also been true."

Other research has also contradicted Vrba's hypothesis. Laura Bishop of Liverpool University in England, for instance, has been studying pig fossils from several East African sites. Some of those fossil animals, she has found, had limbs that were adapted not for open habitats but for heavy woods. Peter de Menocal of Columbia University's Lamont-Doherty Earth Observatory has been looking at long-term climate patterns in Africa by measuring the concentration of dust in ocean sediments. Over the past 5 million years, he has found, the African climate has cycled back and forth between dry and wet climates, but the pattern became dramatic only 2.8 million years ago, when Africa became particularly arid. Such a change could have played a role in the dawn of *Homo*, but it came over a million years too late to have had a hand in australopithecines' becoming bipedal.

WHAT MAY finally kill the savanna hypothesis—or save it—are the hominids themselves. More than anything else, walking on two feet is what makes a hominid a hominid. If those first bipedal footsteps were made on savanna, we should find fossils of the first hominids in open habitats. For almost 20 years, the earliest hominid known has been *Australopithecus afarensis*, exemplified by the 3.2-million-year-old skeleton called Lucy. Lucy had a chimp-size skull but an upright posture, which clinches the argument that hominids evolved bipedalism before big brains. But had Lucy completely let go of the trees? It's been a

matter of much debate: oddities such as her curved digits may be the anatomic underpinnings of a partially arboreal life-style or just baggage left over from her tree-climbing ancestry.

Nor did *afarensis* make clear a preference for one sort of habitat over another. Most of the fossils come from two sites: Hadar in Ethiopia, where Lucy was found, and Laetoli in Tanzania, where three hominids presumed to be *afarensis* left their footprints in a layer of newly erupted volcanic ash 3.5 million years ago. Laetoli has been considered one of the driest, barest habitats in the eastern rift and thus has given comfort to the savanna faithful. But at Hadar the *afarensis* fossils appear to have been laid down among woodlands along ancient rivers. Other ambiguous bones that may have belonged to *afarensis* and may have dated back as far as 4 million years ago have been found at nearby East African localities. Environments at these sites run the gamut from arid to lush, suggesting that Lucy and her kin may not have been confined to one particular habitat but rather lived in a broad range of them. So why give some special credit to the savanna for launching our lineage?

"I've always thought that there was scant evidence for the savanna hypothesis, based simply on the fact that hominids are extremely plastic behaviorally," says Bill Kimbel, director of science at the Institute for Human Origins in Berkeley, California.

At least *afarensis* had enough respect for conventional wisdom to stay on the right side of the Rift Valley. Or it did until last year. In November a team led by Michel Brunet of the University of Poitiers in France announced the discovery of a jawbone similar to that of *afarensis* and, at 3 to 3.5 million years old, well within the species' time range. The ecology where Brunet's hominid lived also has a familiar ring: "a vegetational mosaic of gallery forest and wooded savanna with open grassy patches." But in one important respect the fossil is out of left field—Brunet found it in Chad, in north central Africa, more than 1,500 miles from the eastern Rift Valley, where all the other *afarensis* specimens have been found.

This hominid's home is even farther west, in fact, than the dark, humid forests of the western rift, home to the great apes—clearly on the wrong side of the tracks.

In a normal year the Chad fossil would have been the biggest hominid news. But 1995 was anything but a normal year. In August, Meave Leakey of the National Museums of Kenya and her colleagues made public the discovery of a new hominid species, called *Australopithecus anamensis*, even older than *afarensis*. (The fossils were found at two sites near Lake Turkana in Kenya and derive their name from the Turkana word for "lake.") The previous spring, Tim White of the University of California at Berkeley and his associates had named a whole new genus, *Ardipithecus ramidus*, that was older still, represented by fossils found in Aramis, Ethiopia, over the previous three years. The character of *anamensis* and *ramidus* could well have decided the fate of the savanna hypothesis. If they were bipeds living in relatively open territory, they could breathe new life into a hypothesis that's struggling to survive. But if either showed that our ancestors were upright *before* leaving the forest, the idea that has dominated paleoanthropology for a century would be reduced to little more than a historical artifact.

The best hope for the savanna hypothesis rests with Leakey's new species. In its head and neck, *anamensis* shares a number of features with fossil apes, but Leakey's team also found a shinbone that is quite humanlike and emphatically bipedal—in spite of the deep antiquity of *anamensis*: the older of the two sites has been dated to 4.2 million years ago. And the region surrounding the sites was a dry, relatively open bushland.

This is good news for those, such as Peter Wheeler of Liverpool John Moores University in England, whose theories of bipedalism depend on an initial movement out of the closed-canopy forest. Wheeler maintains that standing upright exposes less body surface to the sun, making it possible for proto-hominids to keep cool enough out of the shade to exploit savanna resources. "It's pretty clear that by three and a half million years ago, australopithecines were living in a range of habitats," he says. "But if you look at the oldest evidence for bipedalism—Laetoli, and now the *anamensis* sites—these are actually more open habitats."

However, Alan Walker of Penn State, one of the codiscoverers of *anamensis*, disagrees. Though the regional climate may have been as hot and arid back then as it is now, he says, the local habitat of *anamensis* was probably quite different. Back then the lake was much bigger than it is today and would have supported a massive ring of vegetation. Animal fossils found at the *anamensis* sites—everything from little forest monkeys to grass-eating antelope—were lodged in deltalike sediments that must have been deposited by "monstrous great rivers," says Walker, with gallery forests as much as a mile or two wide on both banks.

Even Laetoli is proving to be a mixed blessing for savanna lovers. In the first studies, researchers focused their attention mainly on the abundance of fossils of arid-adapted antelope at the site. Based on these remains, they concluded that Laetoli was a grassland with scattered trees. But according to Kaye Reed of the Institute of Human Origins, these conclusions are debatable. For one thing, the antelope are gregarious herders, so one should expect to find more of their bones than those of more solitary species. Moreover, the original studies underplayed evidence for a more diverse community, which included woodland-dwelling antelope and an assortment of monkeys. In separate studies, Reed and Peter Andrews of the British Natural History Museum took a more thorough look at Laetoli, and both concluded that the original description was far too bleak. "Monkeys have to have trees to eat and sleep in," says Reed. "I don't want to give the impression that this was some kind of deep forest. But certainly Laetoli was more heavily wooded than we thought."

Ardipithecus ramidus poses a potentially more devastating blow to the savanna hypothesis. The 4.4-million-year-old species has a skull and teeth that are even more primitive and chimplike than *anamensis*. Other traits of its anatomy, how-

ever, align it with later hominids such as *afarensis*. It remains to be seen whether *ramidus* is an early cousin of our direct ancestors or is indeed at the very base of the hominid lineage (its name perhaps reveals the hope of its discoverers, deriving from the Afar word for "root"). What we do know is that the species lived in a densely wooded habitat along with forest-dwelling monkey species and the kudu, an antelope that prefers a bushy habitat.

"We interpret these initial results as evidence that the hominids lived and died in a wooded setting," says Tim White.

A. ramidus would be the final nail in the coffin of the savanna hypothesis, except for one crucial bit of missing information: none of its discoverers will yet say if it was bipedal. Researchers have unearthed a partial skeleton consisting of over 100 fragments of dozens of bones, including hand bones, foot bones, wrist bones—more than enough to determine whether the creature walked like a human or like a chimp. Unfortunately the fragile bones are encased in sediment that must be laboriously chipped away before a proper analysis can begin. Neither White nor any member of his team will comment on what the bones say about locomotion until a thorough study can be completed—which at this point won't be until 1998 at the earliest.

There are plenty of other puzzling bones to ponder in the meantime. Back in South Africa, where Dart found the first australopithecine, researchers have begun analyzing a horde of some 500 new and previously collected specimens from a site called Sterkfontein. Some of them may have a potent impact on the savanna hypothesis.

The most widely publicized is "Little Foot," a tantalizing string of four connected foot bones running from the ankle to the base of the big toe. Little Foot is between 3 and 3.5 million years old, hundreds of thousands of years younger than the *ramidus* and *anamensis* specimens. Yet according to Phillip Tobias and his Witwatersrand colleague Ronald Clarke, the fossil demonstrates that this early species of australopithecine—quite probably *A. africanus*—still spent time in the trees. While the anklebone, Tobias and

Clarke say, is built to take the weight of a bipedal stride, the foot is also surprisingly primitive. This is especially true of the big toe, which they contend splayed out to the side like a chimpanzee's, all the better to grasp tree branches when climbing.

ALTHOUGH not everyone agrees that Little Foot's foot is so apelike, a host of other fossils from Sterkfontein also speak of the trees. Lee Berger, also at Witwatersrand, has analyzed several shoulder girdles from the collection and found them even better suited to climbing and suspending behavior than those of *afarensis*. He and Tobias have analyzed a shinbone from the same site and concluded it was more chimplike than human. And in what is perhaps the most compelling finding so far, Berger and Henry McHenry of the University of California at Davis have analyzed the proportions of arms and legs of the Sterkfontein *africanus* specimens and found that they were closer to chimps than to humans.

The tree-climbing anatomy of *africanus* has prompted Tobias, long a savanna loyalist, to wonder why such an animal would be out on the South African veld, where today there aren't any trees big enough to climb around in. A possible answer came from recent reevaluations of the environments at several *africanus* sites. As in East Africa, the ancient fauna and pollen suggest that conditions there were warmer, wetter, and more wooded than previously thought. Tobias's colleague Marian Bamford has even recovered traces at Sterkfontein of liana vines, which grow primarily in dense forests. The climate at the site did become drier and more open 2.5 million years ago, but by then hominids had been walking on two legs for a least a million and a half years.

If they live up to their discovers' initial claims, the Sterkfontein fossils make hominid history more complicated than we thought. Lucy and her fellow *afarensis* were traipsing through Ethiopia at about the same time Little Foot and the other *africanus* hominids were in South Africa—and the two hominids were using different ways of moving around. While Lucy was more committed to life on the ground, Little Foot went for a mixed strategy, sometimes scuttling up trees. "What this suggests," says Berger, "is that bipedalism may have evolved not once but twice."

And in both cases, the South African researchers argue, bipedalism was not associated with the savanna. "The idea that bipedalism evolved as an adaptation to the savanna," declares Tobias, "can be thrown out the window."

If so, it shall be missed. For all its shortcomings—a shortage of evidence being the first among them—the savanna hypothesis provided a tidy, plausible explanation for a profound mystery: What set human beings apart from the rest of creation? If not the savanna, what did cause the first hominids to become bipedal? Why develop anatomy good only for walking on the ground when you are still living among the trees? For all the effort it has taken to bring down the savanna hypothesis, it will take much more to build up something else in its place.

"I don't really know why we became bipedal," says Andrew Hill. "It's such an unusual thing."

"We're back to square one," says Tobias.

"Square one" is not completely empty. Kevin Hunt of Indiana University, for instance, has recently revived the idea that bipedalism was initially an adaptation for woodland feeding rather than a new way of getting around. Chimps often stand while feeding in small trees and bushes, stabilizing themselves by hanging onto an overhanging branch. Hunt suggests that the earliest australopithecines made an anatomic commitment to this specialized way of obtaining food.

Nina Jablonski of the California Academy of Science in San Francisco sees the beginnings of our bipedalism mirrored instead in the upright threat displays of great apes. Perhaps our ancestors resorted to this behavior more than their ape cousins to maintain the social hierarchy. Originally Jablonski and her colleague George Chaplin of the University

of Western Australia linked the increase in bipedal threat displays to a move to the savanna, where there would be more competition for resources. But in light of the new evidence for wooded habitats, she now concedes that this "essentially primary cause" of bipedalism could have emerged in a forest.

There is a stubborn paradox in such models, based as they are on living primates: since chimps and gorillas have presumably been performing these behaviors for millennia without the evolution of bipedalism, how could the same behaviors have driven just such an evolution in hominids? In the early 1980's, Owen Lovejoy of Kent State in Ohio proposed an elegant explanation for bipedalism that bypassed this logical difficulty and had nothing to do with moving about on the savanna.

"Bipedality is a lousy form of locomotion," says Lovejoy. "It's slower and more awkward, and it puts the animal at greater risk of injury. The advantage must come from some other motivating selective force." To Lovejoy, the force is reproduction itself. In his view, what separated the protohominids from their ape contemporaries was a wholly new reproductive strategy, in which males provided food to females and their mutual offspring. With the males' assistance, the females could forage less and give birth more frequently than their ape counterparts because they could care for more than one child at a time. In return a male gained continual sexual access to a particular female, ensuring that the children he provisioned were most likely his own.

This monogamous arrangement would have provided an enormous evolutionary advantage, says Lovejoy, since it would directly affect the number of offspring an individual female could bear and raise to maturity. But the males would need the anatomic apparatus to carry food back to be shared in the first place. Bipedalism may have been a poor way of getting around, but by freeing the hands for carrying, it would have been an excellent way to bear more offspring.

And in evolution, of course, more offspring is the name of the game.

If the savanna hypothesis is barely kicking, it's worth remembering that it isn't dead. While scientists no longer believe in the classic portrait of protohominids loping about on a treeless gray plain, there's treeless, and then there's treeless. Despite the revisionism of recent years, the fact remains that Africa as a whole has gradually cooled and dried over the past 5 million years. Even if the trees did not disappear completely, early hominids may still have faced sparser forests than their ancestors. Bipedalism may still have been an important part of their adjustment to this new setting.

"We have to be careful about what we call savanna," says Peter Wheeler. "Most savanna is a range of habitats, including bushland and quite dense trees. My arguments for bipedalism being a thermoregulatory adaptation would still apply, unless there was continuous shade cover, as in a closed-canopy forest. Nobody is saying that about *anamensis*."

Unfortunately they *are* saying that about *anamensis's* older cousin, *Ardipithecus ramidus*, in its deep forest home. Which brings us full circle to the business of storytelling. Every reader knows that a good story depends on having strongly drawn characters, ones the author understands well. Rather suddenly, two new protagonists have been added to the opening chapters of human evolution. About *amamensis* we still know very little. But about *ramidus* we know next to nothing. Once upon a time, nearly four and a half million years ago, there was a hominid that lived in the middle of a dark forest. Did it walk on two legs or four? If it walked on two, why?

"The locomotor habit of *ramidus* is crucial," says Lovejoy, who is among those charged with the enviable task of analyzing its bones. "If it is bipedal, then the savanna hypothesis in all its mundane glory would be dead. But if it is quadrupedal, then the old idea, even though I think it is inherently illogical, would not be disproved."

Early Hominid Fossils from Africa

A new species of Australopithecus, *the ancestor of* Homo, *pushes back the origins of bipedalism to some four million years ago*

by Meave Leakey and Alan Walker

The year was 1965. Bryan Patterson, a paleoanthropologist from Harvard University, unearthed a fragment of a fossil arm bone at a site called Kanapoi in northern Kenya. He and his colleagues knew it would be hard to make a great deal of anatomic or evolutionary sense out of a small piece of elbow joint. Nevertheless, they did recognize some features reminiscent of a species of early hominid (a hominid is any upright-walking primate) known as *Australopithecus,* first discovered 40 years earlier in South Africa by Raymond Dart of the University of the Witwatersrand. In most details, however, Patterson and his team considered the fragment of arm bone to be more like those of modern humans than the one other *Australopithecus* humerus known at the time.

The age of the Kanapoi fossil proved somewhat surprising. Although the techniques for dating the rocks where the fossil was uncovered were still fairly rudimentary, the group working in Kenya was able to show that the bone was probably older

than the various *Australopithecus* specimens previously found. Despite this unusual result, however, the significance of Patterson's discovery was not to be confirmed for another 30 years. In the interim, researchers identified the remains of so many important early hominids that the humerus from Kanapoi was rather forgotten.

Yet Patterson's fossil would eventually help establish the existence of a new species of *Australopithecus*—the oldest yet to be identified—and push back the origins of upright walking to more than four million years (Myr) ago. But to see how this happened, we need to trace the steps that paleoanthropologists have taken in constructing an outline for the story of hominid evolution.

Evolving Story of Early Hominids

Scientists classify the immediate ancestors of the genus *Homo* (which includes our own species, *Homo sapiens*) in the genus *Australopithecus.* For several decades, it was believed that

these ancient hominids first inhabited the earth at least three and a half million years ago. The specimens found in South Africa by Dart and others indicated that there were at least two types of *Australopithecus*—*A. africanus* and *A. robustus.* The leg bones of both species suggested that they had the striding, bipedal locomotion that is a hallmark of humans among living mammals. (The upright posture of these creatures was vividly confirmed in 1978 at the Laetoli site in Tanzania, where a team led by archaeologist Mary Leakey discovered a spectacular series of footprints made 3.6 Myr ago by three *Australopithecus* individuals as they walked across wet volcanic ash.) Both *A. africanus* and *A. robustus* were relatively small-brained and had canine teeth that differed from those of modern apes in that they hardly projected past the rest of the tooth row. The younger of the two species, *A. robustus,* had bizarre adaptations for chewing—huge molar and premolar teeth combined with bony crests on the skull where powerful chewing muscles would have been attached.

Paleoanthropologists identified more species of *Australopithecus* over the next several decades. In 1959 Mary Leakey unearthed a skull from yet another East African species closely related to *robustus*. Skulls of these species uncovered during the past 40 years in the northeastern part of Africa, in Ethiopia and Kenya, differed considerably from those found in South Africa; as a result, researchers think that two separate *robustus*-like species—a northern one and a southern one—existed.

In 1978 Donald C. Johanson, now at the Institute of Human Origins in Berkeley, Calif., along with his colleagues, identified still another species of *Australopithecus*. Johanson and his team had been studying a small number of hominid bones and teeth discovered at Laetoli, as well as a large and very important collection of specimens from the Hadar region of Ethiopia (including the famous "Lucy" skeleton). The group named the new species *afarensis*. Radiometric dating revealed that the species had lived between 3.6 and 2.9 Myr ago, making it the oldest *Australopithecus* known at the time.

This early species is probably the best studied of all the *Australopithecus* recognized so far, and it is certainly the one that has generated the most controversy over the past 20 years. The debates have ranged over many issues: whether the *afarensis* fossils were truly distinct from the *africanus* fossils from South Africa; whether there was one or several species at Hadar; whether the Tanzanian and Ethiopian fossils were of the same species; whether the fossils had been dated correctly.

But the most divisive debate concerns the issue of how extensively the bipedal *afarensis* climbed in trees. Fossils of *afarensis* include various bone and joint structures typical of tree climbers. Some scientists argue that such characteristics indicate that these hominids must have spent at least some time in the trees. But others view these features as simply evolutionary baggage, left over from arboreal ancestors. Underlying this discussion is the question of where *Australopithecus* lived—in forests or on the open savanna.

By the beginning of the 1990s, researchers knew a fair amount about the various species of *Australopithecus* and how each had adapted to its environmental niche. A description of any one of the species would mention that the creatures were bipedal and that they had ape-size brains and large, thickly enameled teeth in strong jaws, with nonprojecting canines. Males were typically larger than females, and individuals grew and matured rapidly. But the origins of *Australopithecus* were only hinted at, because the gap between the earliest well-known species in the group (*afarensis,* from about 3.6 Myr ago) and the postulated time of the last common ancestor of chimpanzees and humans (between 5 and 6 Myr ago) was still very great. Fossil hunters had unearthed only a few older fragments of bone, tooth and jaw from the intervening 1.5 million years to indicate the anatomy and course of evolution of the very earliest hominids.

Filling the Gap

Discoveries in Kenya over the past several years have filled in some of the missing interval between 3.5 and 5 Myr ago. Beginning in 1982, expeditions run by the National Museums of Kenya to the Lake Turkana basin in northern Kenya began finding hominid fossils nearly 4 Myr old. But because these fossils were mainly isolated teeth—no jawbones or skulls were preserved—very little could be said about them except that they resembled the remains of *afarensis* from Laetoli. But our recent excavations at an unusual site, just inland from Allia Bay on the east side of Lake Turkana [*see maps*], yielded more complete fossils.

The site at Allia Bay is a bone bed, where millions of fragments of weathered tooth and bone from a wide variety of animals, including hominids, spill out of the hillside. Exposed at the top of the hill lies a layer of hardened volcanic ash called the Moiti Tuff, which has been dated radiometrically to just over 3.9 Myr old. The fossil fragments lie several meters below the tuff, indicating that the remains are older than the tuff. We do not yet understand fully why so many fossils are concentrated in this spot, but we can be certain that they were deposited by the precursor of the present-day Omo River.

Today the Omo drains the Ethiopian highlands located to the north, emptying into Lake Turkana, which has no outlet. But this has not always been so. Our colleagues Frank Brown of the Univer-

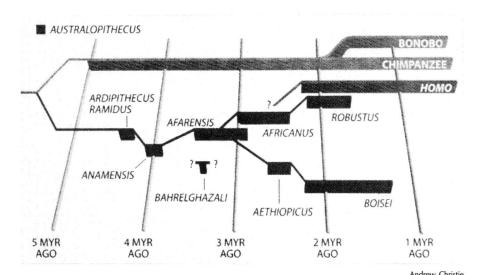

AUSTRALOPITHECUS

BONOBO
CHIMPANZEE
HOMO

ARDIPITHECUS RAMIDUS
AFARENSIS
ANAMENSIS
BAHRELGHAZALI
AFRICANUS
AETHIOPICUS
ROBUSTUS
BOISEI

5 MYR AGO | 4 MYR AGO | 3 MYR AGO | 2 MYR AGO | 1 MYR AGO

Andrew Christie

FAMILY TREE of the hominid species known as *Australopithecus* includes a number of species that lived between roughly 4 and 1.25 Myr ago. Just over 2 Myr ago a new genus, *Homo* (which includes our own species, *Homo sapiens*), evolved from one of the species of *Australopithecus*.

sity of Utah and Craig Feibel of Rutgers University have shown that the ancient Omo River dominated the Turkana area for much of the Pliocene (roughly 5.3 to 1.6 Myr ago) and the early Pleistocene (1.6 to 0.7 Myr ago). Only infrequently was a lake present in the area at all. Instead, for most of the past four million years, an extensive river system flowed across the broad floodplain, proceeding to the Indian Ocean without dumping its sediments into a lake.

The Allia Bay fossils are located in one of the channels of this ancient river system. Most of the fossils collected from Allia Bay are rolled and weathered bones and teeth of aquatic animals—fish, crocodiles, hippopotamuses and the like—that were damaged during transport down the river from some distance away. But some of the fossils are much better preserved; these come from the animals that lived on or near the riverbanks. Among these creatures are several different species of leaf-eating monkeys, related to modern colobus monkeys, as well as antelopes whose living relatives favor closely wooded areas. Reasonably well preserved hominid fossils can also be found here, suggesting that, at least occasionally, early hominids inhabited a riparian habitat.

Where do these *Australopithecus* fossils fit in the evolutionary history of hominids? The jaws and teeth from Allia Bay, as well as a nearly complete radius (the outside bone of the forearm) from the nearby sediments of Sibilot just to the north, show an interesting mixture of characteristics. Some of the traits are primitive ones—that is, they are ancestral features thought to be present before the split occurred between the chimpanzee and human lineages. Yet these bones also share characteristics seen in later hominids and are therefore said to have more advanced features. As our team continues to unearth more bones and teeth at Allia Bay, these new fossils add to our knowledge of the wide range of traits present in early hominids.

Return to Kanapoi

Across Lake Turkana, some 145 kilometers (about 90 miles) south of Allia Bay, lies the site of Kanapoi, where our story began. One of us (Leakey) has mounted expeditions from the National Museums of Kenya to explore the sediments located southwest of Lake Turkana and to document the faunas present during the earliest stages of the basin's history. Kanapoi, virtually unexplored since Patterson's day, has proved to be one of the most rewarding sites in the Turkana region.

A series of deep erosion gullies, known as badlands, has exposed the sediments at Kanapoi. Fossil hunting is difficult here, though, because of a carapace of lava pebbles and gravel that makes it hard to spot small bones and teeth. Studies of the layers of sediment, also carried out by Feibel, reveal that the fossils here have been preserved by deposits from a river ancestral to the present-day Kerio River, which once flowed into the Turkana basin and emptied into an ancient lake we call Lonyumun. This lake reached its maximum size about 4.1 Myr ago and thereafter shrank as it filled with sediments.

Excavations at Kanapoi have primarily yielded the remains of carnivore meals, so the fossils are rather fragmentary. But workers at the site have also recovered two nearly complete lower jaws, one complete upper jaw and lower face, the upper and lower thirds of a tibia (the larger bone of the lower leg), bits of skull and several sets of isolated teeth. After careful study of the fossils from both Allia Bay and Kanapoi—including Patterson's fragment of an arm bone—we felt that in details of anatomy, these specimens were different enough from previously known hominids to warrant designating a new species. So in 1995, in collaboration with both Feibel and Ian McDougall of the Australian National University, we named this new species *Australopithecus anamensis,* drawing on the Turkana word for lake (*anam*) to refer to both the present and ancient lakes.

To establish the age of these fossils, we relied on the extensive efforts of Brown, Feibel and McDougall, who have been investigating the paleogeographic history of the entire lake basin. If there study of the basin's development is correct, the *anamensis* fossils should be between 4.2 and 3.9 Myr old. Currently McDougall is working to determine the age of the so-called Kanapoi Tuff—the layer of volcanic ash that covers most of the fossils at this site. We expect that once McDougall successfully ascertains the age of the tuff, we will be confident in both the age of the fossils and Brown's and Feibel's understanding of the history of the lake basin.

A major question in paleoanthropology today is how the anatomic mosaic of the early hominids evolved. By comparing the nearly contemporaneous Allia Bay and Kanapoi collections of *anamensis,* we can piece together a fairly accurate picture of certain aspects of the species, even though we have not yet uncovered a complete skull.

The jaws of *anamensis* are primitive—the sides sit close together and parallel to each other (as in modern apes), rather than widening at the back of the mouth (as in later hominids, including humans). In its lower jaw, *anamensis* is also chimp-like in terms of the shape of the region where the left and right sides of the jaw meet (technically known as the mandibular symphysis).

Teeth from *anamensis,* however, appear more advanced. The enamel is relatively thick, as it is in all other species of *Australopithecus;* in contrast, the tooth enamel of African great apes is much thinner. The thickened enamel suggests *anamensis* had already adapted to a changed diet—possibly much harder food—even though its jaws and some skull features were still very apelike. We also know that *anamensis* had only a tiny external ear canal. In this regard, it is more like chimpanzees and unlike all later hominids, including humans, which have large external ear canals. (The size of the external canal is unrelated to the size of the fleshy ear.)

The most informative bone of all the ones we have uncovered from this new hominid is the nearly complete tibia—the larger of the two bones in the lower leg. The tibia is revealing because of its important role in weight bearing: the tibia of a biped is distinctly different from the tibia of an animal that walks on all four legs. In size and practically all details of the knee and ankle joints, the tibia found at Kanapoi closely resembles the one from the fully bipedal

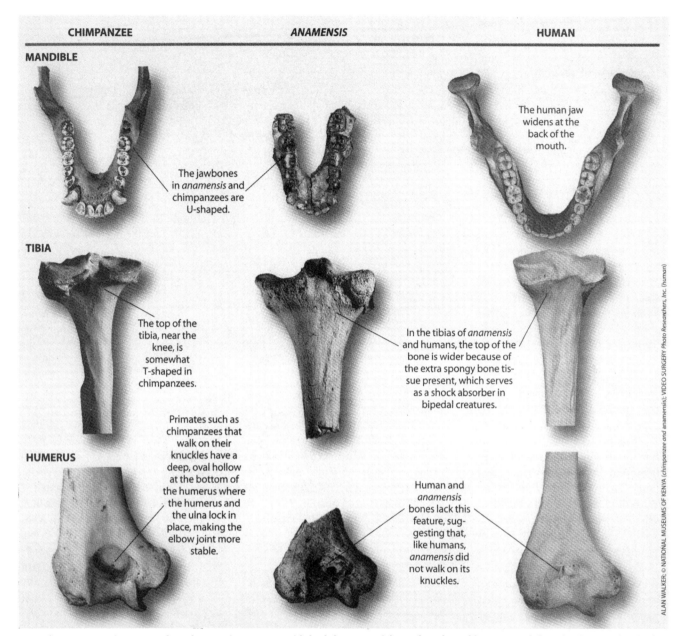

CHIMPANZEE | ***ANAMENSIS*** | **HUMAN**

MANDIBLE

The jawbones in *anamensis* and chimpanzees are U-shaped.

The human jaw widens at the back of the mouth.

TIBIA

The top of the tibia, near the knee, is somewhat T-shaped in chimpanzees.

In the tibias of *anamensis* and humans, the top of the bone is wider because of the extra spongy bone tissue present, which serves as a shock absorber in bipedal creatures.

Primates such as chimpanzees that walk on their knuckles have a deep, oval hollow at the bottom of the humerus where the humerus and the ulna lock in place, making the elbow joint more stable.

HUMERUS

Human and *anamensis* bones lack this feature, suggesting that, like humans, *anamensis* did not walk on its knuckles.

ALAN WALKER; © NATIONAL MUSEUMS OF KENYA (*chimpanzee and anamensis*); VIDEO SURGERY *Photo Researchers, Inc.* (*human*)

FOSSILS from *anamensis (center)* share features in common with both humans (*right*) and modern chimpanzees (*left*). Scientists use the similarities and differences among these species to determine their interrelationships and thereby piece together the course of hominid evolution since the lineages of chimpanzees and humans split some five or six million years ago.

afarensis found at Hadar, even though the latter specimen is nearly a million years younger.

Fossils of other animals collected at Kanapoi point to a somewhat different paleoecological scenario from the setting across the lake at Allia Bay. The channels of the river that laid down the sediments at Kanapoi were probably lined with narrow stretches of forest that grew close to the riverbanks in other-

wise open country. Researchers have recovered the remains of the same spiral-horned antelope found at Allia Bay that very likely lived in dense thickets. But open-country antelopes and hartebeest appear to have lived at Kanapoi as well, suggesting that more open savanna prevailed away from the rivers. These results offer equivocal evidence regarding the preferred habitat of *anamensis:* we know that bushland was present at both

sites that have yielded fossils of the species, but there are clear signs of more diverse habitats at Kanapoi.

An Even Older Hominid?

At about the same time that we were finding new hominids at Allia Bay and Kanapoi, a team led by our colleague Tim D. White of the University of California at Berkeley discovered fossil hominids in Ethiopia that are even older than *anamensis*. In 1992 and 1993 White led an expedition to the Middle Awash area of Ethiopia, where his team uncovered hominid fossils at a site known as Aramis. The group's finds include isolated teeth, a piece of baby's mandible (the lower jaw), fragments from an adult's skull and some arm bones, all of which have been dated to around 4.4 Myr ago. In 1994, together with his colleagues Berhane Asfaw of the Paleoanthropology Laboratory in Addis Ababa and Gen Suwa of the University of Tokyo, White gave these fossils a new name: *Australopithecus ramidus*. In 1995 the group renamed the fossils, moving them to a new genus, *Ardipithecus*. Other fossils buried near the hominids, such as seeds and the bones of forest monkeys and antelopes, strongly imply that these hominids, too, lived in a closed-canopy woodland.

This new species represents the most primitive hominid known—a link between the African apes and *Australopithecus*. Many of the *Ardipithecus ramidus* fossils display similarities to the anatomy of the modern African great apes, such as thin dental enamel and strongly built arm bones. In other features, though—such as the opening at the base of the skull, technically known as the foramen magnum, through which the spinal cord connects to the brain— the fossils resemble later hominids.

Describing early hominids as either primitive or more advanced is a complex issue. Scientists now have almost decisive molecular evidence that humans and chimpanzees once had a common ancestor and that this lineage had previously split from gorillas. This is why we often use the two living species of chimpanzee (*Pan troglodytes* and *P. paniscus*) to illustrate ancestral traits. But we must remember that since their last common ancestor with humans, chimpanzees have had exactly the same amount of time to evolve as humans have. Determining which features were present in the last common ancestor of humans and chimpanzees is not easy.

But *Ardipithecus,* with its numerous chimplike features, appears to have taken the human fossil record back close

to the time of the chimp-human split. More recently, White and his group have found parts of a single *Ardipithecus* skeleton in the Middle Awash region. As White and his team extract these exciting new fossils from the enclosing stone, reconstruct them and prepare them for study, the paleoanthropological community eagerly anticipates the publication of the group's analysis of these astonishing finds.

But even pending White's results, new *Australopithecus* fossil discoveries are offering other surprises, particularly about where these creatures lived. In 1995 a team led by Michel Brunet of the University of Poitiers announced the identification in Chad of *Australopithecus* fossils believed to be about 3.5 Myr old. The new fossils are very fragmentary—only the front part of a lower jaw and isolated tooth. In 1996, however, Brunet and his colleagues designated a new species for their specimen: *A. bahrelghazali*. Surprisingly, these fossils were recovered far from either eastern or southern Africa, the only areas where *Australopithecus* had been found until now. The site, in the Bahr el Ghazal region of Chad, lies 2,500 kilometers west of the western part of the Rift Valley, thus extending the range of *Australopithecus* well into the center of Africa.

The *bahrelghazali* fossils debunk a hypothesis about human evolution postulated in the pages of *Scientific American* by Yves Coppens of the College of France [see "East Side Story: The Origin of Humankind," May 1994]; ironically, Coppens is now a member of Brunet's team. Coppens's article proposed that the formation of Africa's Rift Valley subdivided a single ancient species, isolating the ancestors of hominids on the east side from the ancestors of modern apes on the west side. In general, scientists believe such geographical isolation can foster the development of new species by prohibiting continued interbreeding among the original populations. But the new Chad fossils show that early hominids did live west of the Rift Valley. The geographical separation of apes and hominids previously apparent in the fossil record may be more the result

of accidental circumstances of geology and discovery than the species' actual ranges.

The fossils of *anamensis* that we have identified should also provide some answers in the long-standing debate over whether early *Australopithecus* species lived in wooded areas or on the open savanna. The outcome of this discussion has important implications: for many years, paleoanthropologists have accepted that upright-walking behavior originated on the savanna, where it most likely provided benefits such as keeping the hot sun off the back or freeing hands for carrying food. Yet our evidence suggests that the earliest bipedal hominid known to date lived at least part of the time in wooded areas. The discoveries of the past several years represent a remarkable spurt in the sometimes painfully slow process of uncovering human evolutionary past. But clearly there is still much more to learn.

The Authors

MEAVE LEAKEY and ALAN WALKER, together with Leakey's husband, Richard, have collaborated for many years on the discovery and analysis of early hominid fossils from Kenya. Leakey is head of the division of paleontology at the National Museums of Kenya in Nairobi. Walker is Distinguished Professor of anthropology and biology at Pennsylvania State University. He is a MacArthur Fellow and a member of the American Academy of Arts and Sciences.

Further Reading

AUSTRALOPITHECUS RAMIDUS, A NEW SPECIES OF EARLY HOMINID FROM ARAMIS, ETHIOPIA. Tim D. White, Gen Suwa and Berhane Asfaw in *Nature,* Vol. 371, pages 306–312; September 22, 1994.

NEW FOUR-MILLION-YEAR-OLD HOMINID SPECIES FROM KANAPOI AND ALLIA BAY, KENYA. Meave G. Leakey, Craig S. Feibel, Ian McDougall and Alan Walker in *Nature,* Vol. 376, pages 565–571; August 17, 1995.

FROM LUCY TO LANGUAGE. Donald C. Johanson and Blake Edgar. Peter Nevraumont, Simon & Schuster, 1996.

RECONSTRUCTING HUMAN ORIGINS: A MODERN SYNTHESIS. Glenn C. Conroy. W. W. Norton, 1997.

A New Human Ancestor?

Ethiopian fossils reveal a new branch on the hominid family tree: a small-brained hominid that is a candidate for the ancestor of our lineage

Elizabeth Culotta

About two and half million years ago, on a grassy plain bordering a shallow lake in what is now eastern Ethiopia, a humanlike creature began dismembering an antelope carcass. Nothing remains of the hominid, but the antelope bones show that it wrenched a leg off the carcass, then used a stone tool to slice off the meat and smash the bone. After several tries, it managed to break off both ends of the bone and scrape out the juicy marrow inside.

At just about the same time, two other hominids died near the lake. One, perhaps 1.4 meters tall, had long legs and a human gait but long, apelike forearms. The other, a male, lay some distance away. His limb bones are gone, but the remains of his skull show he had a small brain, big teeth, and an apelike face.

These new fossils give different glimpses of each hominid, and no one can be sure all three belonged to the same species. But even if not, their details are starting to fill in a mysterious chapter of human prehistory. According to the international team that made all three discov-

eries, the big-toothed skull represents an unusual new species that is the best candidate for the ancestor of our own genus, *Homo*. Not everyone in the contentious field of paleoanthropology agrees, but the

SOURCE: ADAPTED FROM T. WHITE; PHOTOS: DAVID L. BRILL

Untangling the family tree. Fossil finds suggest an outline of our evolutionary history, but the ancestor of our genus remains in question.

new species, which Ethiopian anthropologist Berhane Asfaw and his colleagues have named *Australopithecus garhi* (*garhi* means "surprise" in the language spoken by the local Afar people), is certain to shake up views of the tran-

sition from the apelike australopithecines to humankind. And the scored bones from the first hominid's feast are the earliest recorded evidence of hominids butchering animals, bolstering the notion that meat eating was important in human evolution.

"They've put together a whole package here, so that you can say a fair amount about a time we don't know much about," says anthropologist F. Clark Howell of the University of California (UC), Berkeley. With its surprising mix of traits—primitive face and unusually big teeth—the new australopithecine doesn't match the profile many researchers expected for a human ancestor at this stage. "It's very exciting," says paleoanthropologist Alan Walker of Pennsylvania State University in University Park. "Until now it's all been just scraps of teeth and bits of mandible from this time. And this [morphology] is a surprise."

But this rare glimpse of a murky period in human evolution raises as many questions as it answers. *A. garhi* has few

CREDIT: DAVID L. BRILL

Treasure site. A garhi's skull was found on this Ethiopian desert slope.

traits that definitively link it to *Homo*, and like other hominids from the same period, it may simply be an evolutionary dead end that brings us only slightly closer to understanding our own ancestors, says paleoanthropologist Bernard Wood of George Washington University in Washington, D.C. The debate is complicated by the fact that paleoanthropologists are deeply divided over who the first humans, or members of *Homo*, were, and indeed what makes a human. "These are magnificent fossils," says Wood, but he's not ready to admit *A. garhi* into the gallery of our ancestors. "At this point it's impossible to tell what's ancestral to what," he says. "This won't be the last 'surprise.'"

Anthropologists have long been itching to know just what East African hominids were doing between 2 million and 3 million years ago, says one of the team's leaders, paleoanthropologist Tim White of UC Berkeley. Decades of fieldwork and analysis have allowed researchers to identify many characters in the human evolutionary story (see diagram), starting with apelike species such as the 4.2-million-year-old *A. anamensis*. Next in line, known from 3.7 million to 3.0 million years ago, is *A. afarensis*, best known for the famed "Lucy" skeleton: a meter-tall, small-brained, upright hominid that retained apelike limb proportions and a protruding lower face.

More than a million years separate Lucy from the first specimens usually considered to be part of our own genus,

which appear in East Africa around 2 million years ago and tend to have larger brains and a more human face, although they are highly variable. In the interim, the South African fossil record is diverse and confusing, and the East African record has been sparse. The period includes three species that fall into the "robust" australopithecine group— heavy-jawed hominids with skull crests and large back teeth, perhaps for eating hard roots and tubers—that are not part of our own lineage. More promising for those seeking a human ancestor is *A. africanus*, known from South Africa starting at around 2.8 million years ago, which has a more humanlike face than the Lucy species.

But the *A. africanus* fossils were found half a continent away from the East African cradle of *Homo*, and some anthropologists have been hoping for a stronger candidate for the root of our lineage. "After the split with the robust lineage, we have very little evidence," says Walker. That's why White and his team zeroed in on sediments in the desert of Ethiopia's Afar depression. They struck gold with three separate discoveries, all dated securely to 2.5 million years ago by radiometric techniques on an underlying volcanic rock layer.

One dramatic find came in 1997, when El Nio-driven rains washed away stones and dirt on steep slopes near the village of Bouri. Berkeley graduate student Yohannes Haile-Selassie spotted fragments of the skull—the color and thickness of a coconut shell—on the surface. A closer look revealed teeth poking out of the ground. Much of the rest of the skull had washed down the hill, so the team, which includes 40 members from 13 countries, took the slope apart. They dug tons of material from the hill, then sieved it and picked through it for bone—twice. "It was probably the most difficult fossil recovery we've ever

done," says White. "We spent 7 weeks on that slope." Although the delicate bones of the middle face were gone for good, the team found many more skull fragments.

After reconstructing the skull, the researchers were confronted with a face that is apelike in the lower part, with a protruding jaw resembling that of *A. afarensis*. The large size of the palate and teeth suggests that it is a male, with a small braincase of about 450 cubic centimeters. (A modern human brain is about 1400 cubic centimeters.) It is like no other hominid species and is clearly not a robust form. And in a few dental traits, such as the shape of the premolar and the size ratio of the canine teeth to the molars, *A. garhi* resembles specimens of early *Homo*. But its molars are huge—the second molar is 17.7 millimeters across, even larger than the *A. robustus* average. "Selection was driving bigger teeth in both lineages—that's a big surprise," says Walker.

The other dramatic skeletal find had come a year earlier: leg and arm bones of a single ancient hominid individual, found together. The new hominid femur or upper leg bone is relatively long, like that of modern humans. But the forearm is long too, a condition found in apes and other australopithecines but not in humans. The fossils show that human proportions evolved in steps, with the legs lengthening before the forearms shortened, says co-author Owen Lovejoy of Kent State University in Ohio.

The third major find, at the same stratigraphic level and only a meter away from the skeletal bones, preserves dramatic evidence of hominid behavior: bones of antelopes, horses, and other animals bearing cut marks, suggesting that butchery may be the oldest human profession. One antelope bone, described by a team including archaeologist J. Desmond Clark of UC Berkeley and White, records a failed hammerstone blow, which scratched the bone slightly and caused a bone flake to fly off; a second blow was struck from exactly the same angle. Both ends of the bone were broken off, presumably to get at the marrow.

Similarly, an antelope jawbone bears three successive curved marks, apparently made as a hominid sliced out the tongue. In cross section under the microscope, these marks show a parallel series of ragged V-shaped striations with rough inner walls—the telltale signature of a stone tool rather than a predator's teeth, says White. Marks on the leg bone of a three-toed horse show that hominids dismembered the animal and filleted the meat from the bone.

The Bouri sites yielded few of the stone tools the hominids must have used, perhaps because there is no local source of stone. The hominids "must have brought flakes and cobbles in from some distance, so that obviously shows quite a bit of forethought," says Clark. Tool use by this point is no surprise: At other sites, anthropologists have found tools dated to 2.6 million years ago. But there had been little hard evidence of what the oldest tools were used for. The new find shows that tools enabled hominids to get at "a whole new world of food"—bone marrow, says White.

Marrow is rich in fat, and few animals other than humans and hyenas can get at it. Anthropologists have theorized that just such a dietary breakthrough allowed the dramatic increase in brain size (*Science,* 29 May 1998, p. 1345), to perhaps 650 cc or larger, that took place in the *Homo* lineage by 2 million years ago. Two researchers recently proposed that cooked tubers were the crucial new food source (*Science,* 26 March, p. 2004), but most others have assumed it

was meat. The cut marks present convincing evidence that they were right, says Yale University anthropologist Andrew Hill.

Whether or not the three finds can be connected, *A. garhi,* as based on the new skull, is now a prime candidate as an ancestor of our genus. The species is in the right place—East Africa—and the right time—between the time of *A. afarensis* and that of early *Homo*—says White. But making the link to the human lineage isn't easy, in part because the nature of "early *Homo*" is itself something of a mystery. White notes that some of the early *Homo* specimens have large teeth, and that in the teeth "there's not much change at all from *A. garhi* to those specimens of early *Homo.*"

The link between *A. garhi* and *Homo* would be strengthened, of course, if researchers could show that the humanlike long bones come from *A. garhi* rather than from some other humanlike hominid. For now White is willing only to "make up a hypothesis to be tested": *A. garhi,* a small-brained, big-toothed hominid with humanlike leg proportions, began butchering animals by 2.5 million years ago. Thanks in part to the better diet, brain size rapidly increased to that seen in early *Homo,* and the trend toward large back teeth reversed—changes that quickly transformed other parts of the skull as well, such as flattening the protruding jaw.

But some other researchers don't buy that as a likely scenario. There's no reason to expect that every new branch on the hominid tree is our ancestor, says George Washington's Wood. He adds that he is not surprised by *A. garhi's* mix of humanlike and robust features, because, given that climate was changing, "we should expect a variety of creatures with mixtures of adaptations at this time." Other researchers note that the dental data linking the species to *Homo* are weak. "Nothing here aligns garhi closely with *Homo,*" says paleoanthropologist Fred Grine of the State University of New York, Stony Brook. "It's a possible candidate [for *Homo* ancestry], but no better than africanus."

Some anthropologists also say that there may not have been enough time for evolution to have transformed *A. garhi* into *Homo.* The oldest known specimen assigned to *Homo,* a 2.33-million-year-old palate from Hadar, Ethiopia, is more humanlike than *A. garhi,* with smaller teeth. That requires either a burst of evolution or some other explanation, such as sexual dimorphism, if *A. garhi* is to be considered part of our lineage, notes paleoanthropologist Juan Luis Arsuaga of the Universidad Complutense de Madrid in Spain. White says that only further discoveries and analysis will show just where the hominids of that long-vanished plain stand in relation to our own species: "*A. garhi* isn't the end; it's the first step."

Asian Hominids Grow Older

Fossils from China could alter the picture of human dispersal and evolution—and they're just one of several findings, described on the following pages, that challenge the textbooks.

Wanderlust has been a potent factor in human evolution, spurring early members of our lineage to leave their African homes and spread throughout the world. But exactly when this itinerant urge struck has become a hotly debated issue, especially since it has major consequences for scenarios of later human evolution. For years, the majority view held that the first footloose hominid was *Homo erectus,* thought to have left Africa about 1 million years ago. In the past year, however, new data from Java and the republic of Georgia have suggested that *H. erectus* was already present in those Asian locales as early as 1.8 million years ago.

Now, in this week's issue of *Nature,* the idea of an earlier migration gets additional support from a team of Chinese and Western scientists. Based on a three-part package of hominid fossils, dating methods, and primitive tools, they argue that early *Homo* reached central China between 1.7 million and 1.9 million years ago—nearly 800,000 years earlier than had been thought.

Because the field sporadically reverses over time, researchers can date fossils by tying them to a particular period of normal or reversed field.

And the team's claims go beyond dating. They suggest that the ancient wanderer was not *H. erectus* itself, but an even earlier hominid with ties to more primitive African forms. "Our work shows that there was an early dispersal of [primitive] hominids with basic stone tools out of Africa," says paleoanthropologist Russell Ciochon of the University of Iowa, who led the collaboration with the Chinese. Their results could strengthen a minority view that *H. erectus* evolved not in Africa but in Asia, from primitive hominids like the newly

reported Chinese finds. If so, *H. erectus* could be an Asian side branch of the hominid evolutionary tree, rather than

Researchers used a relatively new method called electron spin resonance (ESR), which measures the electric charges induced in tooth enamel over time by naturally occurring radioactive materials in the surrounding sediments.

part of the African lineage that led to modern humans. But judging from early reaction, the fragmentary new evidence

African Origins: West Side Story

Ask most paleoanthropologists where an ancestral ape took its first humanlike steps, and they're likely to point to East Africa. After all, the oldest known bipedal hominid, 4.1-million-year-old *Australopithecine anamensis,* was found in Kenya, while the slightly younger *Australopithecus afarensis,* typified by the famous skeleton "Lucy," was found in Ethiopia. But the discovery of a 3- to 3.5-million-year-old australopithecine fossil in Chad, some 5400 kilometers to the west in the heart of the African continent, has upset that East African–centric view. "Human origins is not just an east-side story," says Michel Brunet, a paleoanthropologist at the University of Poitiers, who found the partial lower jawbone in January. "It's a west-side story, too."

Brunet's find, preliminarily assigned to *Australopithecus afarensis*—although Brunet himself thinks it may be a new species—is reported in the 16 November issue of *Nature.* It already has scientists backpedaling about previous declarations labeling East Africa the cradle of humankind. "I think that's been a very naive view," says Alan Walker, a paleoanthropologist at Pennsylvania State University, "and so we're going to have to rethink things, which is good for the field."

One idea being heavily rethought is the notion that East Africa's long Rift Valley acted as a geographical barrier to ape populations in the late Miocene, 5 million to 7 million years ago, separating those that became hominids in the savannas of the east from forest-dwelling apes in the west. "Now we have early australopithecines all around Africa," says Brunet, "which makes it impossible to tell the exact place of origin."

Or the cause of that origin. Previously, scientists such as Yves Coppens, a paleoanthropologist at the College de France in Paris and co-author of the new paper, had suggested that hominids had evolved in the eastern part of the continent because of habitat changes associated with the development of the Rift Valley. "The rise of the western Rift has been linked to the development of the more open savanna country one finds in East Africa," explains David Pilbeam, director of Harvard University's Peabody Museum and another co-author of the new paper. Open country, in theory, created selective pressure driving ape-like creatures out of the trees and onto the ground. "And that, in turn, was seen as causing the origin of the hominids. But I don't think the Rift Valley was the mechanism," he says. The habitat of the Chad hominids seems to have been a dry, grassy woodland, according to animal fossils from the site. "We have rhinoceroses, giraffe, and hipparion [horse], which suggest grasslands, and pigs and elephants, which are more adapted to woodlands," Brunet says.

The finds will focus more attention on Central and West Africa as potential hotbeds of hominid activity. "We've always thought of the [current] West and Central African tropical rain forest as being around forever, while the east became a savanna," explains Rick Potts, a paleoanthropologist at the Smithsonian Institution's Natural History Museum. "That ecological distinction was thought to be the critical marker of the human-ape split. Now it's clear that we don't really know a single thing about what was going on in West Africa at that time."

Brunet hopes that future discoveries at his site will give scientists a clearer view of ancient West Africa—and help him nail down the precise species of the Chad specimen. Currently, Brunet says, the two australopithecines known from that period are *A. afarensis* and *A. anamensis.* "I think there were more than just two australopithecines 3.5 million years ago—it was more complicated than that, as we know now their origins were, too."

—Virginia Morell

may not be enough to sway researchers who have long held a more classical view of human evolution and dispersal.

The provocative new fossils and stone tools were unearthed in the late 1980s by Chinese scientists, led by Huang Wanpo of the Institute of Vertebrate Paleontology and Paleoanthropology in Beijing. They excavated Longgupo Cave in Sichuan province, known as Dragon Bone Slope in Chinese because the cave's roof and walls have collapsed. They found a rich collection of bones, including prized evidence of hominids: a jaw fragment with two teeth, an upper incisor, and two crude stone tools. The Chinese scientists also analyzed traces of Earth's ancient magnetic field left in sediments associated with the fossils.

Because the field sporadically reverses over time, researchers can date fossils by tying them to a particular period of normal or reversed field. The hominid fossils were determined to have been deposited during a period of normal magnetic polarity, and the Chinese correlated this to a normal polarity event dated at Africa's Olduvai Gorge to 1.77 million to 1.95 million years ago. But their work was not widely known because it was published in a Chinese journal.

Then in 1992 the Chinese invited Ciochon and his colleagues to visit the site to explore its geology and confirm the dates. For this, the researchers used a relatively new method called electron spin resonance (ESR), which measures the electric charges induced in tooth enamel over time by naturally occurring radioactive materials in the surrounding sediments. They weren't able to excavate the cave and so couldn't date the hominid levels directly. But Henry Schwarcz of McMaster University in Hamilton, Ontario, applied ESR dating to a deer tooth from one of the cave's upper levels; he estimated a minimum age of 750,000 years and a most likely age of 1 million years. Together with associated animal fossils, the ESR date "calibrates and constrains the paleomagnetics," indirectly confirming the hominid ages, says Ciochon.

The Chinese fossils may have a significance even beyond their advanced age: their primitive form. Ciochon says that characters such as a double-rooted premolar and the pattern of cusps on the molar resemble those of early African *Homo* species that predated *H. erectus.* He points to either *H. habilis,* the most ancient member of our genus, known

---

The Longgupo Cave fossils provide a link between Asian and early Africa forms, the new Asian finds resemble 1.6 million-year-old fossils found in East Africa.

—**Bernard Wood**

---

only from Africa, or *H. ergaster,* a species recognized by some researchers as the precursor to *H. erectus* in Africa. The Longgupo Cave fossils provide a link between Asian and early African forms, agrees paleoanthropologist Bernard Wood of the University of Liver-

pool, noting that the new Asian finds resemble 1.6 million-year-old fossils found in East Africa.

The tools—rounded pieces of igneous rock that show signs of repeated battering—provide additional support for the idea that the fossils belong to a primitive *Homo.* These crude implements recall the basic choppers found with early hominids at Olduvai Gorge, rather than the more complicated tools associated with *H. erectus,* says the team's archaeologist, Roy Larick of the University of Massachusetts, Amherst.

All this adds up to a coherent picture of a pre-*erectus* hominid that left Africa

---

Simple stone tools show that hominids were able to conquer new territory before they developed the more complex hand axes once thought to be a prerequisite for long-distance dispersal.

---

perhaps 2 million years ago, says Ciochon. And the simple stone tools show that hominids were able to conquer new territory before they developed the more complex hand axes once thought to be a prerequisite for long-distance dispersal. "This shows that very soon after the origin of *Homo,* hominids became mobile and were able to disperse rapidly over huge distances," agrees Peter Andrews of the Natural History Museum in London. Furthermore, in this scenario, these first travelers evolved into *H. erectus* while in Asia. And because everyone agrees that our own species arose in Africa, this implies that *erectus* itself was an Asian creature and an evolutionary side branch not directly ancestral to modern humans.

Parts of this theory are extremely controversial, but the early dates are in

---

The geology of cave deposits such as Longgupo is notoriously complex, because material falling from above may become jumbled with rocks of difference ages.

—**Philip Rightmire**

---

accord with two recent observations. Last year, Carl Swisher of the Berkeley Geochronology Center and colleagues redated *H. erectus* skulls from Java to 1.6 million and 1.8 million years old (*Science,* 25 February 1994, pp. 1087 and 1118). And earlier this year, scientists in the Republic of Georgia published an *H. erectus* jawbone, estimated to be 1.6 million to 1.8 million years old, from the site of Dmanisi, Georgia.

But many researchers remain skeptical of all three of the earlier dates. The geology of cave deposits such as Longgupo is notoriously complex, because material falling from above may become jumbled with rocks of different ages, says hominid expert Philip Rightmire of the State University of New York, Binghamton. He's not convinced that the hominids are truly older than the deer tooth dated by ESR. And the Chinese team remains leery of the Java dates. The problem, says Ciochon, is that the Chinese hominid looks more primitive than the Javanese ones. Unless there were two ancient hominids in Asia, it doesn't make sense to have a pre-*erectus* hominid in China at the same time as true *erectus* in Java. Meanwhile, Swisher and Georgian colleagues are now redoing the paleomagnetics for the Dmanisi site.

But even more contentious than the date is the notion that the travelers were "pre-*erectus.*" That conclusion is based on "pretty scrappy evidence," says Rightmire. The hominid fossils are incomplete, and the stone tools are so simple that Rightmire and others wonder if they are really artifacts. "This is not the material on which I'd choose to erect bold new scenarios of Chinese prehistory," he says. As F. Clark Howell of the University of California, Berkeley, points out, partial jawbones of early hominids are difficult to classify. To paleoanthropologist Alan Walker of Pennsylvania State University, who supports the more classical idea that *H. erectus* led to *H. sapiens,* the Chinese hominid is "just early *erectus.*" If so, *H. erectus* could have evolved in Africa, then dispersed to Asia, albeit earlier than had been thought. But the link between *erectus* and *H. sapiens* would be intact.

Other anthropologists have more fundamental concerns about the fossils' identity. Milford Wolpoff of the Univer-

Yet despite the murmurs of doubt, the evidence is mounting in favor of an early excursion out of Africa, accomplished with only stone technology.

sity of Michigan, who saw the specimens on a trip to China several years ago, isn't even convinced that the partial jaw is a hominid. "I believe it is a piece of an orangutan or other *Pongo,*" he

says. He bases that conclusion on a wear facet on the preserved premolar, which to him suggests that the missing neighboring tooth is shaped more like an orang's than a human's.

Yet despite the murmurs of doubt, the evidence is mounting in favor of an early excursion out of Africa, accomplished with only crude stone technology. Whether the first travelers are properly called *H. erectus* or something else, the newest work all points to the same conclusion: The urge to wander is an ancient trait that evolved near the dawn of our lineage.

—Elizabeth Culotta

Scavenger Hunt

As paleoanthropologists close in on their quarry, it may turn out to be a different beast from what they imaged

Pat Shipman

Pat Shipman is an assistant professor in the Department of Cell Biology and Anatomy at The Johns Hopkins University School of Medicine.

In both textbooks and films, ancestral humans (hominids) have been portrayed as hunters. Small-brained, big-browed, upright, and usually mildly furry, early hominid males gaze with keen eyes across the gold savanna, searching for prey. Skillfully wielding a few crude stone tools, they kill and dismember everything from small gazelles to elephants, while females care for young and gather roots, tubers, and berries. The food is shared by group members at temporary camps. This familiar image of Man the Hunter has been bolstered by the finding of stone tools in association with fossil animal bones. But the role of hunting in early hominid life cannot be determined in the absence of more direct evidence.

I discovered one means of testing the hunting hypothesis almost by accident. In 1978, I began documenting the microscopic damage produced on bones by different events. I hoped to develop a diagnostic key for identifying the postmortem history of specific fossil bones, useful for understanding how fossil assemblages were formed. Using a scanning electron microscope (SEM) because of its excellent resolution and superb depth of field, I inspected high-fidelity replicas of modern bones that had been subjected to known events or conditions. (I had to use replicas, rather than real bones, because specimens must fit into the SEM's small vacuum chamber.) I soon established that such common events as weathering, root etching, sedimentary abrasion, and carnivore chewing produced microscopically distinctive features.

In 1980, my SEM study took an unexpected turn. Richard Potts (now of Yale University), Henry Bunn (now of the University of Wisconsin at Madison), and I almost simultaneously found what appeared to be stone-tool cut marks on fossils from Olduvai Gorge, Tanzania, and Koobi Fora, Kenya. We were working almost side by side at the National Museums of Kenya, in Nairobi, where the fossils are stored. The possibility of cut marks was exciting, since both sites preserve some of the oldest known archaeological materials. Potts and I returned to the United States, manufactured some stone tools, and started "butchering" bones and joints begged from our local butchers. Under the SEM, replicas of these cut marks looked very different from replicas of carnivore tooth scratches, regardless of the species of carnivore or the type of tool involved. By comparing the marks on the fossils with our hundreds of modern bones of known history, we were able to demonstrate convincingly that hominids using stone tools had processed carcasses of many different animals nearly two million years ago. For the first time, there was a firm link between stone tools and at least some of the early fossil animal bones.

This initial discovery persuaded some paleoanthropologists that the hominid hunter scenario was correct. Potts and I were not so sure. Our study had shown that many of the cut-marked fossils also bore carnivore tooth marks and that some of the cut marks were in places we hadn't expected—on bones that bore little meat in life. More work was needed.

In addition to more data about the Olduvai cut marks and tooth marks, I needed specific information about the patterns of cut marks left by known hunters performing typical activities associated with hunting. If similar patterns occurred on the fossils, then the early hominids probably behaved similarly to more modern hunters; if the patterns were different, then the behavior was probably also different. Three activities related to hunting occur often enough in peoples around the world and leave consistent enough traces to be used for such a test.

First, human hunters systematically disarticulate their kills, unless the animals are small enough to be eaten on the spot. Disarticulation leaves cut marks in a predictable pattern on the skeleton. Such marks cluster near the major joints of the limbs: shoulder, elbow, carpal joint (wrist), hip, knee, and hock (ankle). Taking a carcass apart at the joints is much easier than breaking or cutting through bones. Disarticulation enables hunters to carry food back to a central place or camp, so that they can share it with others or cook it or even store it by placing portions in trees, away from the reach of carnivores. If early hominids were hunters who transported and shared their kills, disarticulation marks would occur near joints in frequencies comparable to those produced by modern human hunters.

Second, human hunters often butcher carcasses, in the sense of removing meat from the bones. Butchery marks are usually found on the shafts of bones from the upper part of the front or hind limb, since this is where the big muscle masses lie. Butchery may be carried out at the kill site—especially if the animal is very large and its bones very heavy—or it may take place at the base camp, during the process of sharing food with others. Compared with disarticulation, butchery leaves relatively few marks. It is hard for a hunter to locate an animal's joints without leaving cut marks on the bone. In contrast, it is easier to cut the meat away from the midshaft of the bone without making such marks. If early hominids shared their food, however, there ought to be a number of cut marks located on the midshaft of some fossil bones.

Finally, human hunters often remove skin or tendons from carcasses, to be used for clothing, bags, thongs, and so on. Hide or tendon must be separated from the bones in many areas where there is little flesh, such as the lower limb bones of pigs, giraffes, antelopes, and zebras. In such cases, it is difficult to cut the skin without leaving a cut mark on the bone. Therefore, one expects to find many more cut marks on such bones than on the flesh-covered bones of the upper part of the limbs.

Unfortunately, although accounts of butchery and disarticulation by modern human hunters are remarkably consistent, quantitative studies are rare. Further, virtually all modern hunter-gatherers use metal tools, which leave more cut marks than stone tools. For these reasons I hesitated to compare the fossil evidence with data on modern hunters. Fortunately, Diane Gifford of the University of California, Santa Cruz, and her colleagues had recently completed a quantitative study of marks and damage on thousands of antelope bones processed by Neolithic (Stone Age) hunters in Kenya some 2,300 years ago. The data from Prolonged Drift, as the site is called, were perfect for comparison with the Olduvai material.

Assisted by my technician, Jennie Rose, I carefully inspected more than 2,500 antelope bones from Bed I at Olduvai Gorge, which is dated to between 1.9 and 1.7 million years ago. We made high-fidelity replicas of every mark that we thought might be either a cut mark or a carnivore tooth mark. Back in the United States, we used the SEM to make positive identifications of the marks. (The replication and SEM inspection was time consuming, but necessary: only about half of the marks were correctly identified by eye or by light microscope.) I then compared the patterns of cut mark and tooth mark distributions on Olduvai fossils with those made by Stone Age hunters at Prolonged Drift.

By their location, I identified marks caused either by disarticulation or meat removal and then compared their frequencies with those from Prolonged Drift. More than 90 percent of the Neolithic marks in these two categories were from disarticulation, but to my surprise, only about 45 percent of the corresponding Olduvai cut marks were from disarticulation. This difference is too great to have occurred by chance; the Olduvai bones did not show the predicted pattern. In fact, the Olduvai cut marks attributable to meat removal and disarticulation showed essentially the same pattern of distribution as the carnivore tooth marks. Apparently, the early hominids were not regularly disarticulating carcasses. This finding casts serious doubt on the idea that early hominids carried their kills back to camp to share with others, since both transport and sharing are difficult unless carcasses are cut up.

When I looked for cut marks attributable to skinning or tendon removal, a more modern pattern emerged. On both the Neolithic and Olduvai bones, nearly 75 percent of all cut marks occurred on bones that bore little meat; these cut marks probably came from skinning. Carnivore tooth marks were much less common on such bones. Hominids were using carcasses as a source of skin and tendon. This made it seem more surprising that they disarticulated carcasses so rarely.

A third line of evidence provided the most tantalizing clue. Occasionally, sets of overlapping marks occur on the Olduvai fossils. Sometimes, these sets include both cut marks and carnivore tooth marks. Still more rarely, I could see under the SEM which mark had been made first, because its features were overlaid by those of the later mark, in much the same way as old tire tracks on a dirt road are obscured by fresh ones. Although only thirteen such sets of marks were found, in eight cases the hominids made the cut marks after the carnivores made their tooth marks. This finding suggested a new hypothesis. Instead of hunting for prey and leaving the remains behind for carnivores to scavenge, perhaps hominids were scavenging from the carnivores. This might explain the hominids' apparently unsystematic use of carcasses: they took what they could get, be it skin, tendon, or meat.

Man the Scavenger is not nearly as attractive an image as Man the Hunter, but it is worth examining. Actually, although hunting and scavenging are different ecological strategies, many mammals do both. The only pure scavengers alive in Africa today are vultures; not one of the modern African mammalian carnivores is a pure scavenger. Even spotted hyenas, which have massive, bone-crushing teeth well adapted for eating the bones left behind by others, only scavenge about 33 percent of their food. Other carnivores that scavenge when there are enough carcasses around include lions, leopards, striped hyenas, and jackals. Long-term behavioral studies suggest that these carnivores scavenge when they can and kill when they must. There are only two nearly pure predators, or hunters—the cheetah and the wild dog—that rarely, if ever, scavenge.

What are the costs and benefits of scavenging compared with those of predation? First of all, the scavenger avoids the task of making sure its meal is dead: a predator has already endured the energetically costly business of chasing or stalking animal after animal until one is killed. But while scavenging may be cheap, it's risky. Predators rarely give up their prey to scavengers without defending it. In such disputes, the larger animal, whether a scavenger or a predator, usually wins, although smaller animals in a pack may defeat a lone, larger animal. Both predators and scavengers suffer the dangers inherent in fighting for possession of a carcass. Smaller scavengers such as jackals or striped hyenas avoid disputes to some extent by specializing in darting in and removing a piece of a carcass without trying to take possession of the whole thing. These two strategies can be characterized as that of the bully or that of the sneak: bullies need to be large to be successful, sneaks need to be small and quick.

Because carcasses are almost always much rarer than live prey, the major cost peculiar to scavenging is that scavengers must survey much larger areas than predators to find food. They can travel slowly, since their "prey" is already dead, but endurance is important. Many predators specialize in speed at the expense of endurance, while scavengers do the opposite.

The more committed predators among the East African carnivores (wild dogs and cheetahs) can achieve great top speeds when running, although not for long. Perhaps as a consequence, these "pure" hunters enjoy a much higher success rate in hunting (about three-fourths of their chases end in kills) than any of the scavenger-hunters do (less than half of their chases are successful). Wild dogs and cheetahs are efficient hunters, but they are neither big enough nor efficient enough in their locomotion to make good scavengers. In fact, the cheetah's teeth are so specialized for meat slicing that they probably cannot withstand the stresses of bone crunching and carcass dismembering carried out by scavengers. Other carnivores are less

successful at hunting, but have specializations of size, endurance, or (in the case of the hyenas) dentition that make successful scavenging possible. The small carnivores seem to have a somewhat higher hunting success rate than the large ones, which balances out their difficulties in asserting possession of carcasses.

In addition to endurance, scavengers need an efficient means of locating carcasses, which, unlike live animals, don't move or make noises. Vultures, for example, solve both problems by flying. The soaring, gliding flight of vultures expends much less energy than walking or cantering as performed by the part-time mammalian scavengers. Flight enables vultures to maintain a foraging radius two to three times larger than that of spotted hyenas, while providing a better vantage point. This explains why vultures can scavenge all of their food in the same habitat in which it is impossible for any mammal to be a pure scavenger. (In fact, many mammals learn where carcasses are located from the presence of vultures.)

Since mammals can't succeed as full-time scavengers, they must have another source of food to provide the bulk of their diet. The large carnivores rely on hunting large animals to obtain food when scavenging doesn't work. Their size enables them to defend a carcass against others. Since the small carnivores—jackals and striped hyenas—often can't defend carcasses successfully, most of their diet is composed of fruit and insects. When they do hunt, they usually prey on very small animals, such as rats or hares, that can be consumed in their entirety before the larger competitors arrive.

The ancient habitat associated with the fossils of Olduvai and Koobi Fora would have supported many herbivores and carnivores. Among the latter were two species of large saber-toothed cats, whose teeth show extreme adaptations for meat slicing. These were predators with primary access to carcasses. Since their teeth were unsuitable for bone crushing, the saber-toothed cats must have left behind many bones covered with scraps of meat, skin, and tendon.

Were early hominids among the scavengers that exploited such carcasses?

All three hominid species that were present in Bed I times (*Homo habilis, Australopithecus africanus, A. robustus*) were adapted for habitual, upright bipedalism. Many anatomists see evidence that these hominids were agile tree climbers as well. Although upright bipedalism is a notoriously peculiar mode of locomotion, the adaptive value of which has been argued for years (See Matt Cartmill's article, "Four Legs Good, Two Legs Bad," *Natural History,* November 1983), there are three general points of agreement.

First, bipedal running is nether fast nor efficient compared to quadrupedal gaits. However, at moderate speeds of 2.5 to 3.5 miles per hour, bipedal *walking* is more energetically efficient than quadrupedal walking. Thus, bipedal walking is an excellent means of covering large areas slowly, making it an unlikely adaptation for a hunter but an appropriate and useful adaptation for a scavenger. Second, bipedalism elevates the head, thus improving the hominid's ability to spot items on the ground—an advantage both to scavengers and to those trying to avoid becoming a carcass. Combining bipedalism with agile tree climbing improves the vantage point still further. Third, bipedalism frees the hands from locomotive duties, making it possible to carry items. What would early hominids have carried? Meat makes a nutritious, easy-to-carry package; the problem is that carrying meat attracts scavengers. Richard Potts suggests that carrying stone tools or unworked stones for toolmaking to caches would be a more efficient and less dangerous activity under many circumstances.

In short, bipedalism is compatible with a scavenging strategy. I am tempted to argue that bipedalism evolved because it provided a substantial advantage to scavenging hominids. But I doubt hominids could scavenge effectively without tools, and bipedalism predates the oldest known stone tools by more than a million years.

Is there evidence that, like modern mammalian scavengers, early hominids had an alternative food source, such as either hunting or eating fruits and insects? My husband, Alan Walker, has shown that the microscopic wear on an animal's teeth reflects its diet. Early hominid teeth wear more like that of chimpanzees and other modern fruit eaters than that of carnivores. Apparently, early hominids ate mostly fruit, as the smaller, modern scavengers do. This accords with the estimated body weight of early hominids, which was only about forty to eighty pounds—less than that of any of the modern carnivores that combine scavenging and hunting but comparable to the striped hyena, which eats fruits and insects as well as meat.

Would early hominids have been able to compete for carcasses with other carnivores? They were too small to use a bully strategy, but if they scavenged in groups, a combined bully-sneak strategy might have been possible. Perhaps they were able to drive off a primary predator long enough to grab some meat, skin, or marrow-filled bone before relinquishing the carcass. The effectiveness of this strategy would have been vastly improved by using tools to remove meat or parts of limbs, a task at which hominid teeth are poor. As agile climbers, early hominids may have retreated into the trees to eat their scavenged trophies, thus avoiding competition from large terrestrial carnivores.

In sum, the evidence on cut marks, tooth wear, and bipedalism, together with our knowledge of scavenger adaptation in general, is consistent with the hypothesis that two million years ago hominids were scavengers rather than accomplished hunters. Animal carcasses, which contributed relatively little to the hominid diet, were not systematically cut up and transported for sharing at base camps. Man the Hunter may not have appeared until 1.5 to 0.7 million years ago, when we do see a shift toward omnivory, with a greater proportion of meat in the diet. This more heroic ancestor may have been *Homo erectus,* equipped with Acheulean-style stone tools and, increasingly, fire. If we wish to look further back, we may have to become accustomed to a less flattering image of our heritage.

New Clues to the History Of Male and Female Size Differences

By John Noble Wilford

AMONG other differences more obvious and beguiling, women on average are shorter and weigh less than men, as the most casual admirer must have observed. It is a striking example of what scientists call sexual dimorphism, the phenomenon in many species of male-female physical differences that go beyond those directly linked to reproduction.

Dimorphism in primates is especially pronounced in gorillas and orangutans; the males are almost twice the size of females. Male chimpanzees are about 35 percent larger than females, which may also have been the size difference among Lucy and her kind, the early human ancestors known as australopithecines.

A discovery in Spain sheds light on an old question.

The celebrated Lucy, a fossil female from 3.2 million years ago, was a diminutive adult, 3 feet 7 inches tall and no more than 60 pounds. Another skeleton found in related African fossil beds, presumably that of a male, measured 5 feet 3 inches and 110 pounds. By contrast, modern humans are not only bigger, but their body-size dimorphism has declined. On the whole, men today are only about 15 to 20 percent heavier and 5 to 12 percent taller than women.

That raises a question that has troubled paleoanthropologists for a long time: When and why did sexual dimorphism in humans diminish to the present level? No one knows the answer, but a new study of Spanish fossils shows that the change occurred much earlier than once supposed, well before the emergence of modern Homo sapiens or even the Neanderthals.

In an article in the current issue of the journal Science, a team of Spanish paleontologists led by Dr. Juan Luis Arsuaga of the Complutense University of Madrid reported findings showing that the ratio of male–female sizes of Neanderthal ancestors 300,000 years ago was no different from what it is among modern humans today.

"We know for the first time with any confidence the degree of sexual dimorphism in one fossil hominid species, compared with modern humans," Dr. Arsuaga said in a telephone interview. "Before, it was not known because we lacked large enough samples that we were sure belonged to a single population."

The research was based on an examination of a rich fossil lode at the Sima de los Huesos site in the Atapuerca Mountains of northern Spain, near Burgos. It was the second major fossil discovery in the region in the last two years. In 1995, another group of Spanish paleontologists found fossils and stone tools from human ancestors who inhabited the Gran Dolina cavern at least 780,000 years ago, the earliest evidence of European colonization by human ancestors.

Dr. Arsuaga's team collected fossils of at least 32 individuals who occupied another cave there in the Middle Pleistocene geological period. These were the remains of Homo heidelbergensis. Scientists consider the species to be an-cestors of Neanderthals, the prototypical "cave men" who became extinct about 30,000 years ago, and to share a common ancestor with Homo sapiens some half a million years ago.

Many scientists had assumed that Neanderthals and other prehistoric members of the human family tree in Europe had a greater sexual dimorphism than modern humans. After studying the craniums and skeletal bones, the Spanish paleontologists determined that the fossils did not "show an unusual size variation compared with the distribution of samples of the same size randomly generated from large samples of modern humans."

In search of the origins of human monogamy.

Dr. Erik Trinkaus, a paleoanthropologist at Washington University in St. Louis, said the findings appeared to be valid, though not surprising. His own research has established, he said, no difference in sexual dimorphism between Neanderthals and modern humans. The new study extends that conclusion back to a somewhat earlier time and a different species, H. heidelbergensis.

Why does it matter, knowing the degree of sexual dimorphism in human evolutionary history?

Dr. Bernard A. Wood, a specialist in human evolution at George Washington University in Washington, said that a better understanding of body-size vari-

VIVE LA DIFFERENCE, MORE OR LESS

The existence of a size difference between the sexes in a species, called sexual dimorphism, can be relatively small or pronounced. Now scientists have evidence that the small difference of today has existed much longer than previously supposed.

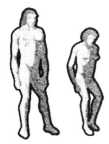

LUCY (3.2 million years ago)

Lucy was a member of the Australopithecus afarensis branch of the human family tree, which flourished from 4 million to 3 million years ago. Males were about 35 percent larger than females in her day, about the same dimorphism that exists in chimpanzees today.

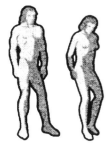

HEIDELBERG (300,000 years ago)

The species Homo heidelbergensis was ancestral to the Neanderthals. New findings in Spain show that, on average, males were about 15 percent larger than females. This deviation is comparable to the male-female size differences of today.

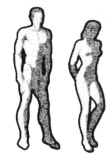

MODERN HUMANS

Men are 15 percent larger than women, on average. Scientists still don't know when dimorphism began to diminish from its early high level, which could indicate the timing of important shifts in early sexual behavior and family life.

Source: Dr. Juan Luis Arsuaga/Complutense University
The New York Times

ations in human ancestral species should help fossil hunters interpret their finds. Are the differences in the size of skeletons within the range of sexual dimorphism? If so, in the absence of any clearer evidence like pelvic bones, the large adults can be assumed to be male and the smaller ones, female. But if the size differences exceed the determined limits of intraspecies variation, then they are more likely to represent differences in species, not males and females.

Such a controversy has swirled around the Lucy skeleton. She was so small compared with many of the other skeletons lumped in the same species, Australopithecus afarensis, which lived about four million to three million years ago that some paleontologists argued that the fossils may represent several species, not one.

As it is, Dr. Trinkaus said, scientists are not sure just how dimorphic A. afarensis was. As much as chimpanzees—35 percent? So it seems, but the estimate could be skewed by a case of circular reasoning. If they cannot tell otherwise, paleontologists classify all large skeletons as male and smaller ones as female, which could shift estimates of size variation to a higher level of sexual dimorphism.

Because the level of dimorphism has implications for behavior, the knowledge can be the basis for making careful inferences about the social life of ancient ancestors. In living nonhuman primates, for example, those that are highly dimorphic, like gorillas, tend to be polygynous species, not monogamous.

Males fight with each other for sexual access to females, and the larger, stronger ones would presumably have an advantage and thus pass on more of their genes to succeeding generations.

While this would have favored the continuation of large males, it presumably would also have tended to produce larger females over time, which has been the long-term trend. Scientists note that the gap in male-female sizes in the human lineage has been closing less as a result of the slight increases in male stature than as a result of the tremendous leap in female size.

"By determining the degree of sexual dimorphism," Dr. Arsuaga said, "we can possibly infer that social aspects of life 300,000 years ago would be more similar to modern humans than we used to think."

Likewise, if scientists could identify the time when dimorphism diminished toward current levels, they might be able to mark the fateful transition in human sexual and family life, when humans began to bond, as a rule, with only one partner. Perhaps then large size ceased to be such an overwhelming advantage for males, and the level of sexual dimorphism began gradually to decline.

Dr. Arsuaga speculated that the transition probably occurred about two million years ago with the first Homo species that began migrating out of Africa. They were evolving larger brains and making more stone tools. But the few scraps of fossil bones from that time are insufficient to indicate any shift in dimorphism—nothing to match the finds at Sima de los Huesos.

The Spanish paleontologists said they have just begun to investigate the fossil riches there, with further explorations next summer likely to produce more discoveries.

"The fossil record here is so spectacular," Dr. Arsuaga said. "There are probably many more individual skeletons, so many that we can conduct even more detailed aspects of these peoples' lives, their growth, brain development, biomechanics, pathology and nutrition as a species. The sample is so great, it's almost like being able to study a living people."

Unit 5

Unit Selections

Key Points to Consider

❖ When, where, and why did *Homo erectus* evolve? Is it one species or two? Explain your reasoning.

❖ What evidence is there for hard times among the Neanderthals?

❖ What were Cro-Magnons trying to say or do with their cave art?

❖ When did language ability arise in our ancestry and why?

❖ How do we measure evolutionary time?

❖ Explain whether the Cro-Magnons were the first or the last modern humans to appear on Earth.

❖ What are the strengths and weaknesses of the "Eve hypothesis"?

❖ Are Neanderthals part of our ancestry?

❖ How would you draw the late hominid family tree?

❖ How do you assess the evidence for cannibalism?

 **Links**   **www.dushkin.com/online/**

25. **Archaeology Links (NC)**
    http://www.arch.dcr.state.nc.us/links.htm#stuff/
26. **Human Prehistory**
    http://users.hol.gr/~dilos/prehis.htm

These sites are annotated on pages 4 and 5.

The most important aspect of human evolution is also the most difficult to decipher from the fossil evidence: our development as sentient, social beings, capable of communicating by means of language.

We detect hints of incipient humanity in the form of crudely chipped tools, the telltale signs of a home base, or the artistic achievements of ornaments and cave art. Yet none of these indicators of a distinctly hominid way of life can provide us with the nuances of the everyday lives of these creatures, their social relations, or their supernatural beliefs, if any. Most of what remains is the rubble of bones and stones from which we interpret what we can of their lifestyle, thought processes, and ability to communicate. Our ability to glean from the fossil record is not completely without hope, however. In fact, informed speculation is what makes possible such essays as "Hard Times among the Neanderthals" by Erik Trinkaus, "Old Masters" by Pat Shipman and "The Gift of Gab" by Matt Cartmill. Each is a fine example of careful, systematic, and thought-provoking work that is based upon an increased understanding of hominid fossil sites as well as the more general environmental circumstances in which our predecessors lived.

Beyond the technological and anatomical adaptations, questions have arisen as to how our hominid forebears organized themselves socially and whether modern-day human behavior is inherited as a legacy of our evolutionary past or is a learned product of contemporary circumstances. Attempts to address these questions have given rise to the technique referred to as the "ethnographic analogy."

This is a method whereby anthropologists use "ethnographies" or field studies of modern-day hunters and gatherers whose lives we take to be the best approximations we have to what life might have been like for our ancestors. Granted, these contemporary foragers have been living under conditions of environmental and social change just as industrial peoples have. Nevertheless, it seems that, at least in some aspects of their lives, they have not changed as much as we have. So, if we are to make any enlightened assessments of prehistoric behavior patterns, we are better off looking at them than at ourselves.

As if to show that controversial interpretations of the evidence are not limited to the earlier hominid period (see unit 2), in this unit we see how the question of cannibalism has arisen anew ("Archaeologists Rediscover Cannibals" by Ann Gibbons) and how long-held beliefs about *Homo erectus* are being threatened by new fossil evidence (see James Shreeve's essay "*Erectus* Rising"). We also consider new evidence bearing upon the "Eve hypothesis," as addressed in "The Dating Game" and "The Neanderthal Peace," both articles by James Shreeve, and "Learning to Love Neanderthals" by Robert Kunzig. In the case of the "Eve hypothesis," the issue of when and where the family tree of modern humans actually began has pitted the bone experts, on the one hand, against a new type of anthropologist who specializes in molecular biology, on the other. For some scientists, the new evidence fits in quite comfortably with previously held positions; for others it seems that reputations, as well as theories, are at stake.

Late Hominid Evolution

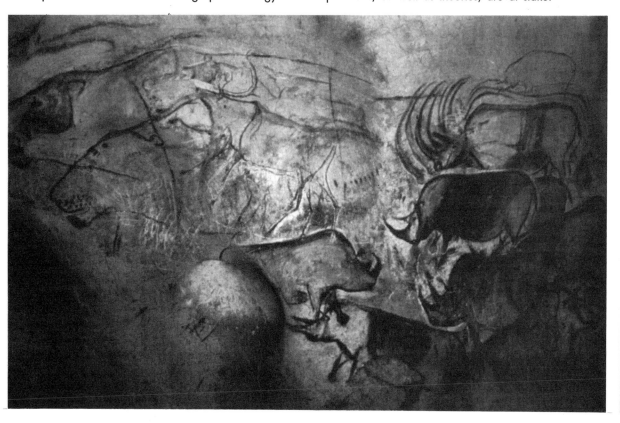

Erectus Rising

Oh No. Not This. The Hominids Are Acting Up Again . . .

James Shreeve

James Shreeve is the coauthor, with anthropologist Donald Johanson, of Lucy's Child: The Discovery of a Human Ancestor. *His book,* The Neandertal Enigma: Solving the Mystery of Modern Human Origins, *was published in 1995, and he is at work on a novel that a reliable source calls "a murder thriller about the species question."*

Just when it seemed that the recent monumental fuss over the origins of modern human beings was beginning to quiet down, an ancient ancestor is once more running wild. Trampling on theories. Appearing in odd places, way ahead of schedule. Demanding new explanations. And shamelessly flaunting its contempt for conventional wisdom in the public press.

The uppity ancestor this time is *Homo erectus*—alias Java man, alias Peking man, alias a mouthful of formal names known only to the paleontological cognoscenti. Whatever you call it, *erectus* has traditionally been a quiet, average sort of hominid: low of brow, thick of bone, endowed with a brain larger than that of previous hominids but smaller than those that followed, a face less apelike and projecting than that of its ancestors but decidedly more simian than its descendants'. In most scenarios of human evolution, *erectus*'s role was essentially to mark time—a million and a half years of it—between its obscure, presumed origins in East Africa just under 2 million years ago and its much more recent evolution into something deserving the name *sapiens*.

Erectus accomplished only two noteworthy deeds during its long tenure on Earth. First, some 1.5 million years ago, it developed what is known as the Acheulean stone tool culture, a technology exemplified by large, carefully crafted tear-shaped hand axes that were much more advanced than the bashed rocks that had passed for tools in the hands of earlier hominids. Then, half a million years later, and aided by those Acheulean tools, the species carved its way out of Africa and established a human presence in other parts of the Old World. But most of the time, *Homo erectus* merely existed, banging out the same stone tools millennium after millennium, over a time span that one archeologist has called "a period of unimaginable monotony."

Or so read the old script. These days, *erectus* has begun to ad-lib a more vigorous, controversial identity for itself. Research within the past year has revealed that rather than being 1 million years old, several *erectus* fossils from Southeast Asia are in fact almost 2 million years old. That is as old as the oldest African members of the species, and it would mean that *erectus* emerged from its home continent much earlier than has been thought—in fact, almost immediately after it first appeared. There's also a jawbone, found in 1991 near the Georgian city of Tbilisi, that resembles *erectus* fossils from Africa and may be as old as 1.8 million years, though that age is still in doubt. These new dates—and the debates they've engendered—have shaken *Homo erectus* out of its interpretive stupor, bringing into sharp relief just how little agreement there is on the rise and demise of the last human species on Earth, save one.

"Everything now is in flux," says Carl Swisher of the Berkeley Geochronology Center, one of the prime movers behind the redating of *erectus* outside Africa. "It's all a mess."

Asian and African fossils were lumped into one farflung taxon, a creature not quite like us but human enough to be welcomed into our genus: Homo erectus.

The focal point for the flux is the locale where the species was first found: Java. The rich but frustration-soaked history of paleoanthropology on that tropical island began just over 100 years ago, when a young Dutch anatomy professor named Eugène Dubois conceived the idée fixe that the "missing link" between ape and man was to be found in the jungled remoteness of the Dutch East Indies. Dubois had never left Holland, much less traveled to the Dutch East Indies, and his pick for the spot on Earth where humankind first arose owed as much to a large part of the Indonesian archipelago's being a Dutch colony as it did to any scientific evidence. He nevertheless found this missing link—the top of an oddly thick skull with massive browridges—in 1891 on the banks of the Solo River, near a community called Trinil in central Java. About a year later a thighbone that Dubois thought might belong to the same individual was found nearby; it looked so much like a modern human thighbone that Dubois assumed this ancient primate had walked upright. He christened the creature *Pithecanthropus erectus*—"erect ape-man"—and returned home in triumph.

Finding the fossil proved to be the easy part. Though Dubois won popular acclaim, neither he nor his "Java man" received the full approbation of the anatomists of the day, who considered his ape-man either merely an ape or merely a man. In an apparent pique, Dubois cloistered away the fossils for a quarter-century, refusing others the chance to view his prized possessions. Later, other similarly primitive human remains began to turn up in China and East Africa. All shared a collection of anatomical traits, including a long, low braincase with prominent browridges and a flattened forehead; a sharp angle to the back of the skull when viewed in profile; and a deep, robustly built jaw showing no hint of a chin. Though initially given separate regional names, the fossils were eventually lumped together into one far-flung taxon, a creature not quite like us but human enough to be welcomed into our genus: *Homo erectus*.

Over the decades the most generous source of new *erectus* fossils has been the sites on or near the Solo River in Java. The harvest continues: two more skulls, including one of the most complete *erectus* skulls yet known, were found at a famous fossil site called Sangiran just in the past year. Though the Javan yield of ancient humans has been rich, something has always been missing—the crucial element of time. Unless the age of a fossil can be determined, it hangs in limbo, its importance and place in the larger scheme of human evolution forever undercut with doubt. Until researchers can devise better methods for dating bone directly—right now there are no techniques that can reliably date fossilized, calcified bone more than 50,000 years old—a specimen's age has to be inferred from the geology that surrounds it. Unfortunately, most of the discoveries made on the densely populated and cultivated island of Java have been made not by trained excavators but by sharp-eyed local farmers who spot the bones as they wash out with the annual rains and later sell them. As a result, the original location of many a prized specimen, and thus all hopes of knowing its age, are a matter of memory and word of mouth.

Despite the problems, scientists continue to try to pin down dates for Java's fossils. Most have come up with an upper limit of around 1 million years. Along with the dates for the Peking man skulls found in China and the Acheulean tools from Europe, the Javan evidence has come to be seen as confirmation that *erectus* first left Africa at about that time.

There are those, however, who have wondered about these dates for quite some time. Chief among them is Garniss Curtis, the founder of the Berkeley Geochronology Center. In 1971 Curtis, who was then at the University of California

By the early 1970s most paleontologists were firmly wedded to the idea that Africa was the only human-inhabited part of the world until one million years ago.

at Berkeley, attempted to determine the age of a child's skull from a site called Mojokerto, in eastern Java, by using the potassium-argon method to date volcanic minerals in the sediments from which the skull was purportedly removed. Potassium-argon dating had been in use since the 1950s, and Curtis had been enormously successful with it in dating ancient African hominids—including Louis Leakey's famous hominid finds at Olduvai Gorge in Tanzania. The method takes advantage of the fact that a radioactive isotope of potassium found in volcanic ash slowly and predictably decays over time into argon gas, which becomes trapped in the crystalline structure of the mineral. The amount of argon contained in a given sample, measured against the amount of the potassium isotope, serves as a kind of clock that tells how much time has passed since a volcano exploded and its ash fell to earth and buried the bone in question.

Applying the technique to the volcanic pumice associated with the skull from Mojokerto, Curtis got an extraordinary age of 1.9 million years. The wildly anomalous date was all too easy to dismiss, however. Unlike the ash deposits of East Africa, the volcanic pumices in Java are poor in potassium. Also, not unexpectedly, a heavy veil of uncertainty obscured the collector's memories of precisely where he had found the fossil some 35 years earlier. Besides, most paleontologists were by this time firmly wedded to the idea that Africa was the only human-inhabited part of the world until 1 million years ago. Curtis's date was thus deemed wrong for the most stubbornly cherished of reasons: because it couldn't possibly be right.

In 1992 Curtis—under the auspices of the Institute for Human Origins in Berkeley—returned to Java with his colleague Carl Swisher. This time he was backed up by far more sensitive equipment and a powerful refinement in the dating technique. In conventional potassium-argon dating, several grams' worth of volcanic crystals gleaned from a site are needed to run a single experiment. While the bulk of these crystals are probably from the eruption that covered the fossil, there's always the possibility that other materials, from volcanoes millions of years older, have gotten mixed in and will thus make the fossil appear to be much older than it actually is. The potassium-argon method also requires that the researcher divide the sample of crystals in two. One half is dissolved in acid and passed through a flame; the wavelengths of light emitted tell how much potassium is in the sample. The other half is used to measure the amount of argon gas that's released when the crystals are heated. This two-step process further increases the chance of error, simply by giving the experiment twice as much opportunity to go wrong.

The refined technique, called argon-argon dating, neatly sidesteps most of these difficulties. The volcanic crystals are first placed in a reactor and bombarded with neutrons; when one of these neutrons penetrates the potassium nucleus, it displaces a proton, converting the potassium into an isotope of argon that doesn't occur in nature. Then the artificially created argon and the naturally occurring argon are measured in a single experiment. Because the equipment used to measure the isotopes can look for both types of argon at the same time, there's no need to divide the sample, and so the argon-argon method can produce clear results from tiny amounts of material.

In some cases—when the volcanic material is fairly rich in potassium—all the atoms of argon from a single volcanic crystal can be quick-released by the heat from a laser beam and then counted. By doing a number of such single-crystal experiments, the researchers can easily pick out and discard any data from older, contaminant crystals. But even when the researchers are forced to sample more than one potassium-poor crystal to get any reading at all—as was the case at Mojokerto—the argon-argon method can still produce a highly reliable age. In this case, the researchers carefully heat a few crystals at a time to higher and higher temperatures, using a precisely controlled laser. If all the crystals in a sample are the same age, then the amount of argon released at each temperature will be the same. But if contaminants are mixed in, or if severe weathering has altered the crystal's chemical composition, the argon measurements will be erratic, and the researchers will know to throw out the results.

Curtis and Swisher knew that in the argon-argon step-heating method they had the technical means to date the potassium-poor deposits at Mojokerto accurately. But they had no way to prove that those deposits were the ones in which the skull had been buried: all they had was the word of the local man who had found it. Then, during a visit to the museum in the regional capital, where the fossil was being housed, Swisher noticed something odd. The hardened sediments that filled the inside of the fossil's

braincase looked black. But back at the site, the deposits of volcanic pumice that had supposedly sheltered the infant's skull were whitish in color. How could a skull come to be filled with black sediments if it had been buried in white ones? Was it possible that the site and the skull had nothing to do with each other after all? Swisher suspected something was wrong. He borrowed a penknife, picked up the precious skull, and nicked off a bit of the matrix inside.

"I almost got kicked out of the country at that point," he says. "These fossils in Java are like the crown jewels."

Luckily, his impulsiveness paid off. The knife's nick revealed white pumice under a thin skin of dark pigment: years earlier, someone had apparently painted the surface of the hardened sediments black. Since there were no other deposits within miles of the purported site that contained a white pumice visually or chemically resembling the matrix in the skull, its tie to the site was suddenly much stronger. Curtis and Swisher returned to Berkeley with pumice from that site and within a few weeks proclaimed the fossil to be 1.8 million years old, give or take some 40,000 years. At the same time, the geochronologists ran tests on pumice from the lower part of the Sangiran area, where erectus facial and cranial bone fragments had been found. The tests yielded an age of around 1.6 million years. Both numbers obviously shatter the 1-million-year barrier for erectus outside Africa, and they are a stunning vindication of Curtis's work at Mojokerto 20 years ago. "That was very rewarding," he says, "after having been told what a fool I was by my colleagues."

While no one takes Curtis or Swisher for a fool now, some of their colleagues won't be fully convinced by the new dates until the matrix inside the Mojokerto skull itself can be tested. Even then, the possibility will remain that the skull may have drifted down over the years into deposits containing older volcanic crystals that have nothing to do with its original burial site, or that it was carried by a river to another, older site. But Swisher contends that the chance of such an occurrence is remote: it would have to have happened at both Mojokerto and Sangiran for the

fossils' ages to be refuted. "I feel really good about the dates," he says. "But it has taken me a while to understand their implications."

The implications that can be spun out from the Javan dates depend on how one chooses to interpret the body of fossil evidence commonly embraced under the name Homo erectus. The earliest African fossils traditionally attributed to erectus are two nearly complete skulls from the site of Koobi Fora in Kenya, dated between 1.8 and 1.7 million years old. In the conventional view, these early specimens evolved from a more primitive, smaller-brained ancestor called Homo habilis, well represented by bones from Koobi Fora, Olduvai Gorge, and sites in South Africa.

If this conventional view is correct, then the new dates mean that erectus must have migrated out of Africa very soon after it evolved, quickly reaching deep into the farthest corner of Southeast Asia. This is certainly possible: at the time, Indonesia was connected to Asia by lower sea levels—thus providing an overland route from Africa—and Java is just 10,000 to 15,000 miles from Kenya, depending on the route. Even if erectus traveled just one mile a year, it would still take no more than 15,000 years to reach Java—a negligible amount of evolutionary time.

If erectus did indeed reach Asia almost a million years earlier than thought, then other, more controversial theories become much more plausible. Although many anthropologists believe that the African and Asian erectus fossils all represent a single species, other investigators have recently argued that the two groups are too different to be so casually lumped together. According to paleoanthropologist Ian Tattersall of the American Museum of Natural History in New York, the African skulls traditionally assigned to erectus often lack many of the specialized traits that were originally used to define the species in Asia, including the long, low cranial structure, thick skull bones, and robustly built faces. In his view, the African group deserves to be placed in a separate species, which he calls Homo ergaster.

MEANWHILE, IN SIBERIA...

The presence of *Homo erectus* in Asia twice as long ago as previously thought has some people asking whether the human lineage might have originated in Asia instead of Africa. This long-dormant theory runs contrary to all current thinking about human evolution and lacks an important element: evidence. Although the new Javan dates do place the species in Asia at around the same time it evolved in Africa, all confirmed specimens of other, earlier hominids—the first members of the genus *Homo,* for instance, and the australopithecines, like Lucy—have been found exclusively in Africa. Given such an overwhelming argument, most investigators continue to believe that the hominid line began in Africa.

Most, but not all. Some have begun to cock an ear to the claims of Russian archeologist Yuri Mochanov. For over a decade Mochanov has been excavating a huge site on the Lena River in eastern Siberia—far from Africa, Java, or anywhere else on Earth an ancient hominid bone has ever turned up. Though he hasn't found any hominid fossils in Siberia, he stubbornly believes he's uncovered the next best thing: a trove of some 4,000 stone artifacts—crudely made flaked tools, but tools nonetheless—that he maintains are at least 2 million years old, and possibly 3 million. This, he says, would mean that the human lineage arose not in tropical Africa but in the cold northern latitudes of Asia.

"For evolutionary progress to occur, there had to be the appearance of new conditions: winter, snow, and, accompanying them, hunger," writes Mochanov. "[The ancestral primates] had to learn to walk on the ground, to change their carriage, and to become accustomed to meat—that is, to become 'clever animals of prey.'" And to become clever animals of prey, they'd need tools.

Although he is a well-respected investigator, Mochanov has been unable to convince either Western anthropologists or his Russian colleagues of the age of his site. Until recently the chipped rocks he was holding up as human artifacts were simply dismissed as stones broken by natural processes, or else his estimate of the age of the site was thought to be wincingly wrong. After all, no other signs of human occupation of Siberia appear until some 35,000 years ago.

But after a lecture swing through the United States earlier this year—in which he brought more data and a few prime examples of the tools for people to examine and pass around—many archeologists concede that it is difficult to explain the particular pattern of breakage of the rocks by any known natural process. "Everything I have heard or seen about the context of these things suggests that they are most likely tools," says anthropologist Rick Potts of the Smithsonian Institution, which was host to Mochanov last January.

They're even willing to concede that the site might be considerably older than they'd thought, though not nearly as old as Mochanov estimates. (To date the site, Mochanov compared the tools with artifacts found early in Africa; he also em-

ployed an arcane dating technique little known outside Russia.) Preliminary results from an experimental dating technique performed on soil samples from the site by Michael Waters of Texas A&M and Steve Forman of Ohio State suggest that the layer of sediment bearing the artifacts is some 400,000 years old. That's a long way from 2 million, certainly, but it's still vastly older than anything else found in Siberia—and the site is 1,500 miles farther north than the famous Peking man site in China, previously considered the most northerly home of *erectus.*

"If this does turn out to be 400,000 years old, it's very exciting," says Waters. "If people were able to cope and survive in such a rigorous Arctic environment at such an early time, we would have to completely change our perception of the evolution of human adaptation."

"I have no problem with hominids being almost anywhere at that age—they were certainly traveling around," says Potts. "But the environment is the critical thing. If it was really cold up there"—temperatures in the region now often reach −50 degrees in deep winter—"we'd all have to scratch our heads over how these early hominids were making it in Siberia. There is no evidence that Neanderthals, who were better equipped for cold than anyone, were living in such climates. But who knows? Maybe a population got trapped up there, went extinct, and Mochanov managed to find it." He shrugs. "But that's just arm waving."

—*J. S.*

Most anthropologists believe that the only way to distinguish between species in the fossil record is to look at the similarities and differences between bones; the age of the fossil should not play a part. But age is often hard to ignore, and Tattersall believes that the new evidence for what he sees as two distinct populations living at the same time in widely separate parts of the Old World is highly suggestive. "The new dates help confirm that these were indeed two different species," he says. "In my view, *erectus* is a separate variant that evolved only in Asia."

Other investigators still contend that the differences between the African and Asian forms of *erectus* are too minimal to merit placing them in separate species. But if Tattersall is right, his theory

raises the question of who the original emigrant out of Africa really was. *Homo ergaster* may have been the one to make the trek, evolving into *erectus* once it was established in Asia. Or perhaps a population of some even more primitive, as-yet-unidentified common ancestor ventured forth, giving rise to *erectus* in Asia while a sister population evolved into *ergaster* on the home continent.

Furthermore, no matter who left Africa first, there's the question of what precipitated the migration, a question made even more confounding by the new dates. The old explanation, that the primal human expansion across the hem of the Old World was triggered by the sophisticated Acheulean tools, is no longer tenable with these dates, simply

because the tools had not yet been invented when the earliest populations would have moved out. In hindsight, that notion seems a bit shopworn anyway. Acheulean tools first appear in Africa around 1.5 million years ago, and soon after at a site in the nearby Middle East. But they've never been found in the Far East, in spite of the abundant fossil evidence for *Homo erectus* in the region.

Until now, that absence has best been explained by the "bamboo line." According to paleoanthropologist Geoffrey Pope of William Paterson College in New Jersey, *erectus* populations venturing from Africa into the Far East found the land rich in bamboo, a raw material more easily worked into cutting and butchering tools than recalcitrant stone.

Sensibly, they abandoned their less efficient stone industry for one based on the pliable plant, which leaves no trace of itself in the archeological record. This is still a viable theory, but the new dates from Java add an even simpler dimension to it: there are no Acheulean tools in the Far East because the first wave of *erectus* to leave Africa didn't have any to bring with them.

So what *did* fuel the quick-step migration out of Africa? Some researchers say the crucial development was not cultural but physical. Earlier hominids like *Homo habilis* were small-bodied creatures with more ape-like limb proportions, notes paleoanthropologist Bernard Wood of the University of Liverpool, while African *erectus* was built along more modern lines. Tall, relatively slender, with long legs better able to range over distance and a body better able to dissipate heat, the species was endowed with the physiology needed to free it from the tropical shaded woodlands of Africa that sheltered earlier hominids. In fact, the larger-bodied *erectus* would have required a bigger feeding range to sustain itself, so it makes perfect sense that the expansion out of Africa should begin soon after the species appeared. "Until now, one was always having to account for what kept *erectus* in Africa so long after it evolved," says Wood. "So rather than raising a problem, in some ways the new dates in Java solve one."

Of course, if those dates are right, the accepted time frame for human evolution outside the home continent is nearly doubled, and that has implications for the ongoing debate over the origins of modern human beings. There are two opposing theories. The "out of Africa" hypothesis says that *Homo sapiens* evolved from *erectus* in Africa, and then—sometime in the last 100,000 years—spread out and replaced the more archaic residents of Eurasia. The "multiregional continuity" hypothesis says that modern humans evolved from *erectus* stock in various parts of the Old World, more or less simultaneously and independently. According to this scenario, living peoples outside Africa should look for their most recent ancestors not in African fossils but in the anatomy of ancient fossils within their own region of origin.

As it happens, the multiregionalists have long claimed that the best evidence for their theory lies in Australia, which is generally thought to have become inhabited around 50,000 years ago, by humans crossing over from Indonesia. There are certain facial and cranial characteristics in modern Australian aborigines, the multiregionalists say, that can be traced all the way back to the earliest specimens of *erectus* at Sangiran—characteristics that differ from and precede those of any more recent, *Homo sapiens* arrival from Africa. But if the new Javan dates are right, then these unique characteristics, and thus the aborigines' Asian *erectus* ancestors, must have been evolving separately from the rest of humankind for almost 2 million years. Many anthropologists, already skeptical of the multiregionalists' potential 1-million-year-long isolation for Asian *erectus,* find a 2-million-year-long isolation exceedingly difficult to swallow. "Can anyone seriously propose that the lineage of Australian aborigines could go back that far?" wonders paleoanthropologist Chris Stringer of the Natural History Museum in London, a leading advocate of the out-of-Africa theory.

The multiregionalists counter that they've never argued for *complete* isolation—that there's always been some flow of genes between populations, enough interbreeding to ensure that clearly beneficial *sapiens* characteristics would quickly be conferred on peoples throughout the Old World. "Just as genes flow now from Johannesburg to Beijing and from Melbourne to Paris, they have been flowing that way ever since humanity evolved," says Alan Thorne of the Australian National University in Canberra, an outspoken multiregionalist.

Stanford archeologist Richard Klein, another out-of-Africa supporter, believes the evidence actually *does* point to just such a long, deep isolation of Asian populations from African ones. The fossil record, he says, shows that while archaic forms of *Homo sapiens* were developing in Africa, *erectus* was remaining much the same in Asia. In fact, if some *erectus* fossils from a site called Ngandong in Java turn out to be as young as 100,000 years, as some researchers believe, then *erectus* was still alive on Java at the same time that fully modern human beings were living in Africa and the Middle East. Even more important, Klein says, is the cultural evidence. That Acheulean tools never reached East Asia, even after their invention in Africa, could mean the inventors never reached East Asia either. "You could argue that the new dates show that until very recently there was a long biological and cultural division between Asia on one hand, and Africa and Europe on the other," says Klein. In other words, there must have been two separate lineages of *erectus,* and since there aren't two separate lineages of modern humans, one of those must have gone extinct: presumably the Asian lineage, hastened into oblivion by the arrival of the more culturally adept, tool-laden *Homo sapiens.*

Naturally this argument is anathema to the multiregionalists. But this tenacious debate is unlikely to be resolved without basketfuls of new fossils, new ways of interpreting old ones—and new dates. In Berkeley, Curtis and Swisher are already busy applying the argon-argon method to the Ngandong fossils, which could represent some of the last surviving *Homo erectus* populations on Earth. They also hope to work their radiometric magic on a key *erectus* skull from Olduvai Gorge. In the meantime, at least one thing has become clear: *Homo erectus,* for so long the humdrum hominid, is just as fascinating, contentious, and elusive a character as any other in the human evolutionary story.

FURTHER READING

Eugène Dubois & the Ape-Man from Java. Bert Theunissen. Kluwer Academic, 1989. When a Dutch army surgeon, determined to prove Darwin right, traveled to Java in search of the missing link between apes and humans, he inadvertently opened a paleontological Pandora's box. This is Dubois's story, the story of the discovery of *Homo erectus.*

Hard Times Among the Neanderthals

*Although life was difficult, these prehistoric people may not
have been as exclusively brutish as usually supposed*

Erik Trinkaus

Throughout the century that followed the discovery in 1856 of the first recognized human fossil remains in the Neander Valley (*Neanderthal* in German) near Düsseldorf, Germany, the field of human paleontology has been beset with controversies. This has been especially true of interpretations of the Neanderthals, those frequently maligned people who occupied Europe and the Near East from about 100,000 years ago until the appearance of anatomically modern humans about 35,000 years ago.

During the last two decades, however, a number of fossil discoveries, new analyses of previously known remains, and more sophisticated models for interpreting subtle anatomical differences have led to a reevaluation of the Neanderthals and their place in human evolution.

This recent work has shown that the often quoted reconstruction of the Neanderthals as semierect, lumbering caricatures of humanity is inaccurate. It was based on faulty anatomical interpretations that were reinforced by the intellectual biases of the turn of the century. Detailed comparisons of Neanderthal skeletal remains with those of modern humans have shown that there is nothing in Neanderthal anatomy that conclusively indicates locomotor, manipulative, intellectual, or linguistic abilities

inferior to those of modern humans. Neanderthals have therefore been added to the same species as ourselves—*Homo sapiens*—although they are usually placed in their own subspecies, *Homo sapiens neanderthalensis.*

Despite these revisions, it is apparent that there are significant anatomical differences between the Neanderthals and present-day humans. If we are to understand the Neanderthals, we must formulate hypotheses as to why they evolved from earlier humans about 100,000 years ago in Europe and the Near East, and why they were suddenly replaced about 35,000 years ago by peoples largely indistinguishable from ourselves. We must determine, therefore, the behavioral significance of the anatomical differences between the Neanderthals and other human groups, since it is patterns of successful behavior that dictate the direction of natural selection for a species.

In the past, behavioral reconstructions of the Neanderthals and other prehistoric humans have been based largely on archeological data. Research has now reached the stage at which behavioral interpretations from the archeological record can be significantly supplemented by analyses of the fossils themselves. These analyses promise to tell us

a considerable amount about the ways of the Neanderthals and may eventually help us to determine their evolutionary fate.

One of the most characteristic features of the Neanderthals is the exaggerated massiveness of their trunk and limb bones. All of the preserved bones suggest a strength seldom attained by modern humans. Furthermore, not only is this robustness present among the adult males, as one might expect, but it is also evident in the adult females, adolescents, and even children. The bones themselves reflect this hardiness in several ways.

First, the muscle and ligament attachment areas are consistently enlarged and strongly marked. This implies large, highly developed muscles and ligaments capable of generating and sustaining great mechanical stress. Secondly, since the skeleton must be capable of supporting these levels of stress, which are frequently several times as great as body weight, the enlarged attachments for muscles and ligaments are associated with arm and leg bone shafts that have been reinforced. The shafts of all of the arm and leg bones are modified tubular structures that have to absorb stress from bending and twisting without fracturing. When the habitual load on a bone

Reprinted with permission from *Natural History*, December 1978. © 1978 by the American Museum of Natural History.

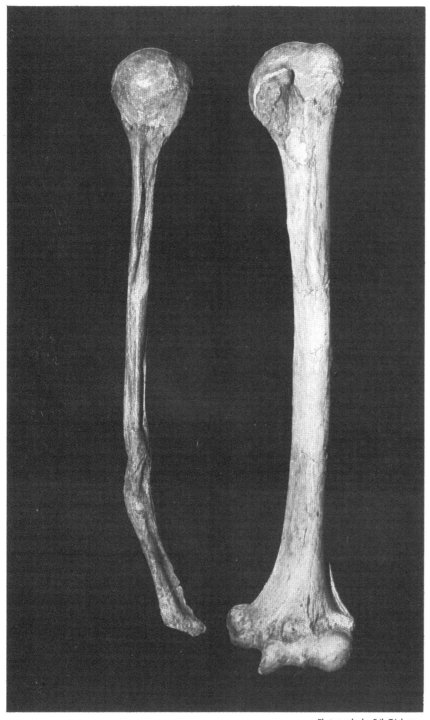

Photographs by Erik Trinkaus

Diagonal lines on these two arm bones from Shanidar 1 are healed fractures. The bone on the right is normal. That on the left is atrophied and has a pathological tip, caused by either amputation or an improperly healed elbow fracture.

the force per unit area of cartilage is reduced, decreasing the pressure on the cartilage.

Most of the robustness of Neanderthal arm bones is seen in muscle and ligament attachments. All of the muscles that go from the trunk or the shoulder blade to the upper end of the arm show massive development. This applies in particular to the muscles responsible for powerful downward movements of the arm and, to a lesser extent, to muscles that stabilize the shoulder during vigorous movements.

Virtually every major muscle or ligament attachment on the hand bones is clearly marked by a large roughened area or a crest, especially the muscles used in grasping objects. In fact, Neanderthal hand bones frequently have clear bony crests, where on modern human ones it is barely possible to discern the attachment of the muscle on the dried bone.

In addition, the flattened areas on the ends of the fingers, which provide support for the nail and the pulp of the finger tip, are enormous among the Neanderthals. These areas on the thumb and the index and middle fingers are usually two to three times as large as those of similarly sized modern human hands. The overall impression is one of arms to rival those of the mightiest blacksmith.

Neanderthal legs are equally massive; their strength is best illustrated in the development of the shafts of the leg bones. Modern human thigh and shin bones possess characteristic shaft shapes adapted to the habitual levels and directions of the stresses acting upon them. The shaft shapes of the Neanderthals are similar to those in modern humans, but the cross-sectional areas of the shafts are much greater. This implies significantly higher levels of stress.

Further evidence of the massiveness of Neanderthal lower limbs is provided by the dimensions of their knee and ankle joints. All of these are larger than in modern humans, especially with respect to the overall lengths of the bones.

The development of their limb bones suggests that the Neanderthals frequently generated high levels of mechanical stress in their limbs. Since most mechanical stress in the body is pro-

increases, the bone responds by laying down more bone in those areas under the greatest stress.

In addition, musculature and body momentum generate large forces across the joints. The cartilage, which covers joint surfaces, can be relatively easily overworked to the point where it degenerates, as is indicated by the prevalence of arthritis in joints subjected to significant wear and tear over the years. When the surface area of a joint is increased,

duced by body momentum and muscular contraction, it appears that the Neanderthals led extremely active lives. It is hard to conceive of what could have required such exertion, especially since the maintenance of vigorous muscular activity would have required considerable expenditure of energy. That level of energy expenditure would undoubtedly have been maladaptive had it not been necessary for survival.

The available evidence from the archeological material associated with the Neanderthals is equivocal on this matter. Most of the archeological evidence at Middle Paleolithic sites concerns stone

The ankle and big toe of Shanidar 1's right foot show evidence of arthritis, which suggests an injury to those parts. The left foot is normal though incomplete.

tool technology and hunting activities. After relatively little change in technology during the Middle Paleolithic (from about 100,000 years to 35,000 years before the present), the advent of the Upper Paleolithic appears to have brought significant technological advances. This transition about 35,000 years ago is approximately coincident with the replacement of the Neanderthals by the earliest anatomically modern humans. However, the evidence for a significant change in hunting patterns is not evident in the animal remains left behind. Yet even if a correlation between the robustness of body build and the level of hunting ef-

ficiency could be demonstrated, it would only explain the ruggedness of the Neanderthal males. Since hunting is exclusively or at least predominantly a male activity among humans, and since Neanderthal females were in all respects as strongly built as the males, an alternative explanation is required for the females.

Some insight into why the Neanderthals consistently possessed such massiveness is provided by a series of partial skeletons of Neanderthals from the Shanidar Cave in northern Iraq. These fossils were excavated between 1953 and 1960 by anthropologist Ralph Solecki of Columbia University and have been studied principally by T. Dale Stewart, an anthropologist at the Smithsonian Institution, and myself. The most remarkable aspect of these skeletons is the number of healed injuries they contain. Four of the six reasonably complete adult skeletons show evidence of trauma during life.

The identification of traumatic injury in human fossil remains has plagued paleontologists for years. There has been a tendency to consider any form of damage to a fossil as conclusive evidence of prehistoric violence between humans if it resembles the breakage patterns caused by a direct blow with a heavy object. Hence a jaw with the teeth pushed in or a skull with a depressed fracture of the vault would be construed to indicate blows to the head.

The central problem with these interpretations is that they ignore the possibility of damage after death. Bone is relatively fragile, especially as compared with the rock and other sediment in which it is buried during fossilization. Therefore when several feet of sediment caused compression around fossil remains, the fossils will almost always break. In fact, among the innumerable cases of suggested violence between humans cited over the years, there are only a few exceptional examples that cannot be readily explained as the result of natural geologic forces acting after the death and burial of the individual.

One of these examples is the trauma of the left ninth rib of the skeleton of Shanidar 3, a partially healed wound inflicted by a sharp object. The implement

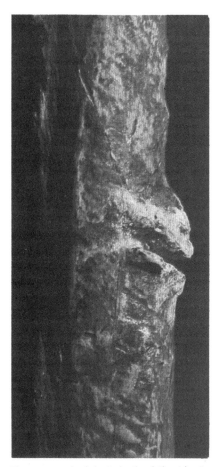

The scar on the left ninth rib of Shanidar 3 is a partially healed wound inflicted by a sharp object. This wound is one of the few examples of trauma caused by violence.

cut obliquely across the top of the ninth rib and probably pierced the underlying lung. Shanidar 3 almost certainly suffered a collapsed left lung and died several days or weeks later, probably as a result of secondary complications. This is deduced from the presence of bony spurs and increased density of the bone around the cut.

The position of the wound on the rib, the angle of the incision, and the cleanness of the cut make it highly unlikely that the injury was accidentally inflicted. In fact, the incision is almost exactly what would have resulted if Shanidar 3 had been stabbed in the side by a right-handed adversary in face-to-face conflict. This would therefore provide conclusive evidence of violence between humans, the *only* evidence so far found of such violence among the Neanderthals.

In most cases, however, it is impossible to determine from fossilized re-

mains the cause of an individual's death. The instances that can be positively identified as prehistoric traumatic injury are those in which the injury was inflicted prior to death and some healing took place. Shortly after an injury to bone, whether a cut or a fracture, the damaged bone tissue is resorbed by the body and new bone tissue is laid down around the injured area. As long as irritation persists, new bone is deposited, creating a bulge or spurs of irregular bone extending into the soft tissue. If the irritation ceases, the bone will slowly re-form so as to approximate its previous, normal condition. However, except for superficial injuries or those sustained during early childhood, some trace of damage persists for the life of the individual.

In terms of trauma, the most impressive of the Shanidar Neanderthals is the first adult discovered, known as Shanidar 1. This individual suffered a number of injuries, some of which may be related. On the right forehead there are scars from minor surface injuries, probably superficial scalp cuts. The outside of the left eye socket sustained a major blow that partially collapsed that part of the bony cavity, giving it a flat rather than a rounded contour. This injury possibly caused loss of sight in the left eye and pathological alterations of the right side of the body.

Shanidar 1's left arm is largely preserved and fully normal. The right arm, however, consists of a highly atrophied but otherwise normal collarbone and shoulder blade and a highly abnormal upper arm bone shaft. That shaft is atrophied to a fraction of the diameter of the left one but retains most of its original length. Furthermore, the lower end of the right arm bone has a healed fracture of the atrophied shaft and an irregular, pathological tip. The arm was apparently either intentionally amputated just above the elbow or fractured at the elbow and never healed.

This abnormal condition of the right arm does not appear to be a congenital malformation, since the length of the bone is close to the estimated length of the normal left upper arm bone. If, however, the injury to the left eye socket also affected the left side of the brain,

directly or indirectly, by disrupting the blood supply to part of the brain, the result could have been partial paralysis of the right side. Motor and sensory control areas for the right side are located on the left side of the brain, slightly behind the left eye socket. This would explain the atrophy of the whole right arm since loss of nervous stimulation will rapidly lead to atrophy of the affected muscles and bone.

The abnormality of the right arm of Shanidar 1 is paralleled to a lesser extent in the right foot. The right ankle joint shows extensive arthritic degeneration, and one of the major joints of the inner arch of the right foot has been completely reworked by arthritis. The left foot, however, is totally free of pathology. Arthritis from normal stress usually affects both lower limbs equally; this degeneration therefore suggests that the arthritis in the right foot is a secondary result of an injury, perhaps a sprain, that would not otherwise be evident on skeletal remains. This conclusion is supported by a healed fracture of the right fifth instep bone, which makes up a major portion of the outer arch of the foot. These foot pathologies may be tied into the damage to the left side of the skull; partial paralysis of the right side would certainly weaken the leg and make it more susceptible to injury.

The trauma evident on the other Shanidar Neanderthals is relatively minor by comparison. Shanidar 3, the individual who died of the rib wound, suffered debilitating arthritis of the right ankle and neighboring foot joints, but lacks any evidence of pathology on the left foot; this suggests a superficial injury similar to the one sustained by Shanidar 1. Shanidar 4 had a healed broken rib. Shanidar 5 received a transverse blow across the left forehead that left a large scar on the bone but does not appear to have affected the brain.

None of these injuries necessarily provides evidence of deliberate violence among the Neanderthals; all of them could have been accidentally self-inflicted or accidentally caused by another individual. In either case, the impression gained of the Shanidar Neanderthals is of a group of invalids. The crucial variable, however, appears to be age. All four of these individuals died at relatively advanced ages, probably between 40 and 60 years (estimating the age at death for Neanderthals beyond the age of 25 is extremely difficult); they therefore had considerable time to accumulate the scars of past injuries. Shanidar 2 and 6, the other reasonably complete Shanidar adults, lack evidence of trauma, but they both died young, probably before reaching 30.

Other Neanderthal remains, all from Europe, exhibit the same pattern. Every fairly complete skeleton of an elderly adult shows evidence of traumatic injuries. The original male skeleton from the Neander Valley had a fracture just below the elbow of the left arm, which probably limited movement of that arm for life. The "old man" from La Chapelle-aux-Saints, France, on whom most traditional reconstructions of the Neanderthals have been based, suffered a broken rib. La Ferrassi 1, the old adult male from La Ferrassie, France, sustained a severe injury to the right hip, which may have impaired his mobility.

In addition, several younger specimens and ones of uncertain age show traces of trauma. La Quina 5, the young adult female from La Quina, France, was wounded on her right upper arm. A young adult from Sala, Czechoslovakia, was superficially wounded on the right forehead just above the brow. And an individual of unknown age and sex from the site of Krapina, Yugoslavia, suffered a broken forearm, in which the bones never reunited after the fracture.

The evidence suggests several things. First, life for the Neanderthals was rigorous. If they lived through childhood and early adulthood, they did so bearing the scars of a harsh and dangerous life. Furthermore, this incident of trauma correlates with the massiveness of the Neanderthals; a life style that so consistently involved injury would have required considerable strength and fortitude for survival.

There is, however, another, more optimistic side to this. The presence of so many injuries in a prehistoric human group, many of which were debilitating

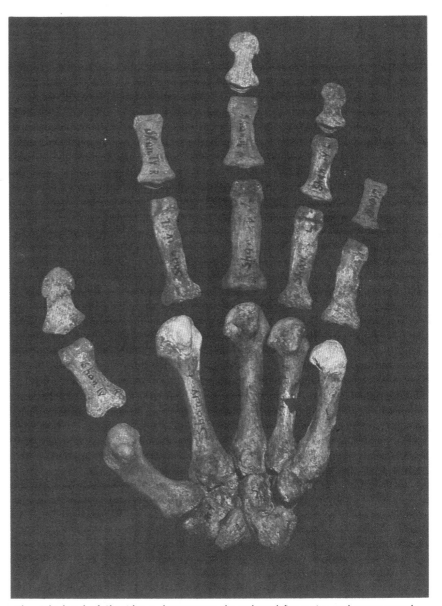

The right hand of Shanidar 4 demonstrates the enlarged finger tips and strong muscle markings characteristic of Neanderthal hands.

and sustained years before death, shows that individuals were taken care of long after their economic usefulness to the social group had ceased. It is perhaps no accident that among the Neanderthals, for the first time in human history, people lived to a comparatively old age. We also find among the Neanderthals the first intentional burials of the dead, some of which involved offerings. Despite the hardships of their life style, the Neanderthals apparently had a deep-seated respect and concern for each other.

Taken together, these different pieces of information paint a picture of life among the Neanderthals that, while harsh and dangerous, was not without personal security. Certainly the hardships the Neanderthals endured were beyond those commonly experienced in the prehistoric record of human caring and respect as well as of violence between individuals. Perhaps for these reasons, despite their physical appearance, the Neanderthals should be considered the first modern humans.

OLD MASTERS

Brilliant paintings brightened the caves of our early ancestors. But were the artists picturing their mythic beliefs or simply showing what they ate for dinner?

Pat Shipman

Pat Shipman wrote about killer bamboo in [Discover,] *February [1990].*

Fifty years ago, in a green valley of the Dordogne region of southwest France, a group of teenage boys made the first claustrophobic descent into the labyrinthine caverns of Lascaux. When they reached the main chamber and held their lamps aloft, the sight that flickered into view astonished them. There were animals everywhere. A frieze of wild horses, with chunky bodies and fuzzy, crew-cut manes, galloped across the domed walls and ceiling past the massive figure of a white bull-like creature (the extinct aurochs). Running helter-skelter in the opposing direction were three little stags with delicately drawn antlers. They were followed by more bulls, cows, and calves rounding the corner of the chamber.

Thousands have since admired these paintings in Lascaux's Hall of the Bulls, probably the most magnificent example of Ice Age art known to us today. In fact, by 1963 so many tourists wanted to view the cave that officials were forced to close Lascaux to the general public; the paintings were being threatened as the huge influx of visitors warmed the air in the cave and brought in corrosive algae and pollen. (Fortunately, a nearby exhibit called Lascaux II faithfully reproduces the paintings.) After I first saw these powerful images, they haunted me for several months. I had looked at photographs of Lascaux in books, of course, so I knew that the paintings were beautiful; but what I

didn't know was that they would reach across 17,000 years to grab my soul.

Lascaux is not an Ice Age anomaly. Other animal paintings, many exquisitely crafted, adorn hundreds of caves throughout the Dordogne and the French Pyrenees and the region known as Cantabria on the northern coast of Spain. All these images were created by the people we commonly call Cro-Magnons, who lived during the Upper Paleolithic Period, between 10,000 and 30,000 years ago, when Europe lay in the harsh grip of the Ice Age.

The image of Paleolithic humans moving by flickering lamps, singing, chanting, and drawing their knowledge of their world is hard to resist.

What did this wonderful art mean, and what does it tell us about the prehistoric humans who created it? These questions have been asked since the turn of the century, when cave paintings in Spain were first definitively attributed to Paleolithic humans. Until recently the dominant answers were based on rather sweeping symbolic interpretations—attempts, as it were, to read the Paleolithic psyche.

These days some anthropologists are adopting a more literal-minded approach. They are not trying to empathize with the artists' collective soul—a perilous exercise in imagination, consider-

ing how remote Cro-Magnon life must have been from ours. Rather, armed with the tools of the late twentieth century—statistics, maps, computer analyses of the art's distribution patterns—the researchers are trying to make sense of the paintings by piecing together their cultural context.

This is a far cry from earlier attempts at interpretation. At the beginning of the century, with little more to go on than his intuition, the French amateur archeologist Abbé Henri Breuil suggested that the pictures were a form of hunting magic. Painting animals, in other words, was a magical way of capturing them, in the hope that it would make the beasts vulnerable to hunters. Abstract symbols painted on the walls were interpreted as hunting paraphernalia. Straight lines drawn to the animals' sides represented spears, and V and O shapes on their hides were seen as wounds. Rectangular grids, some observers thought, might have been fences or animal traps.

In the 1960s this view was brushed aside for a much more complex, somewhat Freudian approach that was brought into fashion by anthropologist André Leroi-Gourhan. He saw the cave paintings as a series of mythograms, or symbolic depictions, of how Paleolithic people viewed their world— a world split between things male and female. Femaleness was represented by animals such as the bison and aurochs (which were sometimes juxtaposed with human female figures in the paintings), and maleness was embodied by such animals as the horse and ibex (which, when accompanied by human figures, were shown only with males). Female images, Leroi-Gourhan suggested, were clustered in the central parts of the dark, womblike caves, while male images either consorted with the female ones or encircled them in the more peripheral areas.

Leroi-Gourhan also ascribed sex to the geometric designs on the cave walls. Thin shapes such as straight lines, which often make up barbed, arrowlike structures, were seen as male (phallic) signs. Full shapes such as ovals, V shapes, triangles, and rectangles were female (vulval) symbols. Thus, an arrow stuck into a V-shaped wound on an animal's hide was a male symbol entering a complementary female one.

Leroi-Gourhan was the first to look for structure in the paintings systematically, and his work reinforced the notion that these cave paintings had underlying designs and were not simply idle graffiti or random doodles. Still, some scholars considered his *"perspective sexomaniaque"* rather farfetched; eventually even he played down some of the sexual interpretations. However, a far bigger problem with both his theory and Breuil's was their sheer monolithic scope: a single explanation was assumed to account for 20,000 years of paintings produced by quite widely scattered groups of people.

. . . animals may have been depicted more or less frequently depending on how aggressive they were to humans.

Yet it is at least as likely that the paintings carried a number of different messages. The images' meaning may have varied depending on who painted them and where. Increasingly, therefore, researchers have tried to relate the content of the paintings to their context—their distribution within a particular cave, the cave's location within a particular region, and the presence of other nearby dwelling sites, tools, and animal bones in the area.

Anthropologist Patricia Rice and sociologist Ann Paterson, both from West Virginia University, made good use of this principle in their study of a single river valley in the Dordogne region, an area that yielded 90 different caves containing 1,955 animal portrayals and 151 dwelling sites with animal bone deposits. They wanted to find out whether the number of times an animal was painted simply reflected how common it was or whether it revealed further information about the animals or the human artists.

By comparing the bone counts of the various animals—horses, reindeer, red deer, ibex, mammoths, bison, and aurochs—Rice and Paterson were able to score the animals according to their abundance. When they related this number to the number of times a species turned up in the art, they found an interesting relationship: Pictures of the smaller animals, such as deer, were proportionate to their bone counts. But the bigger species, such as horses and bison, were portrayed more often than you'd expect from the faunal remains. In fact, it turned out that to predict how often an animal would appear, you had to factor in not just its relative abundance but its weight as well.

A commonsense explanation of this finding was that an animal was depicted according to its usefulness as food, with the larger, meatier animals shown more often. This "grocery store" explanation of the art worked well, except for the ibex, which was portrayed as often as the red deer yet was only half its size and, according to the bone counts, not as numerous. The discrepancy led Rice and Paterson to explore the hypothesis that animals may have also been depicted more or less frequently depending on how aggressive they were to humans.

To test this idea, the researchers asked wildlife-management specialists to score the animals according to a "danger index." The feisty ibex, like the big animals, was rated as highly aggressive; and like these other dangerous animals, it was painted more often than just the numbers of its remains would suggest. Milder-tempered red deer and reindeer, on the other hand, were painted only about as often as you'd expect from their bones. Rice and Paterson concluded that the local artists may have portrayed the animals for both "gro-

cery store" and "danger index" reasons. Maybe such art was used to impress important information on the minds of young hunters—drawing attention to the animals that were the most worthwhile to kill, yet balancing the rewards of dinner with the risks of attacking a fearsome animal.

One thing is certain: Paleolithic artists knew their animals well. Subtle physical details, characteristic poses, even seasonal changes in coat color or texture, were deftly observed. At Lascaux bison are pictured shedding their dark winter pelts. Five stags are shown swimming across a river, heads held high above the swirling tide. A stallion is depicted with its lip curled back, responding to a mare in heat. The reddish coats, stiff black manes, short legs, and potbellies of the Lascaux horses are so well recorded that they look unmistakably like the modern Przhevalsky's horses from Mongolia.

New findings at Solutré, in east-central France, the most famous horse-hunting site from the Upper Paleolithic, show how intimate knowledge of the animal's habits was used to the early hunter's advantage. The study, by archeologist Sandra Olsen of the Virginia Museum of Natural History, set out to reexamine how vast numbers of horses—from tens to hundreds of thousands, according to fossil records—came to be killed in the same, isolated spot. The archeological deposits at Solutré are 27 feet thick, span 20,000 years, and provide a record of stone tools and artifacts as well as faunal remains.

The traditional interpretation of this site was lots of fun but unlikely. The Roche de Solutré is one of several high limestone ridges running east-west from the Saône River to the Massif Central plateau; narrow valleys run between the ridges. When the piles of bones were discovered, in 1866, it was proposed that the site was a "horse jump" similar to the buffalo jumps in the American West, where whole herds of bison were driven off cliffs to their death. Several nineteenth-century paintings depict Cro-Magnon hunters driving a massive herd of wild horses up and off the steep rock of Solutré. But Olsen's bone analysis has shown that the horse jump scenario is almost certainly wrong.

For one thing, the horse bones are not at the foot of the steep western end of Solutré, but in a natural cul-de-sac along the southern face of the ridge. For the horse jump hypothesis to work, one of two fairly incredible events had to occur. Either the hunters drove the horses off the western end and then dragged all the carcasses around to the southern face to butcher them or the hunters herded the animals up the steep slope and then forced them to veer off the southern side of the ridge. But behavioral studies show that, unlike bison, wild horses travel not in herds but in small, independent bands. So it would have been extremely difficult for our Cro-Magnons on foot to force lots of horses together and persuade them to jump en masse.

Instead the horse behavior studies suggested to Olsen a new hypothesis. Wild horses commonly winter in the lowlands and summer in the highlands. This migration pattern preserves their forage and lets them avoid the lowland's biting flies and heat in summer and the highland's cold and snow in winter. The Solutré horses, then, would likely have wintered in the Saône's floodplain to the east and summered in the mountains to the west, migrating through the valleys between the ridges. The kill site at Solutré, Olsen notes, lies in the widest of these valleys, the one offering the easiest passage to the horses. What's more, from the hunters' point of view the valley has a convenient cul-de-sac running off to one side. The hunters, she proposes, used a drive lane of brush, twigs, and rocks to divert the horses from their migratory path into the cul-de-sac and then speared the animals to death. Indeed, spear points found at the site support this scenario.

The bottom line in all this is that Olsen's detailed studies of this prehistoric hunting site confirm what the cave art implies: these early humans used their understanding of animals' habits, mating, and migration patterns to come up with extremely successful hunting strategies. Obviously this knowledge must have been vital to the survival of the group and essential to hand down to successive generations. Perhaps the animal friezes in the

cave were used as a mnemonic device or as a visual teaching aid in rites of initiation—a means for people to recall or rehearse epic hunts, preserve information, and school their young. The emotional power of the art certainly suggests that this information was crucial to their lives and could not be forgotten.

The transmission of this knowledge may well have been assisted by more than illustration. French researchers Iégor Reznikoff and Michel Dauvois have recently shown that cave art may well have been used in rituals accompanied by songs or chants. The two studied the acoustic resonances of three caves in the French Pyrenees by singing and whistling through almost five octaves as they walked slowly through each cave. At certain points the caves resonated in response to a particular note, and these points were carefully mapped.

When Reznikoff and Dauvois compared their acoustic map with a map of the cave paintings, they found an astonishing relationship. The best resonance points were all well marked with images, while those with poor acoustics had very few pictures. Even if a resonance point offered little room for a full painting, it was marked in some way—by a set of red dots, for example. It remains to be seen if this intriguing correlation holds true for other caves. In the meantime, the image of Paleolithic humans moving by flickering lamps, singing, chanting, and drawing their knowledge of their world indelibly into their memories is so appealing that I find it hard to resist.

Yet the humans in this mental image of mine are shadowy, strangely elusive people. For all the finely observed animal pictures, we catch only the sketchiest glimpses of humans, in the form of stick figures or stylized line drawings. Still, when Rice and Paterson turned to study these human images in French and Spanish caves, a few striking patterns did emerge. Of the 67 images studied, 52 were male and a mere 15 were female. Only men were depicted as engaged in active behavior, a category that included walking, running, carrying spears, being speared, or falling. Females were a picture of passivity; they

stood, sat, or lay prone. Most women were shown in close proximity to another human figure or group of figures, which were always other women. Seldom were men featured in social groups; they were much more likely to be shown facing off with an animal.

For all the finely observed animal pictures, we catch only the sketchiest glimpses of humans, in the form of stick figures or stylized line drawings.

These images offer tantalizing clues to Paleolithic life. They suggest a society where males and females led very separate lives. (Male-female couples do not figure at all in Paleolithic art, for all the sexual obsessions of earlier researchers.) Males carried out the only physical activities—or at least the only ones deemed worthy of recording. Their chief preoccupation was hunting, and from all appearances, what counted most was the moment of truth between man and his prey. What women did in Paleolithic society (other than bear children and gather food) remains more obscure. But whatever they did, they mostly did it in the company of other women, which would seem to imply that social interaction, cooperation, and oral communication played an important role in female lives.

If we could learn the sex of the artists, perhaps interpreting the social significance of the art would be easier. Were women's lives so mysterious because the artists were male and chauvinistically showed only men's activities in their paintings? Or perhaps the artists were all female. Is their passive group activity the recording and encoding of the information vital to the group's survival in paintings and carvings? Did they spend their time with other women, learning the songs and chants and the artistic techniques that transmitted and preserved their knowledge? The art that brightened the caves of the Ice Age endures. But the artists who might shed light on its meaning remain as enigmatic as ever.

The Gift *of* Gab

Grooves and holes in fossil skulls may reveal when our ancestors began to speak. The big question, though, is what drove them to it?

By Matt Cartmill

MATT CARTMILL is a professor at Duke, where he teaches anatomy and anthropology and studies animal locomotion. Cartmill is also the author of numerous articles and books on the evolution of people and other animals, including an award-winning book on hunting. A View to a Death in the Morning. He is the president of the American Association of Physical Anthropologists.

People can talk. Other animals can't. They can all communicate in one way or another—to lure mates, at the very least—but their whinnies and wiggles don't do the jobs that language does. The birds and beasts can use their signals to attract, threaten, or alert each other, but they can't ask questions, strike bargains, tell stories, or lay out a plan of action.

Those skills make *Homo sapiens* a uniquely successful, powerful, and dangerous mammal. Other creatures' signals carry only a few limited kinds of information about what's happening at the moment, but language lets us tell each other in limitless detail about what used to be or will be or might be. Language lets us get vast numbers of big, smart fellow primates all working together on a single task—building the Great Wall of China or fighting World War II or flying to the moon. It lets us construct and communicate the gorgeous fantasies of literature and the profound fables of myth. It lets us cheat death by pouring out our knowledge, dreams, and memories into younger people's minds. And it does powerful things for us inside our own minds because we do a lot of our thinking by talking silently to ourselves. Without language, we would be only a sort of upright chimpanzee with funny feet and clever hands. With it, we are the self-possessed masters of the planet.

How did such a marvelous adaptation get started? And if it's so marvelous, why hasn't any other species come up with anything similar? These may be the most important questions we face in studying human evolution. They are also the least understood. But in the past few years, linguists and anthropologists have been making some breakthroughs, and we are now beginning to have a glimmering of some answers.

COULD NEANDERTHALS talk? They seem to have had nimble tongues, but some scientists think the geometry of their throats prevented them from making many clear vowel sounds.

We can reasonably assume that by at least 30,000 years ago people were talking—at any rate, they were producing carvings, rock paintings, and jewelry, as well as ceremonial graves containing various goods. These tokens of art and religion are high-level forms of symbolic behavior, and they imply that the everyday symbol-handling machinery of human language must have been in place then as well.

Language surely goes back further than that, but archeologists don't agree on just how far. Some think that earlier, more basic human behaviors—hunting in groups, tending fires, making tools—also demanded language. Others think these activities are possible without speech. Chimpanzees, after all, hunt communally, and with human guidance they can learn to tend fires and chip flint.

Paleontologists have pored over the fossil bones of our ancient relatives in search of evidence for speech abilities. Because the most crucial organ for language is the brain, they have looked for signs in the impressions left by the brain on the inner surfaces of fossil skulls, particularly impressions made by parts of the brain called speech areas because damage to them can impair a person's ability to talk or understand language. Unfortunately, it turns out that you can't tell whether a fossil hominid was able to talk simply by looking at brain impressions on the inside of its skull. For one thing, the fit between the brain and the bony braincase is loose in people and other large mammals, and so the impressions we derive from fossil skulls are disappointingly fuzzy. Moreover, we now know that language functions are not tightly localized but spread across many parts of the brain.

Faced with these obstacles, researchers have turned from the brain to other organs used in speech, such as the throat and tongue. Some have measured the fossil skulls and jaws of early hominids, tried to reconstruct the shape of their vocal tracts, and then applied the laws of acoustics to them to see whether they might have been capable of producing human speech.

All mammals produce their vocal noises by contracting muscles that compress the rib cage. The air in the lungs is driven out through the windpipe to the larynx, where it flows between the vocal cords. More like flaps than cords, these structures vibrate in the breeze, producing a buzzing sound that becomes the voice. The human difference lies in what happens to the air after it gets past the vocal cords.

In people, the larynx lies well below the back of the tongue, and most of the air goes out through the mouth when we talk. We make only a few sounds by exhaling through the nose—for instance, nasal consonants like *m* or *n,* or the so-called nasal vowels in words like the French *bon* and *vin.* But in most mammals, including apes, the larynx sticks farther up behind the tongue, into the back of the nose, and most of the exhaled air passes out through the nostrils. Nonhuman mammals make mostly nasal sounds as a result.

At some point in human evolution the larynx must have descended from its previous heights, and this change had some serious drawbacks. It put the opening of the windpipe squarely in the path of descending food, making it dangerously easy for us to choke to death if a chunk of meat goes down the wrong way—something that rarely happens to a dog or a cat. Why has evolution exposed us to this danger?

Some scientists think that the benefits outweighed the risks, because lowering the larynx improved the quality of our vowels and made speech easier to understand. The differences between vowels are produced mainly by changing the size and shape of the airway between the tongue and the roof of the mouth. When the front of the tongue almost touches the palate, you get the *ee* sound

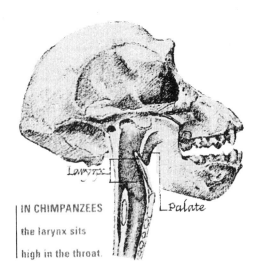

IN CHIMPANZEES the larynx sits high in the throat.

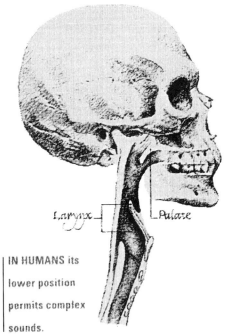

IN HUMANS its lower position permits complex sounds.

Illustrations by Dugald Stermer

in *beet;* when the tongue is humped up high in the back (and the lips are rounded), you get the *oo* sound in *boot,* and so on. We are actually born with a somewhat apelike throat, including a flat tongue and a larynx lying high up in the neck, and this arrangement makes a child's vowels sound less clearly separated from each other than an adult's.

Philip Lieberman of Brown University thinks that an ape-like throat persisted for some time in our hominid ancestors. His studies of fossil jaws and skulls persuade him that a more modern throat didn't evolve until some 500,000 years ago, and that some evolutionary lines in the genus *Homo* never did ac-

quire modern vocal organs. Lieberman concludes that the Neanderthals, who lived in Europe until perhaps 25,000 years ago, belonged to a dead-end lineage that never developed our range of vowels, and that their speech—if they had any at all—would have been harder to understand than ours. Apparently, being easily understood wasn't terribly important to them—not important enough, at any rate, to outweigh the risk of inhaling a chunk of steak into a lowered larynx. This suggests that vocal communication wasn't as central to their lives as it is to ours.

Many paleoanthropologists, especially those who like to see Neanderthals as a separate species, accept this story. Others have their doubts. But the study of other parts of the skeleton in fossil hominids supports some of Lieberman's conclusions. During the 1980s a nearly complete skeleton of a young *Homo* male was recovered from 1.5-million-year-old deposits in northern Kenya. Examining the vertebrae attached to the boy's rib cage, the English anatomist Ann MacLarnon discovered that his spinal cord was proportionately thinner in this region than it is in people today. Since that part of the cord controls most of the muscles that drive air in and out of the lungs, MacLarnon concluded that the youth may not have had the kind of precise neural control over breathing movements that is needed for speech.

This year my colleague Richard Kay, his student Michelle Balow, and I were able to offer some insights from yet another part of the hominid body. The tongue's movements are controlled almost solely by a nerve called the hypoglossal. In its course from the brain to the tongue, this nerve passes through a hole in the skull, and Kay, Balow, and I found that this bony canal is relatively big in modern humans—about twice as big in cross section as that of a like-size chimpanzee. Our larger canal presumably reflects a bigger hypoglossal nerve, giving us the precise control over tongue movements that we need for speech.

We also measured this hole in the skulls of a number of fossil hominids.

Australopithecines have small canals like those of apes, suggesting that they couldn't talk. But later *Homo* skulls, beginning with a 400,000-year-old skull from Zambia, all have big, humanlike hypoglossal canals. These are also the skulls that were the first to house brains as big as our own. On these counts our work supports Lieberman's ideas. We disagree only on the matter of Neanderthals. While he claims their throats couldn't have produced human speech, we find that their skulls also had human-size canals for the hypoglossal nerve, suggesting that they could indeed talk.

THE VERDICT IS STILL out on the language abilities of Neanderthals. I tend to think they must have had fully human language. After all, they had brains larger than those of most humans.

In short, several lines of evidence suggest that neither the australopithecines nor the early, small-brained species of *Homo* could talk. Only around half a million years ago did the first big-brained *Homo* evolve language. The verdict is still out on the language abilities of Neanderthals. I tend to think that they must have had fully human language. After all, they had brains larger than those of most modern humans, made elegant stone tools, and knew how to use fire. But if Lieberman and his friends are right about those vowels, Neanderthals may have sounded something like the Swedish chef on *The Muppet Show*.

We are beginning to get some idea of when human language originated, but the fossils can't tell us how it got started, or what the intermediate stages between animal calls and human language might have been like. When trying to understand the origin of a trait

that doesn't fossilize, it's sometimes useful to look for similar but simpler versions of it in other creatures living today. With luck, you can find a series of forms that suggest how simple primitive makeshifts could have evolved into more complex and elegant versions. This is how Darwin attacked the problem of the evolution of the eye. Earlier biologists had pointed to the human eye as an example of a marvelously perfect organ that must have been specially created all at once in its final form by God. But Darwin pointed out that animal eyes exist in all stages of complexity, from simple skin cells that can detect only the difference between light and darkness, to pits lined with such cells, and so on all the way to the eyes of people and other vertebrates. This series, he argued, shows how the human eye could have evolved from simpler precursors by gradual stages.

Can we look to other animals to find simpler precursors of language? It seems unlikely. Scientists have sought experimental evidence of language in dolphins and chimpanzees, thus far without success. But even if we had no experimental studies, common sense would tell us that the other animals can't have languages like ours. If they had, we would be in big trouble because they would organize against us. They don't. Outside of Gary Larson's *Far Side* cartoons and George Orwell's *Animal Farm*, farmers don't have to watch their backs when they visit the cowshed. There are no conspiracies among cows, or even among dolphins and chimpanzees. Unlike human slaves or prisoners, they never plot rebellions against their oppressors.

Even if language as a whole has no parallels in animal communication, might some of its peculiar properties be foreshadowed among the beasts around us? If so, that might tell us something about how and in what order these properties were acquired. One such property is reference. Most of the units of human languages refer to things—to individuals (like *Fido*), or to types of objects (*dog*), actions (*sit*), or properties (*furry*). Animal signals don't have this kind of referential meaning. Instead, they have what is called instrumental meaning:

this is, they act as stimuli that trigger desired responses from others. A frog's mating croak doesn't *refer* to sex. Its purpose is to get some, not to talk about it. People, too, have signals of this purely animal sort—for example, weeping, laughing, and screaming—but these stand outside language. They have powerful meanings for us but not the kind of meaning that words have.

AUSTRALOPITHECUS africanus and other early hominids couldn't speak.

Some animal signals have a focused meaning that looks a bit like reference. For example, vervet monkeys give different warning calls for different predators. When they hear the "leopard" call, vervets climb trees and anxiously look down; when they hear the "eagle" call, they hide in low bushes or look up. But although the vervets' leopard call is in some sense about leopards, it isn't a word for leopard. Like a frog's croak or human weeping, its meaning is strictly instrumental; it's a stimulus that elicits an automatic response. All a vervet can "say" with it is "*Eeek!* A leopard!"—not "I really hate leopard!" or "No leopards here, thank goodness" or "A leopard ate Alice yesterday."

In these English sentences, such referential words as *leopard* work their magic through an accompanying framework of nonreferential, grammatical words, which set up an empty web of meaning that the referential symbols fill in. When Lewis Carroll tells us in "Jabberwocky" that "the slithy toves did gyre and gimble in the wabe," we have no idea what he is talking about, but we do know certain things—for instance, that all this happened in the past and that there was more than one tove but only one wabe. We know these things because of the grammatical structure of the sentence, a structure that linguists call syntax. Again, there's nothing much like it in any animal signals.

But if there aren't any intermediate stages between animal calls and human speech, then how could language evolve? What was there for it to evolve from? Until recently, linguists have shrugged off these questions—or else concluded that language didn't evolve at all, but just sprang into existence by accident, through some glorious random mutation. This theory drives Darwinians crazy, but the linguists have been content with it because it fits neatly into some key ideas in modern linguistics.

Forty years ago most linguists thought that people learn to talk through the same sort of behavior reinforcement used in training an animal to do tricks: when children use a word correctly or produce a grammatical sentence, they are rewarded. This picture was swept away in the late 1950s by the revolutionary ideas of Noam Chomsky. Chomsky argued that the structures of syntax lie in unconscious linguistic patterns—so-called deep structures—that are very different from the surface strings of words that come out of our mouths. Two sentences that look different on the surface (for instance, "A leopard ate Alice" and "Alice was eaten by a leopard") can mean the same thing because they derive from a single deep structure. Conversely, two sentences with different deep structures and different meanings can look exactly the same on the surface (for example, "Fleeing leopards can be dangerous"). Any models of language learning based strictly on the observable behaviors of language, Chomsky insisted, can't account for these deep-lying patterns of meaning.

Chomsky concluded that the deepest structures of language are innate, not learned. We are all born with the same fundamental grammar hard-wired into our brains, and we are preprogrammed to pick up the additional rules of the lo-cal language, just as baby ducks are hard-wired to follow the first big animal they see when they hatch. Chomsky could see no evidence of other animals' possessing this innate syntax machinery. He concluded that we can't learn anything about the origins of language by studying other animals and they can't

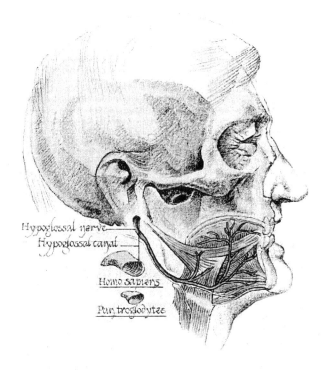

Hypoglossal nerve
Hypoglossal canal

Homo sapiens

Pan troglodytes

THE TONGUE-CONTROLLING

hypoglossal nerve is larger in

humans than in chimps.

learn language from us. If language learning were just a matter of proper training, Chomsky reasoned, we ought to be able to teach English to lab rats, or at least to apes.

As we have seen, apes aren't built to talk. But they can be trained to use sign language or to point to word-symbols on a keyboard. Starting in the 1960s, several experimenters trained chimpanzees and other great apes to use such signs to ask for things and answer questions to get rewards. Linguists, however, were unimpressed. They said that the apes' signs had a purely instrumental meaning: the animals were just doing tricks to get a treat. And there was no trace of syntax in the random-looking jumble of signs the apes produced; an ape that signed "You give me cookie please" one minute might sign "Me cookie please you cookie eat give" the next.

Duane Rumbaugh and Sue Savage-Rumbaugh set to work with chimpanzees at the Yerkes Regional Primate Research Center in Atlanta to try to answer the linguists' criticisms. After many years of mixed results, Sue made a surprising break- through with a young bonobo (or pygmy chimp) named Kanzi. Kanzi had watched his mother, Matata, try to learn signs with little success. When Sue gave up on her and started with Kanzi, she was astonished to discover that he already knew the meaning of 12 of the keyboard symbols. Apparently, he had learned them without any training or rewards. In the years that followed, he learned new symbols quickly and used them referentially, both to answer questions and to "talk" about things that he intended to do or had already done. Still more amazingly, he had a considerable understanding of spoken English—including its syntax. He grasped such grammatical niceties as case structures ("Can you throw a potato to the turtle?") and if-then implication ("You can have some cereal if you give Austin your monster mash to play with"). Upon hearing such sentences, Kanzi behaved appropriately 72 percent of the time—more than a 30-month-old human child given the same tests.

Kanzi is a primatologist's dream and a linguist's nightmare. His language-learning abilities seem inexplicable. He didn't need any rewards to learn language, as the old behaviorists would have predicted; but he also defies the Chomskyan model, which can't explain why a speechless ape would have an innate tendency to learn English. It looks as though some animals can develop linguistic abilities for reasons unrelated to language itself.

Neuroscientist William Calvin of the University of Washington and linguist Derek Bickerton of the University of Hawaii have a suggestion as to what those reasons might be. In their forthcoming book, *Lingua ex Machina,* they argue that the ability to create symbols—signs that refer to things—is potentially present in any animal that can learn to interpret natural signs, such as a trail of footprints. Syntax, meanwhile,

BRAIN ENLARGEMENT in hominids may have been the result of evolutionary pressures that favored intelligence. As a side effect, human evolution crossed a threshold at which language became possible.

emerges from the abstract thought required for a social life. In apes and some other mammals with complex and subtle social relationships, individuals make alliances and act altruistically towards others, with the implicit understanding that their favors will be returned. To succeed in such societies, animals need to choose trustworthy allies and to detect and punish cheaters who take but never give anything in return. This demands fitting a shifting constellation of indi-

viduals into an abstract mental model of social roles (debtors, creditors, allies, and so on) connected by social expectations ("If you scratch my back, I'll scratch yours"). Calvin and Bickerton believe that such abstract models of social obligation furnished the basic pattern for the deep structures of syntax.

These foreshadowings of symbols and syntax, they propose, laid the groundwork for language in a lot of social animals but didn't create language itself. That had to wait until our ancestors evolved brains big enough to handle the large-scale operations needed to generate and process complex strings of signs. Calvin and Bickerton suggest that brain enlargement in our ancestry was the result of evolutionary pressures that favored intelligence and motor coordination for making tools and throwing weapons. As a side effect of these selection pressures, which had nothing to do with communication, human evolution crossed a threshold at which language became possible. Big-brained, nonhuman animals like Kanzi remain just on the verge of language.

This story reconciles natural selection with the linguists' insistence that you can't evolve language out of an animal communication system. It is also consistent with what we know about language from the fossil record. The earliest hominids with modern-size brains also seem to be the first ones with modern-size hypoglossal canals. Lieberman thinks that these are also the first hominids with modern vocal tracts. It may be no coincidence that all three of these changes seem to show up together around half a million years ago. If

Calvin and Bickerton are right, the enlargement of the brain may have abruptly brought language into being at this time, which would have placed new selection pressures on the evolving throat and tongue.

FOSSILS HINT that language dawned 500,000 years ago.

This account may be wrong in some of its details, but the story in its broad outlines solves so many puzzles and ties up so many loose ends that something like it must surely be correct. It also promises to resolve our conflicting views of the boundary between people and animals. To some people, it seems obvious that human beings are utterly different from any beasts. To others, it's just as obvious that many other animals are essentially like us, only with fewer smarts and more fur. Each party finds the other's view of humanity alien and threatening. The story of language origins sketched above suggests that both parties are right: the human difference is real and profound, but it is rooted in aspects of psychology and biology that we share with our close animal relatives. If the growing consensus on the origins of language can join these disparate truths together, it will be a big step forward in the study of human evolution.

The Dating Game

*By tracking changes in ancient atoms, archeologists are establishing
the astonishing antiquity of modern humanity.*

James Shreeve

James Shreeve wrote fiction before turning to science writing. He is the coauthor (with anthropologist Donald Johanson) of Lucy's Child: The Discovery of a Human Ancestor *and the author of* Nature: The Other Earthlings.

Four years ago archeologists Alison Brooks and John Yellen discovered what might be the earliest traces of modern human culture in the world. The only trouble is, nobody believes them. Sometimes they can't quite believe it themselves.

Their discovery came on a sun-soaked hillside called Katanda, in a remote corner of Zaire near the Ugandan border. Thirty yards below, the Semliki River runs so clear and cool the submerged hippos look like giant lumps of jade. But in the excavation itself, the heat is enough to make anyone doubt his eyes.

Katanda is a long way from the plains of Ice Age Europe, which archeologists have long believed to be the setting for the first appearance of truly modern culture: the flourish of new tool technologies, art, and body ornamentation known as the Upper Paleolithic, which began about 40,000 years ago. For several years Brooks, an archeologist at George Washington University had been pursuing the heretical hypothesis that humans in Africa had invented sophisticated technologies even earlier, while their European counterparts were still getting by with the same sorts of tools they'd been using for hundreds of thousands of years. If conclusive evidence hadn't turned up, it was only because nobody had really bothered to look for it.

"In France alone there must be three hundred well-excavated sites dating from the period we call the Middle Paleolithic," Brooks says. "In Africa there are barely two dozen on the whole continent."

One of those two dozen is Katanda. On an afternoon in 1988 John Yellen—archeology program director at the National Science Foundation and Brooks's husband—was digging in a densely packed litter of giant catfish bones, river stones, and Middle Paleolithic stone tools. From the rubble he extricated a beautifully crafted, fossilized bone harpoon point. Eventually two more whole points and fragments of five others turned up, all of them elaborately barbed and polished. A few feet away, the scientists uncovered pieces of an equally well crafted daggerlike tool. In design and workmanship the harpoons were not unlike those at the very end of the Upper Paleolithic, some 14,000 years ago. But there was one important difference. Brooks and Yellen believe the deposits John was standing in were at least five times that old. To put this in perspective, imagine discovering a prototypical Pontiac in Leonardo da Vinci's attic.

"If the site is as old as we think it ·is," says Brooks, "it could clinch the argument that modern humans evolved in Africa."

Ever since the discovery the couple have devoted themselves to chopping away at that stubborn little word *if*. In the face of the entrenched skepticism of their colleagues, it is an uphill task. But they do have some leverage. In those same four years since the first harpoon was found at Katanda, a breakthrough has revived the question of modern human origins. The breakthrough is not some new skeleton pulled out of the ground. Nor is it the highly publicized Eve hypothesis, put forth by geneticists, suggesting that all humans on Earth today share a common female ancestor who lived in Africa 200,000 years ago. The real advance, abiding quietly in the shadows while Eve draws the limelight, is simply a new way of telling time.

To be precise, it is a whole smorgasbord of new ways of telling time. Lately they have all converged on the same exhilarating, mortifying revelation: what little we thought we knew about the origins of our own species was hopelessly wrong. From Africa to the Middle East to Australia, the new dating methods are overturning conventional wisdom with insolent abandon, leaving the anthropological community dazed amid a rubble of collapsed certitudes. It is in this shell-shocked climate that Alison Brooks's Pontiac in Leonardo's attic might actually find a hearing.

"Ten years ago I would have said it was impossible for harpoons like these to be so old," says archeologist Michael Mehlman of the Smithsonian's National Museum of Natural History. "Now I'm reserving judgment. Anything can happen."

An archeologist with a freshly uncovered skull, stone tool, or bone Pontiac in hand can take two general approaches to determine its age. The first is called relative dating. Essentially the archeologist places the find in the context of the surrounding geological deposits. If the new discovery is found in a brown sediment lying beneath a yellowish layer of sand, then, all things being equal, it is older than the yellow sand layer or any other deposit higher up. The fossilized remains of extinct animals found near the object also provide a "biostratigraphic" record that can offer clues to a new find's relative age. (If a stone tool is found alongside an extinct species of horse, then it's a fair bet the tool was made while that kind of horse was still running around.) Sometimes the tools themselves can be used as a guide, if they match up in character and style with tools from other, better-known sites. Relative dating methods like these can tell you whether a find is older or younger than something else, but they cannot pin an age on the object in calendar years.

The most celebrated *absolute* method of telling archeological time, radiocarbon dating, came along in the 1940s. Plants take in carbon from the atmosphere to build tissues, and other organisms take in plants, so carbon ends up in everything from wood to woodchucks. Most carbon exists in the stable form of carbon 12. But some is made up of the unstable, radioactive form carbon 14. When an organism dies, it contains about the same ratio of carbon 12 to carbon 14 that exists in the atmosphere. After death the radioactive carbon 14 atoms begin to decay, changing into stable atoms of nitrogen. The amount of carbon 12, however, stays the same. Scientists can look at the amount of carbon 12 and—based on the ratio—deduce how much carbon 14 was originally present. Since the decay rate of carbon 14 is constant and steady (half of it disappears every 5,730 years), the difference between the amount of carbon 14 originally in a charred bit of wood or bone

and the amount present now can be used as a clock to determine the age of the object.

Conventional radiocarbon dates are extremely accurate up to about 40,000 years. This is far and away the best method to date a find—as long as it is younger than this cutoff point. (In older materials, the amount of carbon 14 still left undecayed is so small that even the slightest amount of contamination in the experimental process leads to highly inaccurate results.) Another dating technique, relying on the decay of radioactive potassium rather than carbon, is available to date volcanic deposits *older* than half a million years. When it was discovered in the late 1950s, radiopotassium dating threw open a window on the emergence of the first members of the human family—the australopithecines, like the famous Lucy, and her more advanced descendants, *Homo habilis* and *Homo erectus*. Until now, however, the period between half a million and 40,000 years—a stretch of time that just happens to embrace the origin of *Homo sapiens*—was practically unknowable by absolute dating techniques. It was as if a geochronological curtain were drawn across the mystery of our species' birth. Behind that curtain the hominid lineage underwent an astonishing metamorphosis, entering the dateless, dark centuries a somewhat precocious bipedal ape and emerging into the range of radiocarbon dating as the culturally resplendent, silver-tongued piece of work we call a modern human being.

Fifteen years ago there was some general agreement about how this change took place. First, what is thought of as an *anatomically* modern human being—with the rounded cranium, vertical forehead, and lightly built skeleton of people today—made its presence known in Europe about 35,000 years ago. Second, along with those first modern-looking people, popularly known as the Cro-Magnons, came the first signs of complex human *behavior*, including tools made of bone and antler as well as of stone, and art, symbolism, social status, ethnic identity, and probably true human language too. Finally, in any one region there was no overlap in time be-

tween the appearance of modern humans and the disappearance of "archaic" humans such as the classic Neanderthals, supporting the idea that one group had evolved from the other.

"Thanks to the efforts of the new dating methods," says Fred Smith, an anthropologist at Northern Illinois University, "we now know that each of these ideas was wrong."

The technique doing the most damage to conventional wisdom is called thermoluminescence, TL for short. (Reader take heed: the terrain of geochronology is full of terms long enough to tie between two trees and trip over, so acronyms are a must.) Unlike radiocarbon dating, which works on organic matter, TL pulls time out of stone.

If you were to pick an ordinary rock up off the ground and try to describe its essential rockness, phrases like "frenetically animated" would probably not leap to mind. But in fact minerals are in a state of constant inner turmoil. Minute amounts of radioactive elements, both within the rock itself and in the surrounding soil and atmosphere, are constantly bombarding its atoms, knocking electrons out of their normal orbits. All this is perfectly normal rock behavior, and after gallivanting around for a hundredth of a second or two, most electrons dutifully return to their normal positions. A few, however, become trapped en route—physically captured within crystal impurities or electronic aberrations in the mineral structure itself. These tiny prisons hold on to their electrons until the mineral is heated, whereupon the traps spring open and the electrons return to their more stable position. As they escape, they release energy in the form of light—a photon for every homeward-bound electron.

Thermoluminescence was observed way back in 1663 by the great English physicist Robert Boyle. One night Boyle took a borrowed diamond to bed with him, for reasons that remain obscure. Resting the diamond "upon a warm part of my Naked Body," Boyle noticed that it soon emitted a warm glow. So taken was he with the responsive gem that the next day he delivered a paper on the subject at the Royal Society, noting his surprise at the glow since his "constitution," he felt, was "not of the hottest."

Three hundred years later another Englishman, Martin Aitken of Oxford University, developed the methods to turn thermoluminescence into a geophysical timepiece. The clock works because the radioactivity bombarding a mineral is fairly constant, so electrons become trapped in those crystalline prisons at a steady rate through time. If you crush the mineral you want to date and heat a few grains to a high enough temperature—about 900 degrees, which is more body heat than Robert Boyle's constitution could ever have produced—all the electron traps will release their captive electrons at once, creating a brilliant puff of light. In a laboratory the intensity of that burst of luminescence can easily be measured with a device called a photomultiplier. The higher the spike of light, the more trapped electrons have accumulated in the sample, and thus the more time has elapsed since it was last exposed to heat. Once a mineral is heated and all the electrons have returned "home," the clock is set back to zero.

Now, our lineage has been making flint tools for hundreds of thousands of years, and somewhere in that long stretch of prehistory we began to use fire as well. Inevitably, some of our less careful ancestors kicked discarded tools into burning hearths, setting their electron clocks back to zero and opening up a ripe opportunity for TL timekeepers in the present. After the fire went out, those flints lay in the ground, pummeled by radioactivity, and each trapped electron was another tick of the clock. Released by laboratory heat, the electrons flash out photons that reveal time gone by.

In the late 1980s Hélène Valladas, an archeologist at the Center for Low-Level Radioactivity of the French Atomic Energy Commission near Paris, along with her father, physicist Georges Valladas, stunned the anthropological community with some TL dates on burned flints taken from two archeological sites in Israel. The first was a cave called Kebara, which had already yielded an astonishingly complete Neanderthal skeleton. Valladas dated flints from the Neanderthal's level at 60,000 years before the present.

In itself this was no surprise, since the date falls well within the known range of the Neanderthals' time on Earth. The shock came a year later, when she used the same technique to pin a date on flints from a nearby cave called Qafzeh, which contained the buried remains of early modern human beings. This time, the spikes of luminescence translated into an age of around 92,000 years. In other words, the more "advanced" human types were a full 30,000 years *older* than the Neanderthals they were supposed to have descended from.

If Valladas's TL dates are accurate, they completely confound the notion that modern humans evolved from Neanderthals in any neat and tidy way. Instead, these two kinds of human, equally endowed culturally but distinctly different in appearance, might have shared the same little nook of the Middle East for tens of thousands of years. To some, this simply does not make sense.

"If these dates are correct, what does this do to what else we know, to the stratigraphy, to fossil man, to the archeology?" worries Anthony Marks, an archeologist at Southern Methodist University. "It's all a mess. Not that the dates are necessarily wrong. But you want to know more about them."

Marks's skepticism is not entirely unfounded. While simple in theory, in practice TL has to overcome some devilish complications. ("If these new techniques were easy, we would have thought of them a long time ago," says geochronologist Gifford Miller of the University of Colorado.) To convert into calendar years the burst of luminescence when a flint is heated, one has to know both the sensitivity of that particular flint to radiation and the dose of radioactive rays it has received each year since it was "zeroed" by fire. The sensitivity of the sample can be determined by assaulting it with artificial radiation in the lab. And the annual dose of radiation received from *within* the sample itself can be calculated fairly easily by measuring how much uranium or other radioactive elements the sample contains. But determining the annual dose from the environment *around* the sample—the radioactivity in the surrounding soil, and cosmic rays from the atmosphere itself—is an iffier proposition. At some sites fluctuations in this environmental dose through the millennia can turn the "absolute" date derived from TL into an absolute nightmare.

Fortunately for Valladas and her colleagues, most of the radiation dose for the Qafzeh flints came from within the flints themselves. The date there of 92,000 years for the modern human skeletons is thus not only the most sensational number so far produced by TL, it is also one of the surest.

"The strong date at Qafzeh was just good luck," says Valladas. "It was just by chance that the internal dose was high and the environmental dose was low."

More recently Valladas and her colleague Norbert Mercier turned their TL techniques to the French site of Saint-Césaire. Last summer they confirmed that a Neanderthal found at Saint-Césaire was only 36,000 years old. This new date, combined with a fresh radiocarbon date of about 40,000 years tagged on some Cro-Magnon sites in northern Spain, strongly suggests that the two types of humans shared the same corner of Europe for several thousand years as the glaciers advanced from the north.

While Valladas has been busy in Europe and the Middle East, other TL timekeepers have produced some astonishing new dates for the first human occupation of Australia. As recently as the 1950s, it was widely believed that Australia had been colonized only some five thousand years ago. The reasoning was typically Eurocentric: since the Australian aborigines were still using stone tools when the first white settlers arrived, they must have just recently developed the capacity to make the difficult sea crossing from Indonesia in the first place. A decade later archeologists grudgingly conceded that the date of first entry might have been closer to the beginning of the Holocene period, 10,000 years ago. In the 1970s radiocarbon dates on human occupation sites pushed the date back again, as far as 32,000 years ago. And now TL studies at two sites in northern Australia drop that first human footstep on the continent—and the sea voyage that preceded it—all the way back to 60,000 years before the present. If these dates stand up, then the once-maligned ancestors of modern aborigines were building ocean-worthy craft some 20,000 years *before* the first signs of sophisticated culture appeared in Europe.

"Luminescence has revolutionized the whole period I work in," says Australian National University archeologist Rhys Jones, a member of the team responsible for the new TL dates. "In effect, we have at our disposal a new machine—a new time machine."

With so much at stake, however, nobody looks to TL—or to any of the other new "time machines"—as a geochronological panacea. Reputations have been too badly singed in the past by dating methods that claimed more than they could deliver. In the 1970s a flush of excitement over a technique called amino acid racemization led many workers to believe that another continent—North America—had been occupied by humans fully 70,000 years ago. Further testing at the same American sites proved that the magical new method was off by one complete goose egg. The real age of the sites was closer to 7,000 years.

"To work with wrong dates is a luxury we cannot afford," British archeologist Paul Mellars intoned ominously earlier this year, at the beginning of a London meeting of the Royal Society to showcase the new dating technologies. "A wrong date does not simply inhibit research. It could conceivably throw it into reverse."

Fear of just such a catastrophe—not to mention the risk that her own reputation could go up in a puff of light—is what keeps Alison Brooks from declaring outright that she has found exquisitely crafted bone harpoons in Zaire that are more than 40,000 years older than such creations are supposed to be. So far the main support for her argument has been her redating of another site, called Ishango, four miles down the Semliki River from the Katanda site. In the 1950s the Belgian geologist Jean de Heinzelin excavated a harpoon-rich "aquatic civilization" at Ishango that he thought was 8,000 years old. Brooks's radiocarbon dating of the site in the mid-1980s pushed the age back to 25,000. By tracing the layers of sediment shared between Ishango and Katanda, Brooks and her colleagues are convinced that Katanda is much farther down in the stratigraphy—twice as old as Ishango, or perhaps even more. But even though Brooks and Yellen talk freely about their harpoons at meetings, they have yet to utter such unbelievable numbers in the unforgiving forum of an academic journal.

"It is precisely because no one believes us that we want to make our case airtight before we publish," says Brooks. "We want dates confirming dates confirming dates."

Soon after the harpoons were discovered, the team went to work with thermoluminescence. Unfortunately, no burned flints have been found at the site. Nevertheless, while TL works best on materials that have been completely zeroed by such extreme heat as a campfire, even a strong dose of sunlight can spring some of the electron traps. Thus even ordinary sediments surrounding an archeological find might harbor a readable clock: bleached out by sunlight when they were on the surface, their TL timers started ticking as soon as they were buried by natural processes. Brooks and Yellen have taken soil samples from Katanda for TL, and so far the results are tantalizing—but that's all.

"At this point we think the site is quite old," says geophysicist Allen Franklin of the University of Maryland, who with his Maryland colleague Bill Hornyak is conducting the work. "But we don't want to put a number on it."

As Franklin explains, the problem with dating sediments with TL is that while some of the electron traps might be quickly bleached out by sunlight, others hold on to their electrons more stubbornly. When the sample is then heated in a conventional TL apparatus, these stubborn traps release electrons that were captured perhaps millions of years before the sediments were last exposed to sunlight-teasing date-hungry archeologists with a deceptively old age for the sample.

Brooks does have other irons in the dating fire. The most promising is called electron spin resonance—or ESR, among friends. Like TL, electron spin resonance fashions a clock out of the steadily accumulating electrons caught in traps. But whereas TL measures that accumulation by the strength of the light given off when the traps open, ESR literally counts the captive electrons themselves while they still rest undisturbed in their prisons.

All electrons "spin" in one of two opposite directions—physicists call them up and down. (Metaphors are a must here because the nature of this "spinning" is quantum mechanical and can be accurately described only in huge mathematical equations.) The spin of each electron creates a tiny magnetic force pointing in one direction, something like a compass needle. Under normal circumstances, the electrons are paired so that their opposing spins and magnetic forces cancel each other out. But trapped electrons are unpaired. By manipulating an external magnetic field placed around the sample to be dated, the captive electrons can be induced to "resonate"—that is, to flip around and spin the other way. When they flip, each electron absorbs a finite amount of energy from a microwave field that is also applied to the sample. This loss of microwave energy can be measured with a detector, and it is a direct count of the number of electrons caught in the traps.

ESR works particularly well on tooth enamel, with an effective range from a thousand to 2 million years. Luckily for Brooks and Yellen, some nice fat hippo teeth have been recovered from Katanda in the layer that also held the harpoons. To date the teeth, they have called in Henry Schwarcz of McMaster University in Ontario, a ubiquitous, veteran geochronologist. In the last ten years Schwarcz has journeyed to some 50 sites throughout Europe, Africa, and western Asia, wherever his precious and arcane services are demanded.

Schwarcz also turned up at the Royal Society meeting, where he explained both the power and the problems of the ESR method. On the plus side is that teeth are hardy remains, found at nearly every archeological site in the world, and that ESR can test a tiny sample again and again—with the luminescence techniques, it's a one-shot deal. ESR can also home in on certain kinds of electron traps, offering some refinement over TL, which lumps them all together.

On the minus side, ESR is subject to the same uncertainties as TL concerning the annual soaking of radiation a sample has received from the environment. What's more, even the radiation from *within* a tooth cannot be relied on to be constant through time. Tooth enamel has the annoying habit of sucking up uranium from its surroundings while it sits in the ground. The more uranium the tooth contains, the more electrons are being bombarded out of their normal positions, and the faster the electron traps will fill up. Remember: you cannot know how old something is by counting filled traps unless you know the rate at which the traps were filled, year by year. If the tooth had a small amount of internal uranium for 50,000 years but took in a big gulp of the hot stuff 10,000 years ago, calculations based on the tooth's current high uranium level would indicate the electron traps were filled at a much faster rate than they really were. "The big question is, When did the uranium get there?" Schwarcz says. "Did the tooth slurp it all up in three days, or did the uranium accumulate gradually through time?"

One factor muddying the "big question" is the amount of moisture present around the sample during its centuries of burial: a wetter tooth will absorb uranium faster. For this reason, the best ESR sites are those where conditions are driest. Middle Eastern and African deserts are good bets. As far as modern human origins go, the technique has already tagged a date of about 100,000 years on some human fossils from an Israeli cave called Skhul, neatly supporting the TL date of 92,000 from Qafzeh, a few miles

away. If a new ESR date from a Neanderthal cave just around the corner from Skhul is right, then Neanderthals were also in the Middle East at about the same time. Meanwhile, in South Africa, a human jawbone from the site of Border Cave—"so modern it boggles the mind," as one researcher puts it—has now been dated with ESR at 60,000 years, nearly twice as old as any fossil like it in Europe.

But what of the cultural change to modern human behavior—such as the sophisticated technological development expressed by the Katanda harpoons? Schwarcz's dating job at Katanda is not yet finished, and given how much is at stake, he too is understandably reluctant to discuss it. "The site has good potential for ESR," he says guardedly. "Let's put it this way: if the initial results had indicated that the harpoons were not very old after all, we would have said 'So what?' to them and backed off. Well, we haven't backed off."

There are other dating techniques being developed that may, in the future, add more certainty to claims of African modernity. One of them, called uranium-series dating, measures the steady decay of uranium into various daughter elements inside anything formed from carbonates (limestone and cave stalactites, for instance). The principle is very similar to radiocarbon dating—the amount of daughter elements in a stalactite, for example, indicates how long that stalactite has been around—with the advantage that uranium-series dates can stretch back half a million years. Even amino acid racemization, scorned for the last 15 years, is making a comeback, thanks to the discovery that the technique, unreliable when applied to porous bone, is quite accurate when used on hard ostrich eggshells.

In the best of all possible worlds, an archeological site will offer an opportu-

nity for two or more of these dating techniques to be called in so they can be tested against each other. When asked to describe the ideal site, Schwarcz gets a dreamy look on his face. "I see a beautiful human skull sandwiched between two layers of very pure flow-stone," he says, imagining uranium-series dating turning those cave limestones into time brackets. "A couple of big, chunky hippo teeth are lying next to it, and a little ways off, a bunch of burned flints."

Even without Schwarcz's dream site, the dating methods used separately are pointing to a common theme: the alarming antiquity of modern human events where they are not supposed to be happening in the first place. Brooks sees suggestive traces of complexity not just at Katanda but scattered all over the African continent, as early as 100,000 years before the present. A classic stone tool type called the blade, long considered a trademark of the European Upper Paleolithic, appears in abundance in some South African sites 40,000 to 50,000 years before the Upper Paleolithic begins. The continent may even harbor the earliest hints of art and a symbolic side to human society: tools designed with stylistic meaning; colorful, incandescent minerals, valueless but for their beauty, found hundreds of miles away from their source. More and more, the Cro-Magnons of Europe are beginning to look like the last modern humans to show themselves and start acting "human" rather than the first.

That's not an easy notion for anthropologists and archeologists to swallow. "It just doesn't fit the pattern that those harpoons of Alison's should be so old," says Richard Klein, a paleoanthropologist at the University of Chicago. Then he shrugs. "Of course, if she's right, she has made a remarkable discovery indeed."

Only time will tell.

The Neanderthal Peace

For perhaps 50,000 years, two radically different types of human lived side by side in the same small land. And for all those millennia, the two apparently had nothing whatsoever to do with each other. Why in the world not?

James Shreeve

James Shreeve is a contributing editor of Discover. *His previous book,* Lucy's Child: The Discovery of a Human Ancestor, *was coauthored with anthropologist Donald Johansen.*

I met my first Neanderthal in a café in Paris, just across the street from the Jussieu metro stop. It was a wet afternoon in May, and I was sitting on a banquette with my back to the window. The café was smoky and charmless. Near the entrance a couple of students were thumping on a pinball machine called Genesis, which beeped approval every time they scored. The place was packed with people—foreign students, professors, young professionals, French workers, Arabs, Africans, and even a couple of Japanese tourists, all thrown together by the rain. Our coffee had just arrived, and I found that if I tucked my elbow down when raising my cup, I could drink it without poking the ribs of a bearded man sitting at the table next to me, who was deep into an argument.

Above the noise of the pinball game and the din of private conversations, a French anthropologist named Jean-Jacques Hublin was telling me about the anatomical unity of man. It was he who had brought along the Neanderthal. When we had come into the café, he had placed an object wrapped in a soft rag on the table and had ignored it ever since. Like anything so carefully neglected, it was beginning to monopolize my attention.

"Perhaps you would be interested in this," he said at last, whisking away the rag. There, amid the clutter of demitasses and empty sugar wrappers, was a large human lower jawbone. The teeth, worn and yellowed by time, were all in place. Around us, I felt the café raise a collective eyebrow. The hubbub of talk sank audibly. The bearded man next to me stopped in midsentence, looked at the jaw, looked at Hublin, and resumed his argument. Hublin gently nudged the fossil to the center of the table and leaned back.

"What is it? I asked.

"It is a Neanderthal from a site called Zafarraya, in the south of Spain," he said. "We have only this mandible and an isolated femur. But as you can see, the jaw is almost complete. We are not sure yet, but it may be that this fossil is only 30,000 years old."

"Only" 30,000 years may seem an odd way of expressing time, but coming from a paleoanthropologist, it is like saying that a professional basketball player is *only* 6 foot 4. Hominids—members of the exclusively human family tree—have been on Earth for at least 4 million years. Measured against the earliest members of our lineage, the mineralized piece of bone on the table was a mewling newborn. Even compared with others of its kind, the jaw was astonishingly young. Neanderthals were supposed to have disappeared fully 5,000 years before this one was born, and I had come to France to find out what might have happened to them.

The Neanderthals are the best known and least understood of all human ancestors. To most people, the name instantly brings to mind the image of a hulking brute, dragging his mate around by her coif. This stereotype, born almost as soon as the first skeleton was found in a German cave in the middle of the last century, has been refluffed in comic books, novels, and movies so often that it has successfully passed from cliché to common parlance. But what actually makes a Neanderthal a Neanderthal is not its size or its strength or any measure of its native intelligence but a suite of exquisitely distinct physical traits, most of them in the face and cranium.

Like all Neanderthal mandibles, for example, the one on the table lacked the bony protrusion on the rim of the jaw called a mental eminence—better known as a chin. The places on the outside of the jaw where chewing muscles had once been attached were grossly enlarged, indicating tremendous torque in the bite. Between the last two molars and the upward thrust of the rear of the jaw, Hublin pointed to gaps of almost a quarter inch, an architectural nicety shifting the business of chewing farther toward the front.

In these and in several other features the jaw was uniquely, quintessentially Neanderthal; no other member of the human family before or since shows the same pattern. With a little instruction the Neanderthal pattern is recognizable even to a layman like me. But unlike Hublin, whose expertise allowed him to sit there

From *Discover* magazine, September 1995, pp. 70-81. Adapted from *The Neanderthal Enigma: Solving the Mystery of Modern Human Origins* by James Shreeve. © 1995 by James Shreeve. Reprinted by permission of William Morrow & Company, Inc.

calmly sipping coffee while the jaw of a 30,000-year-old man rested within biting distance of his free hand, I felt like stooping down and paying homage.

Several years before, based on a comparison of DNA found in the mitochondria of modern human cells, a team of biochemists in Berkeley, California, had concluded that all humans on Earth could trace their ancestry back to a woman who had lived in Africa only 200,000 years earlier. Every living branch and twig of the human family tree had shot up from this "mitochondrial Eve" and spread like kudzu over the face of the globe, binding all humans in an intimate web of relatedness.

To me the Eve hypothesis sounded almost too good to be true. If all living people can be traced back to a common ancestor just 200,000 years ago, then the entire human population of the globe is really just one grand brother-and-sisterhood, despite the confounding embellishments of culture and race. Thus on a May afternoon, a café in Paris could play host to clientele from three or four continents, but the scene still amounted to a sort of ad hoc family reunion.

But Eve bore a darker message too. The Berkeley study suggested that at some point between 100,000 and 50,000 years ago, people from Africa began to disperse across Europe and Asia, eventually populating the Americas as well. These people, and these alone, became the ancestors of all future human generations. When they arrived in Eurasia, however, there were thousands, perhaps millions, of other human beings already living there—including the Neanderthals. What happened to them all? Eve's answer was cruelly unequivocal: the Neanderthals—including the Zafarraya population represented by the jaw on the table—were pushed aside, outcompeted, or otherwise driven extinct by the new arrivals from the south.

What fascinates me about the fate of the Neanderthals is the paradox of their promise. Appearing first in Europe about 150,000 years ago, the Neanderthals flourished throughout the increasing cold of an approaching ice age; by 70,000 years ago they had spread throughout Europe and western Asia. As for Neanderthal appearance, the stereo-

type of a muscled thug is not completely off the mark. Thick-boned, barrelchested, a healthy Neanderthal male could lift an average NFL linebacker over his head and throw him through the goalposts. But despite the Neanderthal's reputation for dim-wittedness, there is nothing that clearly distinguishes its brain from that of a modern human except that, on average, the Neanderthal version was slightly *larger*. There is no trace of the thoughts that animated those brains, so we do not know how much they resembled our own. But a big brain is an expensive piece of adaptive equipment. You don't evolve one if you don't use it. Combining enormous physical strength with manifest intelligence, the Neanderthals appear to have been outfitted to face any obstacle the environment could put in their path. They could not lose.

And then, somehow, they lost. Just when the Neanderthals reached their most advanced expression, they suddenly vanished. Their demise coincides suspiciously with the arrival in western Europe of a new kind of human: taller, thinner, more modern-looking. The collision of these two human populations— us and the other, the destined parvenu and the doomed caretaker of a continent—is as potent and marvelous a part of the human story as anything that has happened since.

By itself, the half-jaw on the table in front of me had its own tale to tell. Hublin had said that it was perhaps as young as 30,000 years old. A few months before, an American archeologist named James Bischoff and his colleagues had also announced astonishing ages for some objects from Spanish caves. After applying a new technique to date some modern human-style artifacts, they declared them to be 40,000 years old. This was 6,000 years *before* there were supposed to have been modern humans in Europe. If both Bischoff's and Hublin's dates were right, it meant that Neanderthals and modern humans had been sharing Spanish soil for 10,000 years. That didn't make sense to me.

"At 30,000 years," I asked Hublin, "wouldn't this jaw be the last Neanderthal known?"

"If we are right about the date, yes," he said. "But there is still much work to be done before we can say how old the jaw is for certain."

"But Bischoff says modern humans were in Spain 10,000 years before then," I persisted. "I can understand how a population with a superior technology might come into an area and quickly dominate a less sophisticated people already there. But 10,000 years doesn't sound very quick even in evolutionary terms. How can two kinds of human being exist side by side for that long without sharing their cultures? Without sharing their genes?"

Hublin shrugged in the classically cryptic French manner that means either "The answer is obvious" or "How should I know?"

Among all the events and transformations in human evolution, the origins of modern humans were, until recently, the easiest to account for. Around 35,000 years ago, signs of a new, explosively energetic culture in Europe marked the beginning of the period known as the Upper Paleolithic. They included a highly sophisticated variety of tools, made out of bone and antler as well as stone. Even more important, the people making these tools—usually known as Cro-Magnons, a name borrowed from a tiny rock shelter in southern France where their skeletons were first found, in 1868—had discovered a symbolic plane of existence, evident in their gorgeously painted caves, carved animal figurines, and the beads and pendants adorning their bodies. The Neanderthals who had inhabited Europe for tens of thousands of years had never produced anything remotely as elaborate. Coinciding with this cultural explosion were the first signs of the kind of anatomy that distinguishes modern human beings: a well-defined chin; a vertical forehead lacking pronounced browridges; a domed braincase; and a slender, lightly built frame, among other, more esoteric features.

The skeletons in the Cro-Magnon cave, believed to be between 32,000 and 30,000 years old, provided an exquisite microcosm of the joined emergence of culture and anatomy. Five skeletons, including one of an infant, were found buried in a communal grave, and all exhibited the anatomical characteristics of modern human beings. Scattered in the grave with them were hundreds of artificially pierced seashells and animal teeth, clearly the vestiges of necklaces, bracelets, and other body ornaments. The nearly simultaneous appearance of modern culture and modern anatomy provided a readymade explanation for the final step in the human journey. Since they happened at the same time, the reasoning went, obviously one had caused the other. It all made good Darwinian sense. A more efficient technology emerged to take over the survival role previously provided by brute strength, relaxing the need for the robust physiques and powerful chewing apparatus of the Neanderthals. Voilà. Suddenly there was clever, slender Cro-Magnon man. That this first truly modern human should be indigenous to Europe tightened the evolutionary narrative: modern man appeared in precisely the region of the world where culture—according to Europeans—later reached its zenith. Prehistory foreshadowed history. The only issue to sort out was whether the Cro-Magnons had come from somewhere else or whether the Neanderthals had evolved into them.

The latter scenario, of course, assumes that modern humans and Neanderthals didn't coexist, at least not for any appreciable amount of time. But the jawbone from Zafarraya challenged that neat supposition. Even more damaging were some strange findings in the Mideast. Recent discoveries there too suggest that Neanderthals and modern humans may have inhabited the same land at the same time, and for far, far longer than in Spain.

In Israel, on the southern edge of the Neanderthal range, a wooded rise of limestone issues abruptly out of the Mediterranean below Haifa, ascending in an undulation of hills. This is the Mount Carmel of the Song of Solomon, where Elijah brought down the false priests of Baal, and Deborah laid rout to the Canaanites. In subsequent centuries, armies, tribes, and whole cultures tramped through its rocky passes and over its fertile flanks, bringing Hittites, Persians, Jews, Romans, Mongols, Muslims, Crusaders, Turks, the modern meddling of Europeans—one people slaughtered or swallowed by the next but somehow springing up again and gaining strength enough to slaughter or swallow in its turn.

The story in the Levant never made sense. Here, how modern a hominid looked said nothing about how modernly it behaved.

My interest here is in more ancient confrontations. Mount Carmel lies in the Levant, a tiny hinge of habitability between the sea and the desert, linking the two great landmasses of Africa and Eurasia. A million years ago a massive radiation of large mammals moved through the Levant from Africa toward the temperate latitudes to the north. Among these mammals were some ancestral humans. Time passed. The humans evolved, diversified. The ones in Europe came to look very different from their now-distant relatives who had remained in Africa. The Europeans became the Neanderthals. Then, still long before history began to scar the Levant with its sieges and slaughters, some Neanderthals from Europe and other humans from Africa wandered into this link between their homelands, leaving their bones on Mount Carmel. What happened when they met? How did two kinds of human respond to each other?

Reaching the Stone Age in Israel is easy; I simply rented a car in Tel Aviv and drove a couple of hours up the coastal road. My destination was the cave of Kebara, an excavation hunched above a banana plantation on the sea-weathered western slope of the mountain.

Inside the cave the present Mideast, with all its political complexities, disappeared—here there was only a cool, sheltered emptiness, greatly enlarged by decades of archeological probing. Scattered through the excavation were a dozen or so scientists and students; an equal number were working at tables along the rim. The atmosphere was one of hushed, almost monkish concentration, like that of a reading room in a great library.

The Kebara excavation began ten years ago, picking up on the previous work of Moshe Stekelis of Hebrew University in the 1950s and early 1960s. Stekelis exposed a sequence of Paleolithic deposits and, before his sudden death, discovered the skeleton of an infant Neanderthal. A greater treasure emerged in 1983. After Stekelis's time, the sharp vertical profiles of the excavation crumbled under the feet of a generation of kibbutz children and assorted other slow ravages. A graduate student named Lynne Schepartz was assigned the mundane task of cleaning up the deteriorated exposures by cutting them a little deeper. One afternoon she noticed what appeared to be a human toe bone peeking out of a fused clod of sediments. The next morning her whisk broom exposed a pearly array of human teeth: the lower jaw of an adult Neanderthal skeleton. Stekelis's team had missed it by two inches.

Lynne Schepartz was no longer a graduate student, but she was still spending her summers at Kebara. I found her and asked her how it felt to uncover the fossil. "Unprintable," she said. "I was jumping up and down and screaming."

She had reason to react unprintably. Her discovery turned out to be not just any Neanderthal but the most complete skeleton ever found: the first complete Neanderthal spinal column, the first complete Neanderthal rib cage, the first complete pelvis of any early hominid known. She showed me a plaster cast of the fossil—affectionately known as Moshe—lying on an adjacent table. The bones were arranged exactly as they had been found. Moshe was resting on his back, his right arm folded over his chest, his left hand on his stomach, in a classic attitude of burial. The only missing parts were the right leg, the extremity of the left, and except for the lower jaw, the skull.

Schepartz led me down ladders to Moshe's burial site, a deep rectangular pit near the center of the excavation. On this July morning, the Neanderthal's grave was occupied by a modern human named Ofer Bar-Yosef, who peered back up at me from behind thick glasses, magnifying my sense that I had disturbed the happy toil of a cavernicolous hobbit. He seemed evolved to the task, nimble and gnomishly compact, the better to fit into cramped quarters.

Bar-Yosef told me that he had directed his first archeological excavation at the age of 11, rounding up a crew of his friends in his Jerusalem neighborhood to help him unearth a Byzantine water system. He had not stopped digging since. Kebara was the latest of three major excavations under his direction. "My daughter has been coming to this site since she was a fetus," he told me. "She used to have a playpen set up right over there."

Throughout his career, Bar-Yosef has dug for answers to two personal obsessions: the origins of Neolithic agricultural societies and—the point where our obsessions converge—the twisting conundrum of modern human origins.

The story in the Levant never really made much sense. In the old days, back when everybody "knew" that modern humans first appeared in western Europe, where the really modern folks still live, you could identify a hominid by the kind of tools he left behind. Bulky Neanderthals made bulky flakes, while svelte Cro-Magnons made slim "blades." Narrowness is, in fact, the very definition of a blade, which in paleoarcheology means nothing more than a stone tool twice as long as it is wide. In Europe a new, efficient way of producing blades from a flint core appeared as part of the "cultural explosion" that coincided with the appearance of the Cro-Magnon people. Here in the Levant, however, the arrival of anatomically modern humans was marked by no fancy new tools not to mention no painted caves, beaded necklaces, or other evidence of exploding Cro-Magnon couture. In this part of the world, how modern a hominid looked in its body said nothing about how modernly it behaved.

Just a couple of bus stops up the coastal road from Kebara is the cave of Tabun, with over 80 vertical feet of deposits spanning more than 100,000 years of human occupation. The treasures of Tabun, like those of Kebara, are Neanderthals. Literally around the corner from Tabun is another cave, called Skhul, where some fairly modern-looking humans were found in the 1930s. And a few miles inland from Kebara on a hill in lower Galilee is Qafzeh, where in 1965 a young French anthropologist named Bernard Vandermeersch found a veritable Middle Paleolithic cemetery of distinctly modern humans. But though the bones in these caves include both Neanderthals and modern humans, the tools found with the bones are all pretty much the same.

In 1982, Arthur Jelinek of the University of Arizona made an inspired attempt to massage some sense into the nagging paradox of Mount Carmel. As in Europe later on, he argued, tools get thinner along with the bodies of the people who make them. Only in this case, the reduction is front to back rather than side to side.

The fattest flakes, he showed, came from a layer near the bottom of Tabun cave, where a partial skeleton of a Neanderthal woman had turned up; if flake thickness was indeed a true measure of time, then she was the oldest in the group. The next oldest would be the Neanderthal infant that Stekelis had found at Kebara. The modern humans of Skhul yielded flake tools that were flatter. And the flattest of all belonged to the moderns of Qafzeh cave. Although the physically modern Skhul-Qafzeh people might not have crossed the line into full-fledged, blade-based humanness, they appeared, as Jelinek wrote, "on the threshold of breaking away."

"Our current evidence from Tabun suggests an orderly and continuous progress of industries in the southern Levant," he went on, "paralleled by a morphological progression from Neanderthal to modern man." According to this scenario, the Neanderthals simply evolved into modern humans. There was no collision of peoples or cultures; two kinds of human never met, because there was really only one kind, changing through time.

If Jelinek's conventional chronology based on slimming tool forms was right, the fossils found in Qafzeh could be "proto-Cro-Magnons," the evolutionary link between a Neanderthal past and a Cro-Magnon future—and thence to the present moment. But the dating methods he used were *relative,* merely inferring an age for the skeletons by where they fell in an overall chronological scheme. What was needed was a new way of measuring time, preferably an *absolute* dating technique that could label the Mount Carmel hominids with an age in actual calendar years.

The most celebrated absolute dating method is radiocarbon dating, which measures time by the constant, steady decay of radioactive carbon atoms. Developed in the 1940s, radiocarbon dating is still one of the most accurate ways to pin an age on a site, so long as it is younger than around 40,000 years. In older materials the amount of radioactive carbon still left undecayed is so small that even the slightest amount of contamination leads to highly inaccurate results. Another technique, relying on the decay of radioactive potassium instead of carbon, has been used since the late 1950s to date volcanic deposits older than half a million years. Radiopotassium was the method of choice for dating the famous East African early hominids like Lucy, as well as the new "root hominid," *Australopithecus ramidus,* announced in 1994. Until recently, though, everything that lived between the ranges of these two techniques—including the moderns at Qafzeh and the Neanderthals at Kebara—fell into a chronological black hole.

In the early 1980s, however, Hélène Valladas, a French archeologist, used a new technique called thermoluminescence, or TL, to date flints from the Kebara and Qafzeh caves. As applied to these flints, the technique is based on the fact that minerals give off a burst of light when heated to about 900 degrees. It is also based on the certainty that past humans, like present ones, were sometimes careless. In the Middle Paleolithic, some flint tools happened to lie around in the path of careless feet, and some tools got kicked into fires, opening up an exquisite opportunity for absolute dating. When a flint tool was heated sufficiently by the fire, it gave up its thermoluminescent energy Over thousands of years, that energy slowly built up again. The dating of fire-charred tools is thus, in principle, straightforward: the brighter a bit of flint glows when heated today, the longer since the time it was last used.

By 1987, Valladas and her physicist father, Georges, had squeezed an age of 60,000 years out of the burnt tools found beside Moshe at Kebara. That number pleased everybody, since it agreed with time schemes arrived at through relative dating methods. The shocker came the following year, when Valladas and her colleagues announced the results of their work at Qafzeh: the "modern" skeletons were 92,000 years old, give or take a few thousand.

Several other Neanderthal and modern human sites have since been dated with TL, and the one at Qafzeh remains not only the most sensational but the surest. Key sites in the Levant have also been dated by a "sister" technique called electron-spin resonance (ESR). Large mammal teeth found near the Qafzeh skeletons came back with an ESR date even older than Valladas's thermoluminescent surprise. The skeletons were at least 100,000 and perhaps 115,000 years old. "People said that TL had too many uncertainties," Bernard Vandermeersch told me. "So we gave them ESR. By now it is very difficult to dispute that the first modern humans in the Levant were here by 100,000 years ago."

Clearly, if modern humans were inhabiting the Levant 40,000 years before the Neanderthals, they could hardly have evolved from them. If the dates are indeed correct, it is hard to see what else one can do with the venerated belief in our Neanderthal ancestry but chuck it, once and for all.

Case closed? On the contrary, the dates only twist the mystery on Mount Carmel even tighter. Presuming that the moderns did not just come for a visit 100,000 years ago and then politely withdraw, they must have been around when the Neanderthals arrived 40,000 years later—if the Neanderthals as well

weren't there to begin with: the latest ESR dates for the Tabun Neanderthal woman place her there 110,000 years ago. Either way, two distinct *kinds* of human were apparently squeezed together in an area not much larger than the state of New Jersey, and for a long time—at least 25,000 years and perhaps 50,000 or more.

Rather than resolving the paradox, the new dating techniques only teased out its riddles. If two kinds of human were behaving the same way in the same place at the same time, how can we call them different? If modern humans did not descend from the Neanderthals but replaced them instead, why did it take them so long to get the job done?

At Kebara, I took the paradox with me to mull over outside, on a still summer afternoon, where the horizon manifested the present moment in the silhouette of an oil tanker, far out at sea. If the names "Neanderthal" and "modern human" are meaningful distinctions, if they have as much reality, say, as the oil tanker pasted onto the horizon, then they cannot be blended, any more than one can blend the sea and the sky. But what if they are mere edges after all, edges that might have had firm content in France and Spain but not here, not in *this* past; edges whose contents spilled ever and leaked into each other so profusely that no true edges can be said to exist at all?

In that case, there would be no more mystery. The Levantine paradox would be a trick knot; pull gently from both ends and it unravels on its own. Think of one end of the rope as cultural. Every species has its own ecological niche, its unique set of adaptations to local habitats. The "principle of competitive exclusion" states that two species cannot squeeze into the same niche: the slightly better adapted one will eventually drive the other one out. Traditionally, the human niche has been defined by culture, so it would be impossible for two kinds of human to coexist using the same stone tools to compete for the same plant and animal resources. One would drive the other into extinction, or never allow it to gain a foothold.

"Competitive exclusion would preclude the coexistence of two different kinds of hominid in a small area over a 40,000- or 50,000-year period unless they had different adaptations," says Geoffrey Clark of Arizona State University. "But as far as we can tell, the adaptations were identical at Kebara and Qafzeh." Clark adds to the list of common adaptations the use of symbols—or lack of it. Perhaps Neanderthals lacked complex social symbols like beads, artwork, and elaborate burial. But so, he believes, did their skinny contemporaries down the road at Qafzeh. If *neither* was littering the landscape with signs of some new mental capacity, by what right do we favor the skinny one with a brilliant future and doom the other to dull extinction?

This leads to the morphological end of the rope. If the two human types cannot be distinguished on the basis of their tools, then the only valid way of telling a Neanderthal from a modern human is to declare that one looks "Neanderthalish" and the other doesn't. If you were to take all the relevant fossils and line them up, could you really separate them into two mutually inclusive groups, with no overlap? A replacement advocate might think so, but a believer in continuity like Geoffrey Clark insists that you could not. He thinks the lineup might better be characterized as one widely variable population, running the gamut from the most Neanderthal to the most modern. The early excavators at Tabun and Skhul saw the fossils there as an intermediate grade between archaic and modern *Homo sapiens*. Perhaps they were right. "The skeletal material is anything but clearly 'Neanderthal' and clearly 'modern,'" Clark maintains, "whatever those terms mean in the first place, which I don't think is much."

This view preserves the traditional idea of continuity but abandons the *process:* there was no evolution from one kind of human to another—from Neanderthal to modern—because there was, in fact, no "other." But for all its appeal, the "oneness" solution to the Levantine paradox is fundamentally flawed. Nobody disputes that the tool kits of the two human types are virtually identical. But it does not logically follow that the toolmakers must be identical as well. Middle Paleolithic tool kits are associated in our minds with Neanderthals because they are the best known human occupants of the Middle Paleolithic. But if people with modern anatomy turn out to have been living back then, too, why *wouldn't* they be using the same culture as the Neanderthals?

"If you ask me, forget about the stone tools," Ofer Bar-Yosef told me. "They can tell you nothing, zero. At most they say something about how they were preparing food. But is what you do in the kitchen all of your life? Of course not. Being positive people, we are not willing to admit that some of the missing evidence might be the crucial evidence we need to solve this problem."

Whatever the tools suggest, the skeletons of moderns and Neanderthals look different, and the pattern of their differences is too consistent to dismiss. As anthropologist Erik Trinkaus of the University of New Mexico has shown, those skeletal differences clearly reflect two distinct patterns of behavior, however alike the archeological leavings may be. Furthermore, the two physical types do not follow one from the other, nor do they meet in a fleeting moment before one triumphs and the other fades. They just keep on going, side by side but never mingling. In his behavioral approach to bones, Trinkaus purposely disregards the features that might best discriminate Neanderthals and moderns from each other genetically. By definition, these traits are poor indicators of the effects of lifestyle on bone, since their shape and size are decided by heredity, not by use. But there is one profoundly important aspect of human life where behavior and heredity converge: the act that allows human lineages to continue in the first place.

Humans love to mate. They mate all the time, by night and by day, through all the phases of the female's reproductive cycle. Given the opportunity, humans throughout the world will mate with any other human. The barriers between races and cultures, so cruelly evident in other respects, melt away when sex is at stake. Cortés began the systematic annihilation of the Aztec people—

but that did not stop him from taking an Aztec princess for his wife. Blacks have been treated with contempt by whites in America since they were first forced into slavery, but some 20 percent of the genes in a typical African American are "white." Consider James Cook's voyages in the Pacific in the eighteenth century. "Cook's men would come to some distant land, and lining the shore were all these very bizarre-looking human beings with spears, long jaws, browridges," archeologist Clive Gamble of Southampton University in England told me. "God, how odd it must have seemed to them. But that didn't stop the Cook crew from making a lot of little Cooklets."

Humans love to mate. The barriers between races, so cruelly evident in other respects, melt away when sex is at stake.

Project this universal human behavior back into the Middle Paleolithic. When Neanderthals and modern humans came into contact in the Levant, they would have interbred, no matter how "strange" they might initially have seemed to each other. If their cohabitation stretched over tens of thousands of years, the fossils should show a convergence through time toward a single morphological pattern, or at least some swapping of traits back and forth.

But the evidence just isn't there, not if the TL and ESR dates are correct. Instead the Neanderthals stay staunchly themselves. In fact, according to some recent ESR dates, the least "Neanderthalish" among them is also the oldest. The full Neanderthal pattern is carved deep at the Kebara cave, around 60,000 years ago. The moderns, meanwhile, arrive very early at Qafzeh and Skhul and never lose their modern aspect. Certainly, it is possible that at any moment new fossils will be revealed that conclusively demonstrate the emergence of a "Neandermod" lineage. From the evi-

dence in hand, however, the most likely conclusion is that Neanderthals and modern humans were not interbreeding in the Levant.

Of course, to interbreed, you first have to meet. Some researchers have contended that the coexistence on the slopes of Mount Carmel for tens of thousands of years is merely an illusion created by the poor archeological record. If moderns and Neanderthals were physically isolated from each other, then there is nothing mysterious about their failure to interbreed. The most obvious form of isolation is geographic. But imagine an isolation in time as well. The climate of the Levant fluctuated throughout the Middle Paleolithic—now warm and dry, now cold and wet. Perhaps modern humans migrated up into the region from Africa during the warm periods, when the climate was better suited to their lighter, taller, warm-adapted physiques. Neanderthals, on the other hand, might have arrived in the Levant only when advancing glaciers cooled their European range more than even their cold-adapted physiques could stand. Then the two did not so much cohabit as "time-share" the same pocket of landscape between their separate continental ranges.

While the solution is intriguing, there are problems with it. Hominids are remarkably adaptable creatures. Even the ancient *Homo erectus*—who lacked the large brain, hafted spear points, and other cultural accoutrements of its descendants—managed to thrive in a range of regions and under diverse climatic conditions. And while hominids adapt quickly, glaciers move very, very slowly, coming and going. Even if one or the other kind of human gained sole possession of the Levant during climatic extremes, what about all those millennia that were neither the hottest nor the coldest? There must have been long stretches of time—perhaps enduring as long as the whole of recorded human history—when the Levant climate was perfectly suited to both Neanderthals and modern humans. What part do these in-between periods play in the time-sharing scenario? It doesn't make sense that one human population should politely vacate Mount Carmel just before the other moved in.

If these humans were isolated in neither space nor time but were truly contemporaneous, then how on earth did they fail to mate? Only one solution to the mystery is left. Neanderthals and moderns did not interbreed in the Levant because they *could* not. They were reproductively incompatible, separate species—equally human, perhaps, but biologically distinct. Two separate species, who both just happened to be human at the same time, in the same place.

Cohabitation in the Levant in the last ice age conjures up a chilling possibility. It forces you to imagine two equally gifted, resourceful, emotionally rich human entities weaving through one tapestry of landscape—yet so different from each other as to make the racial diversity of present-day humans seem like nothing. Take away the sexual bridge and you end up with two fully sentient human species pressed into one place, as mindless of each other as two kinds of bird sharing the same feeder in your backyard.

When paleoanthropologists bicker over whether Neanderthal anatomy is divergent enough to justify calling Neanderthals a separate species from us, they are using a *morphological* definition of a species. This is a useful pretense for the paleoanthropologists, who have nothing but the shapes of bone to work with in the first place. But they admit that in the real, vibrantly unruly natural world, bone morphology is a pitifully poor indication of where one species leaves off and another begins. Ian Tattersall, an evolutionary biologist at the American Museum of Natural History, points out that if you stripped the skin and muscle off 20 New World monkey species, their skeletons would be virtually indistinguishable. Many other species look the same even with their skins still on.

The most common definition of *biological* species, as opposed to the morphological make-believes paleontologists have to work with, is a succinct utterance of the esteemed evolutionary biologist Ernst Mayr: "Species are groups of actually or potentially interbreeding natural populations that are reproductively isolated from other such groups." The key phrase is *reproductively isolated:* a

species is something that doesn't mate with anything but itself. The evolutionary barriers that prevent species from wantonly interbreeding and producing a sort of organismic soup on the landscape are called isolating mechanisms. These can be any obstructions that prevent otherwise closely related species from mating to produce fertile offspring. The obstructions may be anatomical. Two species of hyrax in East Africa share the same sleeping holes, make use of common latrines, and raise their young in communal "play groups." But they cannot interbreed, at least in part because of the radically different shapes of the males' penises. Isolating mechanisms need not be so conspicuous. Two closely related species might have different estrous cycles. Or the barrier might come into play after mating: the chromosomes are incompatible or perhaps recombine into an offspring that is incapable of breeding, an infertile hybrid like a mule.

It is easy to see why paleoanthropologists despair over trying to apply Mayr's biological concept of species to ancient hominids. The characteristics needed to recognize a biological species—the isolating mechanism—are not the kind that usually turn up as fossils. How can an estrous cycle be preserved? What does an infertile hybrid, reduced to a few fragments of its skeleton, look like? How does a chromosomal difference turn into stone?

But there is another way of looking at species that might offer hope. The biological-species concept is a curiously negative one: what makes a species itself is that it doesn't mate with anything else. A few years ago a South African biologist named Hugh Patterson turned the biological-species concept inside out, proposing a view of a species based on not with whom it *doesn't* mate but with whom it *does.* Species, according to Patterson, are groups of individuals in nature that share "a common system of fertilization mechanisms."

With reproduction at its core, Patterson's concept is just as "biological" as Mayr's. But he turns the focus away from barriers preventing interbreeding and throws into relief the adaptations that together ensure the successful meeting of a sperm and an egg. Obviously,

sex and conception are fertilization mechanisms, as is the genetic compatibility of the two parents' chromosomes. But long before a sperm cell gets near a receptive egg, the two sexes must have ways of recognizing each other as potential mates. And therein, perhaps, lies a solution to the mystery of Mount Carmel.

Every mating in nature begins with a message. It may be chemically couched: eggs of the brown alga *Ascophyllum nodosum,* for example, send out a chemical that attracts the sperm of *A. nodosum* and no other. It may be a smell. As any dog owner knows, a bitch in heat lures males from all over the neighborhood. Note that the scent does not draw squirrels, tomcats, or teenage boys. Many birds use vocal signals to attract and recognize the opposite sex, but only of their own species. "A female of one species might *hear* the song of the male of another," explains Judith Masters, a colleague of Patterson's at the University of Witwatersrand, "but she won't make any response. There's no need to talk about what *prevents* her from mating with that male. She just doesn't see what all the fuss is about."

A species' mate-recognition system is extremely stable compared with adaptations to the local habitat. A sparrow born with a slightly too short beak may or may not be able to feed its young as well as another with an average-size beak. But a sparrow who sings an unfamiliar song will not attract a mate and is not going to have any young at all. He will be plucked from the gene pool of the next generation, leaving no evolutionary trace of his idiosyncratic serenade. The same goes, of course, for any sparrow hen who fails to respond to potential mates singing the "correct" tune. With this kind of price for deviance, everybody is a conservative. "The only time a species' mate-recognition system will change is when something really dramatic happens," Masters says.

For the drama to unfold, a population must be geographically isolated from its parent species. If the population is small enough and the habitat radically different from what it was previously, even the powerful evolutionary inertia of the mate-recognition system may be overcome. This change in reproduction may be accompanied by new adaptations to the environment. Or it may not. Either way, the only shift that marks the birth of a new species is the one affecting the recognition of mates. Once the recognition threshold is crossed, there is no going back. Even if individuals from the new population and the old come to live in the same region again—let's say in a well-trafficked corridor of fertile land linking their two continental ranges—they will no longer view each other as potential mates.

The human mate-recognition system is overwhelmingly visual. "Love comes in at the eye," wrote Yeats, and the locus of the human body that lures the eye most of all is the face—a trait our species shares with many other primates. "It is a common Old World anthropoid ploy," says Masters. "Cercopithecoid monkeys have a whole repertoire of eyelid flashes. Forest guenons have brightly painted faces with species-specific patterns, which they wave like flags in the forest gloom. Good old evolution tinkering away, providing new variations on a theme."

Faces are exquisitely expressive instruments. Behind our facial skin lies an intricate web of musculature, concentrated especially around the eyes and mouth, evolved purely for social communication—expressing interest, fear, suspicion, joy, contentment, doubt, surprise, and countless other emotions. Each emotion can be further modified by the raise of an eyebrow or the slight flick of a cheek muscle to express, say, measured surprise, wild surprise, disappointed surprise, feigned surprise, and so on. By one estimate, the 22 expressive muscles on each side of the face can be called on to produce 10,000 different facial actions or expressions.

Among this armory of social signals are stereotyped, formal invitations to potential mates. The mating display we call flirtation plays the same on the face of a New Guinean tribeswoman and a

lycéenne in a Parisian café: a bashful lowering of the gaze to one side and down, followed by a furtive look at the other's face and a coy retreat of the eyes. A host of other sexual signals are communicated facially—the downward tilt of the chin, the glance over the shoulder, the slight parting of the mouth. The importance of the face as an attractant is underscored by the lengths to which humans in various cultures go to embellish what is already there. But the underlying message is communicated by the anatomy of the face itself. "'Tis not a lip, or eye, we beauty call, / but the joint force and full result of all," wrote Alexander Pope. And it is that "joint force"—over generations—that keeps our species so forcefully joined.

This brings us back to the Levant: two human species in a tight space for a long time. The vortex of anatomy where Neanderthals and early moderns differ most emphatically, where a clear line can be drawn between *them* and *us* by even the most rabid advocate of continuity is, of course, the face. The Neanderthal's "classic" facial pattern—the midfacial thrust picked up and amplified by the great projecting nose, the puffed-up cheekbones, the long jaw with its chinless finish, the large, rounded eye sockets, the extra-thick browridges shading it like twin awnings—is usually explained as a complex of modifications relating to a cold climate, or as a support to heavy chewing forces delivered to the front teeth. Either way it is assumed to be an environmental adaptation. But what if these adaptive functions of the face were not the reason they evolved in the first place? What if the peculiarities evolved instead as the underpinnings of totally separate, thoroughly Neanderthal mate-recognition system?

Although it is merely a speculation, the idea fits some of the facts and solves some of the problems. Certainly the Neanderthals' ancestors were geographically cut off from other populations enough to allow some new mate-recog-

nition system to emerge. During glacial periods, contact through Asia was blocked by the polar glaciers and vast uninhabitable tundra. Mountain glaciers between the Black and Caspian Seas all but completed a barrier to the south. "The Neanderthals are a textbook case for how to get a separate species," archeologist John Shea told me. "Isolate them for 100,000 years, then melt the glaciers and let 'em loose."

If mate recognition lay behind a species-level difference between Neanderthals and moderns, the Levantine paradox can finally be put to rest. Their cohabitation with moderns no longer needs explanation. Neanderthals and moderns managed to coexist through long millennia, doing the same human-like things but without interbreeding, simply because the issue never really came up.

The idea seems scarcely imaginable. Continuity believers cannot credit the idea of two human types coexisting in sexual isolation. Replacement advocates cannot conceive of such a long period of coexistence without competition, if not outright violent confrontation. They would rather see Neanderthals and moderns pushing each other in and out of the Levant, in an extended struggle finally won by our own ancestors. Of course, if the Neanderthals were a biologically separate species, something must have happened to cause their extinction. After all, we are still here, and they are not.

Why they faded and we managed to survive is a separate story with its own shocks and surprises. But what happened on Mount Carmel might be more remarkable still. It is something that people today are not prepared to comprehend, especially in places like the Levant. Two human species, with far less in common than any two races or ethnic groups now on the planet, may have shared a small, fertile piece of land for 50,000 years, regarding each other the whole time with steady, untroubled, peaceful indifference.

Learning to Love Neanderthals

Does the 25,000-year-old body of a child found in Portugal make it more likely that they are our ancestors?

By Robert Kunzig

Photographs by Claudio Vazquez

What you want, when you hold a pendant fashioned 35,000 years ago by a Neanderthal—a fox's tooth with a tiny hole for a leather string—what you want is something only the movies can give. A close-up, in the lab's neon light, on the mottled canine between your fingers, the focus so tight you can see the scratches made by the stone tool. The picture fades, and next you see the same tooth in different hands, stronger ones with beefy fingers: the hands of the craftsman. He is piercing the tooth with a sharpened piece of flint. Behind him squats a rough tent of hides stretched over mammoth tusks: behind that the dark mouth of a cave. Before and below him a river meanders lazily between birches and willows. Reindeer graze on the far bank. On an early morning in spring, in northern Burgundy during the Ice Age, the light coming in low over the far bluff catches the craftsman's pale, weathered face. It is a human face. The eyes, under the jutting brow, are human eyes, alive with concentration, with memories of other seasons at this place, with intelligence and hope.

No, hold it: Maybe those Neanderthal eyes are blank as a cat's, all surface, with nothing behind them but dumb instinct and a bit of animal cunning—no memories, no plan, no clue.

Back to spring 1999 and the lab, at a modern campus of the University of Paris. An archaeologist named Dominique Baffier holds the tooth. For the past few days newspapers the world over have been reporting the discovery in Portugal of the skeleton of a 4-year-old child, dead for 25,000 years. The discoverers, led by Portuguese archeologist João Zilhão are making a ground-breaking claim, that the skeleton shows traces of both Neanderthal and modern human ancestry, evidence that modern humans did not simply extinguish the Neanderthals, as many researchers had come to think. Instead the two kinds of human were so alike that in Portugal, at least, they intermingled—and made love—for thousands of years.

After João Zilhão and his colleagues excavated the skeleton, they covered the site with rocks to protect it. More Paleolithic treasure may wait in the dirt.

The claim is controversial. So, too, and for similar reasons, is the fox-tooth Baffier is holding. A collection of such ornaments is arrayed on the table in front of her, along with delicate bone tools—awls for punching through animals hides, needles for sewing or perhaps for pinning up hair. All these artifacts were dug from the mouth of a limestone cave four decades ago at Arcy-sur-Cure, a hundred miles southeast of Paris. Just in the past year, though, the Arcy artifacts have become the subject of heated debate. Zilhão, Baffier, and several French colleagues claim the artifacts show that Neanderthals were not inferior to our ancestors, the Cro-Magnons. Independently, they underwent the same leap into modernity, the same emergence of symbolic thought that millennia later allowed Cro-Magnons to paint on cave walls.

A fox-tooth pendant is not a cave painting, as Baffier well knows, for she studies those paintings too. But it is a symbolic statement. "Oh, it's beautiful," she says quietly turning the Neanderthal pendant in her fingers, peering at it over her glasses. "It's beautiful and it's moving. A 35,000-year-old bijou—isn't that moving?"

João Zilhão director of the Portuguese Institute of Archeology, got the call from his wife and fellow archeologist, Cristina Araújo, while he was at a conference in Japan. She had heard from João Maurício and Pedro Souto, coworkers at a Neanderthal cave site. They had heard from a student named Pedro Ferreira, who had gone looking for rock art in the Lapedo Valley, about 90 miles north of Lisbon, and had found some small paintings.

Last November 28, Maurício and Souto went to the Lapedo Valley—which is really a small, steep ravine, a bit over a mile long and a stone's-throw wide. Olive groves and wildflowers,

Photographs by Claudio Vasquez

The Lapedo Valley kid was lying on its back turned slightly toward the cliffs. The feet are at right, the ribs and vertebrae at left.

vegetable fields and villages sprawl up to the lip of the canyon, but its cool, lush depths are a world apart. The small stream at the bottom, the Caranguejeira, is hidden by reeds and bushes; the canyon walls themselves are practically hidden by a riot of diverse greenery.

Ferreira showed Maurício and Souto his rock art, and they confirmed that it looked man-made and old (CopperAge, it turned out). Then they looked across the ravine to the south side. Above the treetops they could see a limestone wall leaning out over the canyon. Prehistoric humans liked to take shelter under overhanging rocks like that. Maurício and Souto decided to have a look.

They found a mess, construction debris strewn along the base of the cliff, including an abandoned trailer, an old tractor hood, and a giant section of concrete drainpipe. In a fissure in the wall, just above eye level, Maurício and Souto saw sediments laced with stone tools, lots of animal bones, and black flecks of charcoal—the remains of Paleolithic campfires. But all along the wall they could also see the white gouge marks left by the teeth of a steam shovel. To build a road, the owner of the area had created a flat terrace where before there had been a tall slope of sediment—sediment that had washed off the top of the cliff and collected at the base over tens of millennia He had used a steam shovel to dig the dirt from the cliff.

In retrospect the demolition man did archeology a big favor. Otherwise, what is now the base of the cliff at Lapedo would still be buried. Maurício saw a rabbit burrow disappearing under the rock. He reached his hand in and pulled out a radius and an ulna—the bones of a human forearm, though he wasn't sure of that at the time.

Soon, as they gently swept away more dirt, they saw a patch of sediment as red as wine and as large as . . . a small child

---

When the news reached Zilhão, he called Cidália Duarte, a physical anthropologist at the Portuguese Institute of Architectural Heritage. They and Araújo went up to Lapedo the next weekend. While Zilhão examined the stone tools embedded in the cliff face seven feet up, Maurício took the bone lady to the bones in the burrow "I looked at them," Duarte recalls, "and I said, 'Ooh—this is human! This is a kid!'" Meanwhile, Zilhão was looking at the stratigraphic sequence. He began to add it up: "If the kid is down there, and this up here is Solutrean. . ."

Solutrean is the third of four successive cultures—Aurignacian, Gravettian, Solutrean, Magdalenian—of the Upper Paleolithic or Late Stone Age. The Solutrean happened around 20,000 years ago. If the Lapedo Valley kid was seven feet below Solutrean sediments, that suggested he died thousands of years before the Solutrean. Zilhão looked at the bones: They were stained reddish with ochre. Red ochre is one of the things Upper Paleolithic moderns painted caves with, but they also buried their dead with it; the color seems to have had symbolic significance.

"So I immediately recognized that something big was there," says Zilhão. "The question was whether the bulldozing had completely destroyed the burial, and all we had to do was collect the fragments, or if something was still there intact."

The next day, Monday, they went back to their day jobs in Lisbon. The following Friday evening they were back at Lapedo, with Duarte and Araújo digging. "By Sunday evening we were really upset, because all we could find were bits and pieces, fragments of bone, and they didn't even have this reddish color," Duarte recalls. With night falling and spirits crumbling, they started tidying up the dig. Those final offhand brushstrokes did it: The red began to appear. Soon, as Araújo and Duarte gently swept away more dirt, they saw a patch of sediment as red as wine and as large as . . . a small child.

Now they faced a paleontological emergency: The skeleton was almost at the surface, exposed to the elements, which in this case included Boy Scouts. A troop had walked by during the weekend and regarded the diggers with ominous curiosity. That week Duarte signed up for an unscheduled vacation. Zilhão quietly abandoned his airy director's office and his paperwork at the archeological institute. "I just went away without telling anybody," he says, "so as not to run the risk of a leak."

They started digging in earnest. Right away they realized how lucky they had been: In removing tens of feet of dirt, the steam shovel had missed by just a few inches the body of the child. Unfortunately it had not missed the skull— Duarte could only find fragments of that. One of the first, though, was a beauty: the left half of the lower jaw, including teeth. It had a sharply pointed chin, which is just what you would expect from a Cro-Magnon; Neanderthals had weak chins.

The child was lying on its back, with its head and torso tilted a bit to its left, toward the cliff, and its right hand on its pelvis. Its right side was crushed, but the left side was intact. Ribs, vertebrae, pelvis, fingers, toes, the long bones of the arms and legs—all were there. Duarte and Araújo worked steadily through the Christmas holidays, hiding their work every night under the old tractor hood. Soon they had a new problem: up to 500 visitors a day. Zilhão's desire to keep the excavation secret had run into his desire to have it documented. He had asked Portuguese public television to videotape it, which the TV folk were happy to do—provided they could also run the story. On Christmas Day it opened the evening news: 'A Child is Born.'

On Christmas Day itself, Duarte was at the site alone. "I wasn't going to leave it there all exposed," she says. The work that day on the rib cage, was particularly delicate. She was digging with a syringe, squirting acetone around the bones to dissolve the dirt—acetone evaporates quickly so it doesn't soak the bone—and then removing it with a paintbrush and a plastic spoon. She was squatting, kneeling, and sometimes lying on her side next to the Kid. Earlier, near the clavicle, she had found a tiny seashell,

covered with red ochre, with a minute hole. The Kid had worn it as a pendant.

At first Zilhão and Duarte guessed they had excavated a boy. The arms and legs looked robust. But then Erik Trinkaus, a paleoanthropologist at Washington University in St. Louis, had a look at them. He decided the Kid was robust not because he was male, but because he had Neanderthal blood.

Trinkaus is a leading authority on both Neanderthal and early modern human anatomy. When the excavation started, Zilhão let him know right away. "João went out and got a digital camera and started e-mailing me images," Trinkaus says. He got excited too: While Upper Paleolithic skeletons in general are rare, there are no reasonably complete children's skeletons at all. Right after New Year's, while Duarte was still excavating, Trinkaus hopped a plane to Portugal.

He measured all the bones he could, especially the limbs—his specialty. In 1981, Trinkaus published a paper on limb evolution that is still cited. In it he documented a geographic pattern in people today: They get shorter the farther they are from the tropics and the closer they are to the poles. More precisely their extremities get shorter. Inuits and Lapps have shorter forearms relative to their upper arms and shorter shinbones relative to their thighbones than

do the Masai of East Africa. There is a simple explanation: Shorter, stockier bodies fare better in cold climates because they have less surface area to radiate heat. By measuring fossil limbs, Trinkaus showed that Neanderthals, denizens of ice age Europe, were hyperarctic—they had an even smaller shinbone-thighbone ratio than do Lapps. Early moderns from the Near East and Europe, on the other hand, were decidedly tropical in their legginess, like Africans today

This was some of the earliest evidence for a theory of human origins that had not even been formulated then, but has since become orthodoxy. The out-of-Africa theory holds that humans today are descended from a small population of (long-legged) early moderns that walked out of Africa around 100,000 years ago. As they spread all over the world, they replaced whatever archaic humans they met, which in Europe were the Neanderthals. In this view, Neanderthals are a distinct population and maybe even a distinct species that went extinct at the hands of our ancestors, leaving no legacy at all.

Two years ago, when a team led by Svante Pääbo, now of the Max Planck Institute for Evolutionary Anthropology in Leipzig, isolated DNA from a Neanderthal bone, many people thought the case had been clinched. The Neanderthal DNA was different enough that there seemed to be no trace of it in modern DNA; it suggested that Neanderthals and modern humans had evolved separately for half a million years and were unlikely to have interbred. Trinkaus was not moved: "There's a general impression that if something comes out of a million-dollar machine, then it's truth, whereas if something comes out of a bunch of dirty old bones that we clean with paintbrushes, then it's vague and ambiguous."

Based on his analysis of dirty bones from the Czech Republic and Croatia, Trinkaus has long favored a sort of watered-down out of Africa, in which the gene pool of modern humans migrating into Europe was seasoned by interbreeding with Neanderthals. The amount of interbreeding would have varied from place to place. One place Trinkaus

The fragile rib cage of the Kid was excavated as a single block encased in plaster; it is now being cleaned in Cidália Duarte's lab.

didn't expect to see much, though, was on the Iberian peninsula, the last refuge of the Neanderthals.

Cro-Magnons reached northern Spain nearly 40,000 years ago, but for some reason they didn't spread south for another 10,000 years. By the time they crossed the Ebro frontier, as Zilhão has dubbed it after the large river in northern Spain, their kind had already executed striking cave paintings. South of the Ebro they encountered Neanderthals who were still making stone tools in the Middle Paleolithic fashion and not making ornaments at all. The better-armed modern invaders would have been almost as tall as the Masai and maybe as black; the indigenes would have been short and as pale as Lapps. It would be easy to picture the former simply wiping out the latter. But it is from this clash of cultures and anatomies, Trinkaus argues, that the Kid was born.

The sharp point of the Kid's chin screamed Cro-Magnon. So did the relatively small front teeth: Neanderthal front teeth were large compared with their molars. So did the red ochre burial style. And so, finally, did the radiocarbon date: At 24,500 years old, it was much younger than the last signs of Neanderthals.

But when Trinkaus measured the angle between the horizontal tooth line and the vertical line from the frontmost tooth down to the chin, the symphyseal angle, he got a clue that this was a strange Cro-Magnon. Instead of jutting forward of the teeth, the chin retreated a shade behind them. Cro-Magnon chins didn't do that, Trinkaus says, but Neanderthal chins did.

Even more significant, he thinks, are the limb proportions. Trinkaus measured the shinbone and the thighbone and found that the ratio fell way over at the Neanderthal end of the curve. He compared the circumference of the bones with their length, and found that the child had leg bones strong enough to support a stocky Neanderthal body The limb proportions along with the receding chin are enough, Trinkaus says, to prove the child had Neanderthal ancestors as well as Cro-Magnon ones. "It only takes one feature," he says. "We've got two."

But many of his peers are skeptical. Arctic limb proportions don't prove a Neanderthal influence, some argue, since Lapps have them too; maybe the Kid was just an ordinary Cro-Magnon who had adapted to the ice age. And the mere fact that the skeleton is that of a child—whose features were still changing, and for whom no good Cro-Maguon comparisons exist—makes some researchers uneasy "If an adult skeleton had been found, nice and complete, I'm sure we would still have fierce discussions," says Jean-Jacques Hublin of the French National Center for Scientific Research. "But interpreting the remains of a child, of which almost none of the skull is left—that's really a perilous exercise.

This summer, as the dig progresses, Duarte will be looking for more pieces of the child's skull. Finding its two front teeth would be nice (Neanderthals had big ones), or the occipital bone in the rear of the skull (it bulged out in Neanderthals), or even the tiny labyrinth of the inner ear. Hublin has used that feature, and that feature alone, to diagnose a Neanderthal bone at Arcy-sur-Cure.

Duarte also hopes to find the Kid's parents; Upper Paleolithic burials often come in groups. But Trinkaus does not expect she will find a Neanderthal mom and a Cro-Magnon dad. Zilhão's archeological evidence suggests the Kid was born at least two millennia after Neanderthals and Cro-Magnons first met in Portugal. The Kid, Trinkaus argues, must be the product of interbreeding over that entire period, not a one-time hybrid produced by star-crossed lovers. "This is not just two individuals who happened to meet in the bushes," he says.

There is likely to be fierce discussion about that conclusion too. The out-of-Africa model can readily tolerate a little hankypanky between Neanderthals and Cro-Magnons. "In fact I expect it," says Hublin. A few hybrids wouldn't even disprove the view that Neanderthals were a different species. Animals of closely related species can sometimes interbreed, and sometimes the offspring are even fertile.

But if whole populations of Neanderthals and Cro-Magnons were blend-

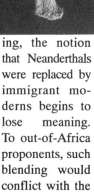

Left: The left arm and hand. Below: The left leg. The thigh bone is eight inches long, the shin just over six, Neanderthal proportions.

ing, the notion that Neanderthals were replaced by immigrant moderns begins to lose meaning. To out-of-Africa proponents, such blending would conflict with the genetic and fossil evidence, and with the simple observation that people today look like Cro-Magnons and not like Neanderthals. To paleontologists who don't believe the out-of-Africa model, however—who think modern humans evolved all over the world from interbreeding populations of archaic humans including Neanderthals—Trinkaus's Lapedo kid is welcome news.

Trinkaus himself has never taken sides in this bitter debate. He sees it now moving onto his middle ground: A migration out of Africa happened, sure, but the migrants also interbred to varying degrees with the people they met along

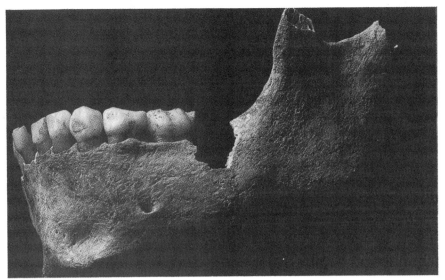

From his Cro-Magnon ancestors, the kid got small front teeth and a pointy chin—but the chin doesn't jut out as a Cro-Magnon's should.

the way. Neanderthals are not us, but neither are they an evolutionary dead end—the Kid, if he is right, puts the truth in the middle. "Trinkaus is in the stratosphere," says Zilhão. "He has believed this for a long time. I couldn't care less—they could just as well be different species as far as I'm concerned. This just comes in handy."

It comes in handy as ammunition in a separate fight that is Zilhão's own. That debate concerns how smart Neanderthals were, and it is centered on the cave digs at Arcy. When French archeologists excavated there in the 1950s, they found dozens of animal-tooth pendants, bone tools, and 40 pounds of red ochre spread over the floor. At other sites, such artifacts have been attributed to modern humans. A few years ago, though, after Hublin CT-scanned a skull fragment found alongside the artifacts, and revealed the inner ear, he convinced most people that the bone was that of a Neanderthal, and so were the artifacts.

The conventional explanation is that the Neanderthal craftsmen at Arcy must have been imitating our ancestors. Modern humans were invading western Europe at around the time—35,000 to 45,000 years ago—when the Neander-

thals were at Arcy. And whereas the few dozen ornaments found there are practically the only ones attributed to Neanderthals, thousands have been found at Cro-Magnon sites. Many researchers say it is common sense to assume that Neanderthals were "acculturated" by Cro-Magnons. Some even argue that the Neanderthals didn't really understand what they were doing: They copied such modern behaviors as wearing pendants, but they couldn't appreciate the symbolic meaning.

'Forty-thousand years ago, people couldn't read or write—are we saying they didn't have the intelligence?'

Lurking under all this is the question of why we survived and the Neanderthals didn't: Was it because their brains were inferior? That idea drives Zilhão up the wall. "What's involved here is not the wiring of the brain cells, it's the wiring of the brains into what we call culture," he says. "Forty thousand years

ago, people couldn't read or write—are we saying they didn't have the intelligence?"

In a controversial paper published last year with Francesco d'Errico of the University of Bordeaux, and Dominique Baffier, Michèle Julien, and Jacques Pelegrin of the University of Paris, Zilhão tried to show that the Neanderthals were intelligent enough to make the Arcy artifacts—and thus the transition to Upper Paleolithic modernity—all by themselves. The Arcy Neanderthals, the researchers argued, made their tools and ornaments using techniques quite unlike those of the moderns—punching a clean hole through a fox tooth, for instance, whereas modern humans did a cruder job of gouging. Zilhão and d'Errico also claim to have proved, through a technical reanalysis of the highly uncertain dates attributed to nearly every relevant site in western Europe, that the Neanderthals couldn't have imitated Cro-Magnons—because they were acting modern thousands of years before any Cro-Magnons were around.

The Lapedo Valley kid drops into the murky waters of this debate like a cannonball. Some researchers say it makes no difference at all if the Kid is a hybrid. But if Trinkaus and Zilhão can prove that modern humans and Neanderthals mixed extensively in Portugal, it would surely affect our view of Neanderthals—by giving us an inkling of the view our ancestors held. Would they really have fraternized with beings who were too dim to understand the purpose of a necklace?

"If you have two populations of hunter-gatherers that are totally different species, that are doing things in very different ways, have different capabilities—they're not going to blend together," Trinkaus says. "They're going to remain separate. So the implication from Portugal is that when these people met, they viewed each other as people. One group may have looked a little funny to the other one—but beyond that they saw each other as human beings. And treated each other as such."

Archaeologists Rediscover Cannibals

*At digs around the world, researchers have unearthed strong
new evidence that people ate their own kind from the
early days of human evolution through recent prehistory*

When Arizona State University bioarchaeologist Christy G. Turner II first looked at the jumbled heap of bones from 30 humans in Arizona in 1967, he was convinced that he was looking at the remains of a feast. The bones of these ancient American Indians had cut marks and burns, just like animal bones that had been roasted and stripped of their flesh. "It just struck me that here was a pile of food refuse," says Turner, who proposed in *American Antiquity* in 1970 that these people from Polacca Wash, Arizona, had been the victims of cannibalism.

But his paper was met with "total disbelief," says Turner. "In the 1960s, the new paradigm about Indians was that they were all peaceful and happy. So, to find something like this was the antithesis of the new way we were supposed to be thinking about Indians"—particularly the Anasazi, thought to be the ancestors of living Pueblo Indians. Not only did Turner's proposal fly in the face of conventional wisdom about the Anasazi culture, but it was also at odds with an emerging consensus that earlier claims of cannibalism in the fossil record rested on shaky evidence. Where earlier generations of archaeologists had seen the remains of cannibalistic feasts, current researchers saw bones scarred by ancient burial practices, war, weathering, or scavenging animals.

To Turner, however, the bones from Polacca Wash told a more disturbing tale, and so he set about studying every prehistoric skeleton he could find in the Southwest and Mexico to see if it was an isolated event. Now, 30 years and 15,000 skeletons later, Turner is putting the final touches on a 1500-page book to be published next year by the University of Utah press in which he says, "Cannibalism was practiced intensively for almost four centuries" in the Four Corners region. The evidence is so strong that Turner says "I would bet a year of my salary on it."

He isn't the only one now betting on cannibalism in prehistory. In the past decade, Turner and other bioarchaeologists have put together a set of clear-cut criteria for distinguishing the marks of

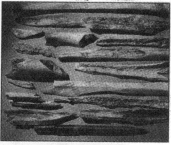

PHOTOS BY: C. TURNER/BERKELEY
Cannibals' house? The Peñasco Blanco great house at Chaco Canyon, New Mexico, where some bones bear cut marks (*upper left*); others were smashed, perhaps to extract marrow.

cannibalism from other kinds of scars. "The analytical rigor has increased across the board," says paleoanthropologist Tim D. White of the University of California, Berkeley. Armed with the new criteria, archaeologists are finding what they say are strong signs of cannibalism throughout the fossil record. This summer, archaeologists are excavating several sites in Europe where the practice may have occurred among our ancestors, perhaps as early as 800,000 years ago. More recently, our brawny cousins, the Neandertals, may have eaten each other. And this behavior wasn't limited to the distant past—strong new evidence suggests that in addition to the Anasazi, the Aztecs of Mexico and the people of Fiji also ate their own kind in the past 2500 years.

These claims imply a disturbing new view of human history, say Turner and others. Although cannibalism is still relatively rare in the fossil record, it is frequent enough to imply that extreme hunger was not the only driving force. Instead of being an aberration, practiced only by a few prehistoric Donner Parties, killing people for food may have been standard human behavior—a means of social control, Turner suspects, or a mob response to stress, or a form of infanticide to thin the ranks of neighboring populations.

Not surprisingly, some find these claims hard to stomach: "These people haven't explored all the alternatives," says archaeologist Paul Bahn, author of the *Cambridge Encyclopedia* entry on cannibalism. "There's no question, for example, that all kinds of weird stuff is done to human remains in mortuary practice"—and in warfare. But even the most prominent skeptic of earlier claims of cannibalism, cultural anthropologist William Arens of the State University of New York, Stony Brook, now admits the case is stronger: "I think the procedures are sounder, and there is more evidence for cannibalism than before."

White learned how weak most earlier scholarship on cannibalism was in 1981, when he first came across what he thought might be a relic of the practice—a massive skull of an early human ancestor from a site called Bodo in Ethiopia. When he got his first look at this 600,000-year-old skull

on a museum table, White noticed that it had a series of fine, deep cuts marks on its cheekbone and inside its eye socket, as if it had been defleshed. To confirm his suspicions, White wanted to compare the marks with a "type collection" for cannibalism—a carefully studied assemblage of bones showing how the signature of cannibalism differs from damage by animal gnawing, trampling, or excavation.

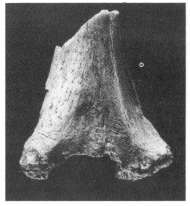

T. D. WHITE/BERKELEY
Unkind cuts. A Neandertal bone from Vindija Cave, Croatia.

"We were naïve at the time," says White, who was working with archaeologist Nicholas Toth of Indiana University in Bloomington. They learned that although the anthropological literature was full of fantastic tales of cannibalistic feasts among early humans at Zhoukoudian in China, Krapina cave in Croatia, and elsewhere, the evidence was weak—or lost.

Indeed, the weakness of the evidence had already opened the way to a backlash, which was led by Arens. He had deconstructed the fossil and historical record for cannibalism in a book called *The Man-Eating Myth: Anthropology and Anthropophagy* (Oxford, 1979). Except for extremely rare cases of starvation or insanity, Arens said, none of the accounts of cannibalism stood up to scrutiny—not even claims that it took place among living tribes in Papua New Guinea (including the Fore, where cannibalism is thought to explain the spread of the degenerative brain disease kuru). There were no reliable eye witnesses for claims of cannibalism, and the archae-

ological evidence was circumstantial. "I didn't deny the existence of cannibalism," he now says, "but I found that there was no good evidence for it. It was bad science."

Physical anthropologists contributed to the backlash when they raised doubts about what little archaeological evidence there was (*Science,* 20 June 1986, p. 1479). Mary Russell, then at Case Western Reserve University in Cleveland, argued, for example, that cut marks on the bones of 20 Neandertals at Krapina Cave could have been left by Neandertal morticians who were cleaning the bones for secondary burial, and the bones would have been smashed when the roof caved in, for example. In his 1992 review in the *Cambridge Encyclopedia,* Bahn concluded that cannibalism's "very existence in prehistory is hard to swallow."

Rising from the ashes

But even as some anthropologists have the ax to Krapina and other notorious cases, a new, more rigorous case for cannibalism in prehistory was emerging, starting in the American Southwest. Turner and his late wife, Jacqueline Turner, had been systematically studying tray after tray of prehistoric bones in museums and private collections in the United States and Mexico. They had identified a pattern of bone processing in several hundred specimens that showed little respect for the dead. "There's no known mortuary practice in the Southwest where the body is dismembered, the head is roasted and dumped into a pit unceremoniously, and other pieces get left all over the floor," says Turner, describing part of the pattern.

White, meanwhile, was identifying other telltale signs. To fill the gap he discovered when he looked for specimens to compare with the Bodo skull, he decided to study in depth one of the bone assemblages the Turners and others had cited. He chose Mancos, a small Anasazi pueblo on the Colorado Plateau from A.D. 1150, where archaeologists had recovered the scattered and broken remains of at least 29 individuals. The project evolved into a landmark book,

Prehistoric Cannibalism at Mancos (Princeton, 1992). While White still doesn't know why the Bodo skull was defleshed—"it's a black box," he says—he extended the blueprint for identifying cannibalism.

In his book, White describes how he painstakingly sifted through 2106 bone fragments, often using an electron microscope to identify cut marks, burn traces, percussion and anvil damage, disarticulations, and breakages. He reviewed how to distinguish marks left by butchering from those left by animal gnawing, trampling, or other wear and tear. He also proposed a new category of bone damage, which he called "pot

T. D. WHITE/BERKELEY

Close shave. Was flesh stripped from the 600,000-year-old Bodo skull in an act of cannibalism?

polish"—shiny abrasions on bone tips that come from being stirred in pots (an idea he tested by stirring deer bones in a replica of an Anasazi pot). And he outlined how to compare the remains of suspected victims with those of ordinary game animals at other sites to see if they were processed the same way.

When he applied these criteria to the Mancos remains, he concluded that they were the leavings of a feast in which 17 adults and 12 children had their heads cut off, roasted, and broken open on rock anvils. Their long bones were broken—he believes for marrow—and their vertebral bodies were missing, perhaps crushed and boiled for oil. Finally, their bones were dumped, like animal bones.

In their forthcoming book, the Turners describe a remarkably similar pattern of bone processing in 300 individuals from 40 different bone assemblages in the Four Corners area of the Southwest, dating from A.D. 900 to A.D. 1700. The strongest case, he says, comes from bones unearthed at the Peñasco Blanco great house at Chaco Canyon in New Mexico, which was the highest center of the Anasazi culture and, he argues, the home of cannibals who terrorized victims within 100 miles of Chaco Canyon, where most of the traumatized bones have been excavated. "Whatever drove the Anasazi to eat people, it happened at Chaco," says Turner.

The case for cannibalism among the Anasazi that Turner and White have put together hasn't swayed all the critics. "These folks have a nice package, but I don't think it proves cannibalism," says Museum of New Mexico archaeologist Peter Bullock. "It's still just a theory."

But even critics like Bullock acknowledge that Turner and White's studies, along with work by the University of Colorado, Boulder's, Paolo Villa and colleagues at another recent site, Fontbrégoua Cave in southeastern France (*Science,* 25 July 1986, p. 431), have raised the standards for how to investigate a case of cannibalism. In fact, White's book has become the unofficial guidebook for the field, says physical anthropologist Carmen Pijoan at the Museum of Anthropology in Mexico City, who has done a systematic review of sites in Mexico where human bones were defleshed. In a forthcoming book chapter, she singles out three sites where she applied diagnostic criteria outlined by Turner, White, and Villa to bones from Aztec and other early cultures and concludes that all "three sites, spread over 2000 years of Mexican prehistory, show a pattern of violence, cannibalism, and sacrifice through time."

White's book "is my bible," agrees paleontologist Yolanda Fernandez-Jalvo of the Museum of Natural History in Madrid, who is analyzing bones that may be the oldest example of cannibalism in the fossil record—the remains of at least six individuals who died 800,000 years ago in an ancient cave at Atapuerca in northern Spain.

Age-old practices

The Spanish fossils have caused considerable excitement because they may represent a new species of human ancestor (*Science,* 30 May, pp. 1331 and 1392). But they also show a pattern familiar from the more recent sites: The bones are highly fragmented and are scored with cut marks, which Fernandez-Jalvo thinks were made when the bodies were decapitated and the bones defleshed. A large femur was also smashed open, perhaps for marrow, says Fernandez-Jalvo, and the whole assemblage had been dumped, like garbage. The treatment was no different from that accorded animal bones at the site. The pattern, says Peter Andrews, a paleoanthropologist at The Natural History Museum, London, is "pretty strong evidence for cannibalism, as opposed to ritual defleshing." He and others note, however, that the small number of individuals at the site and the absence of other sites of similar antiquity to which the bones could be compared leave room for doubt.

A stronger case is emerging at Neandertal sites in Europe, 45,000 to more than 130,000 years old. The new criteria for recognizing cannibalism have not completely vindicated the earlier claims about Krapina Cave, partly because few animal bones are left from the excavation of the site in 1899 to compare with the Neandertal remains. But nearby Vindija Cave, excavated in the 1970s, did yield both animal and human remains. When White and Toth examined the bones recently, they found that both sets showed cut marks, breakage, and disarticulation, and had been dumped on the cave floor. It's the same pattern seen at Krapina, and remarkably similar to that at Mancos, says White, who will publish his conclusions in a forthcoming book with Toth. Marseilles prehistorian Alban DeFleur is finding that Neandertals may also have feasted on their kind in the Moula-Guercy Cave in the Ardeche region of France, where animal and Neandertal bones show similar processing. Taken together, says White, "the evidence from Krapina, Vindija, and Moula is strong."

Not everyone is convinced, however. "White does terrific analysis, but he hasn't proved this is cannibalism," says Bahn. "Frankly, I don't see how he can unless you find a piece of human gut [with human bone or tissue in it]." No matter how close the resemblance to butchered animals, he says, the cut marks and other bone processing could still be the result of mortuary practices. Bullock adds that warfare, not cannibalism, could explain the damage to the bones.

White, however, says such criticism resembles President Clinton's famous claim about marijuana: "Some [although not all] of the Anasazi and Neandertals processed their colleagues. They skinned them, roasted them, cut their muscles off, severed their joints, broke their long bones on anvils with hammerstones, crushed their spongy bones, and put the pieces into pots." Borrowing a line from a review of his book, White says: "To say they didn't eat them is the archaeological equivalent of saying Clinton lit up and didn't inhale."

White's graduate student David DeGusta adds that he has compared human bones at burial sites in Fiji and at a nearby trash midden from the last 2000 years. The intentionally buried bones were less fragmentary and had no bite marks, burns, percussion pits, or other signs of food processing. The human bones in the trash midden, however, were processed like those of pigs. "This site really challenges the claim that these assemblages of bones are the result of mortuary ritual," says DeGusta.

After 30 years of research, Turner says it is a modern bias to insist that cannibalism isn't part of human nature. Many other species eat their own, and our ancestors may have had their own "good" reasons—whether to terrorize subject peoples, limit their neighbors' offspring, or for religious or medicinal purposes. "Today, the only people who eat other people outside of starving are the crazies," says Turner. "We're dealing with a world view that says this is bad and always has been bad. . . . But in the past, that view wasn't necessarily the group view. Cannibalism could have been an adaptive strategy. It has to be entertained."

—Ann Gibbons

Unit 6

Key Points to Consider

❖ Discuss whether the human species can be subdivided into racial categories. Support your position.

❖ How and why did the concept of race develop?

❖ Would you allow archaeologists to study the remains of Kennewick Man, or would you immediately repatriate the bones to Native Americans? Why?

❖ Why is it culture, not race, that best explains human differences in behavior?

❖ To what extent is height a barometer of the health of a society, and why?

 **Links** **www.dushkin.com/online/**

These sites are annotated on pages 4 and 5.

The field of biological anthropology has come a long way since the days when one of its primary concerns was the classification of human beings according to racial type. Although human diversity is still a matter of major interest in terms of how and why we differ from one another, most anthropologists have concluded that human beings cannot be sorted into sharply distinct entities. Without denying the fact of human variation throughout the world, the prevailing view today is that the differences between us exist along geographical gradients, as differences in degree, rather than in terms of the separate and discrete reproductive entities perceived in the past.

One of the old ways of looking at human "races" was that each such group was a subspecies of human that, if left reproductively isolated long enough, would eventually evolve into separate species. While this concept of subspecies, or racial varieties within a species, would seem to apply to some living creatures (such as the dog and wolf or the horse and zebra) and might even be relevant to hominid diversification in the past, the current consensus is that it is not happening today, at least within the human species.

A more recent attempt to salvage the idea of human races has been to perceive them not so much as reproductively isolated entities but as so many clusters of gene frequencies, separable only by the fact that the proportions of traits (such as skin color, hair form, etc.) differ in each artificially constructed group. But, if such "groups" exist only because we say they do and do not have an objective reality of their own, what is to be learned from the exercise? What is its practical value?

It would seem to be more productive to study human traits in terms of their particular adaptiveness (as in "Racial Odyssey" by Boyce Rensberger and "The Tall and the Short of It" by Barry Bogin) or how they help us reconstruct human prehistory (as in "The Lost Man" by Douglas Preston).

Lest anyone think that anthropologists are "in denial" regarding the existence of human races and that the viewpoint expressed here is merely an expression of contemporary political correctness, it should be remembered that serious, scholarly attempts to classify people according to race have been going on now for 200 years and, so far, nothing of value has come of them.

Complicating the matter, as Jonathan Marks elucidates in "Black, White, Other," is that there actually are two concepts of race: the strictly biological one, which was originally set forth by Linnaeus in the 1700s, and the one of popular culture, which has been around since time immemorial. These "two constantly intersecting ways of thinking about the divisions between us," says Marks, have resulted not only in fuzzy thinking about racial biology, but they have also infected the way we think about people and, therefore, the way we treat each other in the social sense. What we should recognize, claim most anthropologists, is that, despite the superficial physical and biological differences between us, when it comes to intelligence, all human beings are basically the same (as discussed by Mark Nathan Cohen in "Culture, Not Race, Explains Human Diversity"). The degrees of variation within our species may be accounted for by the subtle and changing selective forces experienced as one moves from one geographical area to another. However, no matter what the environmental pressures have been, the same intellectual demands have been made upon us all. This is not to say, of course, that we do not vary from each other as individuals. Rather, what is being said is that when we look at these artificially created groups of people called "races," we find the same range of intellectual skills within each group. Indeed, even when we look at traits other than intelligence, we find much greater variation within each group than we find between groups.

It is time, therefore, to put the idea of human races to rest, at least as far as science is concerned. If such notions remain in the realm of social discourse, then so be it. That is where the problems associated with notions of race have to be solved anyway. At least, says Marks, in speaking for the anthropological community: "You may group humans into a small number of races if you want to, but you are denied biology as a support for it."

A REPORTER AT LARGE

The Lost Man

Is it possible that the first Americans weren't who we think they were? And why is the government withholding Kennewick Man, who might turn out to be the most significant archeological find of the decade?

DOUGLAS PRESTON

ON Sunday, July 28, 1996, in the middle of the afternoon, two college students who were watching a hydroplane race on the Columbia River in Kennewick, Washington, decided to take a shortcut along the river's edge. While wading through the shallows, one of them stubbed his toe on a human skull partly buried in the sand. The students picked it up and, thinking it might be that of a murder victim, hid it in some bushes and called the police.

Floyd Johnson, the Benton County coroner, was called in, and the police gave him the skull in a plastic bucket. Late in the afternoon, Johnson called James Chatters, a forensic anthropologist and the owner of a local consulting firm called Applied Paleoscience. "Hey, buddy, I got a skull for you to look at," Johnson said. Chatters had often helped the police identify skeletons and distinguish between those of murder victims and those found in Indian burial sites. He is a small, determined, physically powerful man of forty-eight who used to be a gymnast and a wrestler. His work occasionally involves him in grisly or spectacular murders, where the victims

are difficult to identify, such as burnings and dismemberments.

"When I looked down at the skull," Chatters told me, "right off the bat I saw it had a very large number of Caucasoid features"—in particular, a long, narrow braincase, a narrow face, and a slightly projecting upper jaw. But when Chatters took it out of the bucket and laid it on his worktable he began to see some unusual traits. The crowns of the teeth were worn flat, a common characteristic of prehistoric Indian skulls, and the color of the bone indicated that it was fairly old. The skull sutures had fused, indicating that the individual was past middle age. And, for a prehistoric Indian of what was then an advanced age, he or she was in exceptional health; the skull had, for example, all its teeth and no cavities.

As dusk fell, Chatters and Johnson went out to the site to see if they could find the rest of the skeleton. There, working in the dying light, they found more bones, lying around on sand and mud in about two feet of water. The remains were remarkably complete: only the sternum, a few rib fragments, and

some tiny hand, wrist, and foot bones were missing. The bones had evidently fallen out of a bank during recent flooding of the Columbia River.

The following day, Chatters and Johnson spread the bones out in Chatters's laboratory. In forensic anthropology, the first order of business is to determine sex, age, and race. Determining race was particularly important, because if the skeleton turned out to be Native American it fell under a federal law called the Native American Graves Protection and Repatriation Act, or NAGPRA. Passed in 1990, NAGPRA requires the government—in this case, the Army Corps of Engineers, which controls the stretch of the Columbia River where the bones were found—to ascertain if human remains found on federal lands are Native American and, if they are, to "repatriate" them to the appropriate Indian tribe.

Chatters determined that the skeleton was male, Caucasoid, from an individual between forty and fifty-five years old, and about five feet nine inches tall—much taller than most prehistoric Native Americans in the Northwest. In physical anthropology, the term "Caucasoid" does not necessarily mean "white" or "European"; it is a descriptive term applied to certain biological features of a diverse category that includes, for example, some south-Asian groups as well as Europeans. (In contrast, the term "Caucasian" is a culturally defined racial category.) "I thought maybe we had an early pioneer or fur trapper," he said. As he was cleaning the pelvis, he noticed a gray object embedded in the bone, which had partly healed and fused around it. He took the bone to be X-rayed, but the object did not show up, meaning that it was not made of metal. So he requested a CAT scan. To his surprise, the scan revealed the object to be part of a willow-leaf-shaped spear point, which had been thrust into the bone and broken off. It strongly resembled a Cascade projectile point—an Archaic Indian style in wide use from around nine thousand to forty-five hundred years ago.

The Army Corps of Engineers asked Chatters to get a second opinion. He put the skeleton into his car and drove it a hundred miles to Ellensburg, Washington, where an anthropologist named Catherine J. Mac Millan ran a forensic consulting business called the Bone-Apart Agency. "He didn't say anything," Mac Millan told me. "I examined the bones, and I said 'Male, Caucasian.' He said 'Are you sure?' and I said 'Yeah.' And then he handed me the pelvis and showed me the ancient point embedded in it, and he said, 'What do you think now?' And I said, 'That's extremely interesting, but it still looks Caucasian to me.' "In her report to the Benton County coroner's office she wrote that in her opinion the skeleton was "Caucasian male."

Toward the end of the week, Chatters told Floyd Johnson and the Army Corps of Engineers that he thought they needed to get a radiocarbon date on Kennewick Man. The two parties agreed, so Chatters sent the left fifth metacarpal bone—a tiny bone in the hand—to the University of California at Riverside.

On Friday, August 23rd, Jim Chatters received a telephone call from the radiocarbon lab. The bone was between ninety-three hundred and ninety-six hundred years old. He was astounded. "It was just a phone call," he said. "I thought, Maybe there's been a mistake.

I had to see the report with my own eyes." The report came on Monday, in the form of a fax. "I got very nervous then," Chatters said. He knew that, because of their age, the bones on his worktable had to be one of the most important archeological finds of the decade. "It was just a tremendous responsibility." The following Tuesday, the coroner's office issued a press release on the find, and it was reported in the Seattle *Times,* and other local papers.

Chatters called in a third physical anthropologist, Grover S. Krantz, a professor at Washington State University. Krantz looked at the bones on Friday, August 30th. His report noted some characteristics common to both Europeans and Plains Indians but concluded that "this skeleton cannot be racially or culturally associated with any existing American Indian group." He also wrote, "The Native Repatriation Act has no more applicability to this skeleton than it would if an early Chinese expedition had left one of its members there."

Fifteen minutes after Krantz finished looking at the bones, Chatters received a call from Johnson. Apologetically, the coroner said, "I'm going to have to come over and get the bones." The Army Corps of Engineers had demanded that all study of the bones cease, and had required him to put the skeleton in the county sheriff's evidence locker. On the basis of the carbon date, the Corps had evidently decided that the skeleton was Native American and that it fell under NAGPRA.

"When I heard this, I panicked," Chatters said. "I was the only one who'd recorded any information on it. There were all these things I should have done. I didn't even have photographs of the postcranial skeleton. I thought, Am I going to be the last scientist to see these bones?"

On September 9th, the Umatilla Indians, leading a coalition of five tribes and bands of the Columbia River basin, formally claimed the skeleton under NAGPRA, and the Corps quickly made a preliminary decision to "repatriate" it. The Umatilla Indian Reservation lies just over the border, in northeastern Oregon, and the other tribes live in Washington and Idaho; all consider the Kennewick area part of their traditional

territories. The Umatillas announced that they were going to bury the skeleton in a secret site, where it would never again be available to science.

Three weeks later, the New York *Times* picked up the story, and from there it went to *Time* and on around the world. Television crews from as far away as France and Korea descended on Kennewick. The Corps received more than a dozen other claims for the skeleton, including one from a group known as the Asatru Folk Assembly, the California-based followers of an Old Norse religion, who wanted the bones for their own religious purposes.

On September 2nd, the Corps had directed that the bones be placed in a secure vault at the Pacific Northwest National Laboratory, in Richland, Washington. Nobody outside of the Corps has seen them since. They are now at the center of a legal controversy that will likely determine the course of American archeology.

WHAT was a Caucasoid man doing in the New World more than ninety-three centuries ago? In the reams of press reports last fall, that question never seemed to be dealt with. I called up Douglas Owsley, who is the Division Head for Physical Anthropology at the National Museum of Natural History, Smithsonian Institution, in Washington, D.C., and an expert on Paleo-American remains. I asked him how many well-preserved skeletons that old had been found in North America.

He replied, "Including Kennewick, about seven."

Then I asked if any others had Caucasoid features, and there was a silence that gave me the sense that I was venturing onto controversial ground.

He guardedly replied, "Yes."

"How many?"

"Well," he said, "in varying degrees, all of them."

Kennewick Man's bones are part of a growing quantity of evidence that the earliest inhabitants of the New World may have been a Caucasoid people. Other, tentative evidence suggests that these people may have originally come from Europe. The new evidence is fragmentary, contradictory, and controversial. Critical research remains to be done, and many studies are still unpublished. At the least, the new evidence calls into question the standard Beringian Walk theory, which holds that the first human beings to reach the New World were Asians of Mongoloid stock, who crossed from Siberia to Alaska over a land bridge. The new evidence involves three basic questions. Who were the original Americans? Where did they come from? And what happened to them?

"You're dealing with such a black hole," Owsley told me. "It's hard to draw any firm conclusions from such a small sample of skeletons, and there is more than one group represented. That's why Kennewick is so important."

KENNEWICK MAN made his appearance at the dawn of a new age in physical anthropology. Scientists are now able to extract traces of organic material from a person's bone and perform a succession of powerful biochemical assays which can reveal an astonishing amount of information about the person. In March, for example, scientists at Oxford University announced that they had compared DNA extracted from the molar cavity of a nine-thousand-year-old skeleton known as Cheddar Man to DNA collected from fifteen pupils and five adults from old families in the village of Cheddar, in Somersetshire. They had established a blood tie between Cheddar Man and a schoolteacher who lived just half a mile from the cave where the bones were found.

In the few weeks that Kennewick Man was in the hands of scientists, they discovered a great deal about him. Isotopic-carbon studies of the bones indicate that he had a diet high in marine food—that he may have been a fisherman who ate a lot of salmon. He seems to have been a tall, good-looking man, slender and well proportioned. (Studies have shown that "handsomeness" is largely the result of symmetrical features and good health, both of which Kennewick Man had.) Archeological finds of similar age in the area suggest that he was part of a small band of people who moved about, hunting, fishing, and gathering wild plants. He may have

lived in a simple sewn tent or mat hut that could be disassembled and carried. Some nearby sites contain large numbers of fine bone needles, indicating that a lot of delicate sewing was going on: Kennewick Man may have worn tailored clothing. For a person at that time to live so long in relatively good health indicates that he was clever or lucky, or both, or had family and close friends around him.

He appears to have perished from recurring infections caused by the stone point in his hip. Because of the way his bones were found, and the layer of soil from which they presumably emerged, it may be that he was not deliberately buried but died near the river and was swept away and covered up in a flood. He may have perished alone on a fishing trip, far from his family.

Chatters made a cast of the skull before the skeleton was taken from his office. In the months since, he has been examining it to figure out how Kennewick Man may have looked. He plans to work with physical anthropologists and a forensic sculptor to make a facial reconstruction. "On the physical characteristics alone, he could fit on the streets of Stockholm without causing any kind of notice," Chatters told me. "Or on the streets of Jerusalem or New Delhi, for that matter. I've been looking around for someone who matches this Kennewick gentleman, looking for weeks and weeks at people on the street, thinking, This one's got a little bit here, that one a little bit there. And then, one evening, I turned on the TV, and there was Patrick Stewart"—Captain Picard, of "Star Trek"—"and I said, 'My God, there he is! Kennewick Man!'"

IN September, following the requirements of NAGPRA, the Corps advertised in a local paper its intention of repatriating the skeleton secreted in the laboratory vault. The law mandated a thirty-day waiting period after the advertisements before the Corps could give a skeleton to a tribe.

Physical anthropologists and archeologists around the country were horrified by the seizure of the skeleton. They protested that it was not possible to demonstrate a relationship between

nine-thousand-year-old remains and any modern tribe of the area. "Those tribes are relatively new," says Dennis Stanford, the chairman of the Department of Anthropology at the Smithsonian's National Museum of Natural History. "They pushed out other tribes that were there." Both Owsley and Richard L. Jantz, a biological anthropologist at the University of Tennessee, wrote letters to the Army Corps of Engineers in late September saying that the loss to science would be incalculable if Kennewick Man were to be reburied before being studied. They received no response. Robson Bonnichsen, the director of the Center for the Study of the First Americans, at Oregon State University, also wrote to the Corps and received no reply. Three representatives and a United States senator from the state of Washington got in touch with the Corps, pleading that it allow the skeleton to be studied before reburial, or, at least, refrain from repatriating the skeleton until Congress could take up the issue. The Corps rebuffed them.

The Umatillas themselves issued a statement, which was written by Armand Minthorn, a tribal religious leader. Minthorn, a small, well-spoken young man with long braids, is a member of a new generation of Native American activists, who see religious fundamentalism—in this case, the Washat religion—as a road back to Native American traditions and values:

> Our elders have taught us that once a body goes into the ground, it is meant to stay there until the end of time.... If this individual is truly over 9,000 years old, that only substantiates our belief that he is Native American. From our oral histories, we know that our people have been part of this land since the beginning of time. We do not believe that our people migrated here from another continent, as the scientists do.... Scientists believe that because the individual's head measurement does not match ours, he is not Native American. Our elders have told us that Indian people did not always look the way we look today. Some scientists say that if this individual is not studied further, we, as Indians, will be destroying evidence of our history. We already know our history. It is passed on to us through

our elders and through our religious practices.

Despite the mounting protests, the Corps refused to reconsider its decision to ban scientific study of the Kennewick skeleton. As the thirty-day waiting period came to a close, anthropologists around the country panicked. Just a week before it ended, on October 23rd, a group of eight anthropologists filed suit against the Corps. The plaintiffs included Douglas Owsley, Robson Bonnichsen, and also Dennis Stanford. Stanford, one of the country's top Paleo-Indian experts, is a formidable opponent. While attending graduate school in New Mexico, he roped in local rodeos, and helped support his family by leasing an alfalfa farm. There's still a kind of laconic, frontier toughness about him. "Kennewick Man has the potential to change the way we view the entire peopling of the Americas," he said to me. "We had to act. Otherwise, I might as well retire."

The eight are pursuing the suit as individuals. Their academic institutions are reluctant to get involved in a lawsuit as controversial as this, particularly at a time when most of them are negotiating with tribes over their own collections.

In the suit, the scientists have argued that Kennewick Man may not meet the NAGPRA definition of "Native American" as being "of, or relating to, a tribe, people, or culture that is indigenous to the United States." The judge trying the case has asked both sides to be prepared to define the word "indigenous" as it is used in NAGPRA. This will be an interesting exercise, since no human beings are indigenous to the New World: we are all immigrants.

The scientists have also argued that the Corps had had no evidence to support its claim that the skeleton had a connection to the Umatillas. Alan Schneider, the scientists' attorney, says, "Our analysis of NAGPRA is that first you have to make a determination if the human remains are Native American. And then you get to the question of cultural affiliation. The Army Corps assumed that anyone who died in the continental United States prior to a certain date is automatically Native American."

The NAGPRA law appears to support the scientists' point of view. It says that

when there are no known lineal descendants "cultural affiliation" should be determined using "geographical, kinship, biological, archeological, anthropological, linguistic, folkloric, oral traditional, historical," or other "relevant information or expert opinion" before human remains are repatriated. In other words, human remains must often be studied before anyone can say whom they are related to.

The Corps, represented by the Justice Department, has refused to comment on most aspects of the case. "It's really as if the government didn't want to know the truth about Kennewick Man," Alan Schneider told me in late April. "It seems clear that the government will *never* allow this skeleton to be studied, for any reason, unless it is forced to by the courts."

Preliminary oral arguments in the case were heard on June 2nd in the United States District Court for the District of Oregon. The scientists asked for immediate access to study the bones, and the Corps asked for summary judgment. Judge John Jelderks denied both motions, and said that he would have a list of questions for the Corps that it is to answer within a reasonable time. With the likelihood of appeals, the case could last a couple of years longer, and could ultimately go to the Supreme Court.

Schneider was not surprised that the Corps had sided with the Indians. "It constantly has a variety of issues it has to negotiate with Native American tribes," he told me, and he specified, among others, land issues, water rights, dams, salmon fishing, hydroelectric projects, and toxic-waste dumps. The Corps apparently decided, Schneider speculated, that in this case its political interests would be better served by supporting the tribes than by supporting a disgruntled group of anthropologists with no institutional backing, no money, and no political power. There are large constituencies for the Indians' point of view: fundamentalist Christians and liberal supporters of Indian rights. Fundamentalists of all varieties tend to object to scientific research into the origins of humankind, because the results usually contradict their various creation myths. A novel coalition of conservative Chris-

tians and liberal activists was important in getting NAGPRA through Congress.

KENNEWICK MAN, early as he is, was not one of the first Americans. But he could be their descendant. There is evidence that those mysterious first Americans were a Caucasoid people. They may have come from Europe and may be connected to the Clovis people of America. Kennewick may provide evidence of a connection between the Old World and the New.

The Clovis mammoth hunters were the earliest widespread culture that we know of in the Americas. They appeared abruptly, seemingly out of nowhere, all over North and South America about eleven thousand five hundred years ago—two thousand years before Kennewick. (They were called Clovis after a town in New Mexico near an early site—a campground beside an ancient spring which is littered with projectile points, tools, and the remains of fires.) We have only a few fragments of human bone from the Clovis people and their immediate descendants, the Folsom, and these remains are so damaged that nothing can be learned from them at present.

The oldest bones that scientists have been able to study are the less than a dozen human remains that are contemporaneous with Kennewick. They date from between eight thousand and nearly eleven thousand years ago—the transition period between the Paleo-Indian and the Archaic Indian traditions. Most of the skeletons have been uncovered accidentally in recent years, primarily because of the building boom in the West. (Bones do not survive well in the East: the soil is too wet and acidic.) Some other ancient skeletons, though, have been discovered gathering dust in museum drawers. Among these oldest remains are the Spirit Cave mummy and Wizard's Beach Willie, both from Nevada; the Hourglass Cave and Gordon's Creek skeletons, from Colorado; the Buhl Burial, from Idaho; and remains from Texas, California, and Minnesota.

Douglas Owsley and Richard Jantz made a special study of several of these ancient remains. The best-preserved specimen they looked at was a partial mummy from Spirit Cave, Nevada, which is more than nine thousand years old. Owsley and Jantz compared the Spirit Cave skull with thirty-four population samples from around the world, including ten Native American groups. In an as yet unpublished letter to the Nevada State Museum, they concluded that the Spirit Cave skull was "very different" from any historic-period Native American groups. They wrote, "In terms of its closest classification, it does have a 'European' or 'Archaic Caucasoid' look, because morphometrically it is most similar to the Ainu from Japan and a Medieval period Norse population." Additional early skeletons that they and others have looked at also show Caucasoid-like traits that, in varying degrees, resemble Kennewick Man's. Among these early skeletons, there are no close resemblances to modern Native Americans.

But, even though the skeletons do look Caucasoid, other evidence indicates that the concept of "race" may not be applicable to human beings of ten or fifteen thousand years ago. Recent studies have discovered that all Eurasians may have looked Caucasoid-like in varying degrees. In addition, some researchers believe that the Caucasoid type first emerged in Western Asia or the Middle East, rather than in Europe. The racial differences we see today may be a late (and trivial) development in human evolution. If this is the case, then Kennewick may indeed be a direct ancestor of today's Native Americans—an idea that some preliminary DNA and dental studies seem to support.

Biology, however, isn't the whole story. There is some archeological evidence that the Clovis people of America—Kennewick Man's predecessors—came from Europe, which could account for his Caucasoid features. When the Clovis people appeared in the New World, they possessed an advanced stone and bone technology, and employed it in hunting big game. (It is no small feat to kill a mammoth or a mastodon with a hand-held spear or an atlatl.) If the Clovis people—or their precursors—had migrated to North America from Asia, one would expect to find early forms of their distinctive

tools, of the right age, in Alaska or eastern Siberia. We don't. But we do find such artifacts in Europe and parts of Russia.

Bruce Bradley is the country's leading expert on Paleo-Indian flaked-stone technology. In 1970, as a recent college graduate, Bradley spent time in Europe studying paleolithic artifacts, including those of the Solutrean people, who lived in southwestern France and Spain between twenty thousand and sixteen thousand years ago. During his stay in Europe, Bradley learned how to flake out a decent Solutrean point. Then he came back to America and started studying stone tools made by the Clovis people. He noticed not only striking visual similarities between the Old World Solutrean and the New World Clovis artifacts, as many researchers had before, but also a strikingly similar use of flaking technology.

"The artifacts don't just look identical," Bradley told me. "They are *made* the same way." Both the European Solutrean and the American Clovis stone tools are fashioned using the same complex flaking techniques. He went on, "As far as I know, overshot flaking is unique to Solutrean and Clovis, and the only diving flaking that is more than eleven thousand years old is that of Solutrean and Clovis."

To argue his case, he drove from Cortez, Colorado, where he lives, to see me, in Santa Fe, bringing with him a trunkful of Solutrean artifacts, casts of Clovis bone and stone tools, and detailed illustrations of artifacts, along with a two-pound chunk of gray Texas flint and some flaking tools.

Silently, he laid a piece of felt on my desk and put on it a cast of a Clovis knife from Blackwater Draw, New Mexico. On top of that, he put a broken Solutrean knife from Laugerie-Haute, in France. They matched perfectly—in size, shape, thickness, and pattern of flaking. As we went through the collection, the similarities could be seen again and again. "It isn't just the flaking as you see it on the finished piece that is the same," Bradley explained. "Both cultures had a very specific way of preparing the edge before striking it to get a very specific type of flake. I call these

deep technologies. These are not mere resemblances—they are deep, complex, abstract concepts applied to the stone."

I remarked that according to other archeologists I had talked to the resemblances were coincidental, the result of two cultures—one Old World, the other New World—confronting the same problems and solving them in the same way. "Maybe," Bradley said. "But a lot of older archeologists were not trained with technology in mind. They see convergence of form or look, but they don't see the technology that goes into it. I'm not saying this is the final answer. But there is so much similarity that we cannot say, 'This is just coincidence,' and ignore it."

To show me what he meant, he brought out the piece of gray flint, and we went into the back yard. "This isn't just knocking away at a stone," he said, hefting the piece of flint and examining it from various angles with narrowed eyes. "After the initial flaking of a spear point, there comes a stage where you have almost limitless choices on how to continue." He squatted down and began to work on the flint with hammer stones and antler billets, his hands deftly turning and shaping the material—knapping, chipping, scraping, pressing, flaking. The pile of razor-sharp flakes grew bigger, and the chunk of flint began to take on a definite shape. Over the next ninety minutes, Bradley reduced the nodule to a five-inch-long Clovis spear point.

It was a revelation to see that making a Clovis point was primarily an intellectual process. Sometimes ten minutes would pass while Bradley examined the stone and mapped out the next flaking sequence. "After thirty years of intense practice, I'm still at the level of a mediocre Clovis craftsman," he said, wiping his bloodied hands and reviving himself with a cup of cocoa. "This is as difficult and complex as a game of chess."

Those are not the only similarities between the Old World Solutrean and the New World Clovis cultures. As we went through Bradley's casts of Clovis artifacts, he compared them with pictures of similar Solutrean artifacts. The Clovis people, for example, produced enigmatic bone rods that were bevelled on both ends and crosshatched; so did the Solutrean. The Clovis people fashioned distinctive spear points out of mammoth ivory; so did the Solutrean. Clovis and Solutrean shaft wrenches (tools thought to have been used for straightening spears) look almost identical. At the same time, there are significant differences between the Clovis and the Solutrean tool kits; the Clovis people fluted their spear points, for instance, while the Solutrean did not.

"To get to the kind of complexity you find in Clovis tools, I see a long technological development," Bradley said. "It isn't one person in one place inventing something. But there is no evolution in Clovis technology. It just appears, full blown, all over the New World, around eleven thousand five hundred years ago. Where's the evolution? *Where did that advanced Clovis technology come from?*"

When I mentioned the idea of a possible Old World Solutrean origin for the New World Clovis to Lawrence Straus, a Solutrean expert at the University of New Mexico, he said, "There are two gigantic problems with it—thousands of years of separation and thousands of miles of ocean." Pointing out that the Solutrean technology itself appeared relatively abruptly in the South of France, he added, "I think this is a fairly clear case of mankind's ability to reinvent things."

But Bradley and other archeologists have pointed out that these objections may not be quite so insurmountable. Recently, in southern Virginia, archeologists discovered a layer of non-Clovis artifacts beneath a Clovis site. The layer dates from fifteen thousand years ago—much closer to the late Solutrean period, sixteen thousand five hundred years ago. Only a few rough tools have been found, but if more emerge they might provide evidence for the independent development of Clovis—or provide a link to the Solutrean. The as yet unnamed culture may be precisely the precursor to Clovis which archeologists have been looking for since the nineteen-thirties. The gap between Solutrean and Clovis may be narrowing.

The other problem is how the Solutrean people—if they are indeed the ancestors of the Clovis—might have reached America in the first place. Al-

though most of them lived along riverbanks and on the seacoast of France and Spain, there is no evidence that they had boats. No Paleo-Indian boats have been found on the American side, either, but there is circumstantial evidence that the Paleo-Indians used boats. The ancestors of the Australian aborigines got to Australia in boats from the Indonesian archipelago at least fifty thousand years ago.

But the Solutrean people may not have needed boats at all: sixteen thousand years ago, the North Atlantic was frozen from Norway to Newfoundland. Seasonal pack ice probably extended as far south as Britain and Nova Scotia. William Fitzhugh, the director of the Arctic Studies Center, at the Smithsonian, points out that if human beings had started in France, crossed the English Channel, then hopped along the archipelago from Scotland to the Faeroes to Iceland to Greenland to Newfoundland, and, finally, Nova Scotia, the biggest distance between landfalls would have been about five hundred miles, which Fitzhugh says could have been done on foot over ice. It would still have been a stupendous journey, but perhaps not much more difficult than the Beringian Walk, across thousands of miles of tundra, muskeg, snow, and ice.

"For a long time, most archeologists have been afraid to challenge the Beringian Walk paradigm," Bradley said. "I don't want to try to convince anybody. But I do want to shake the bushes. You could put all the archeological evidence for the Asian-Clovis connection in an envelope and mail it for thirty-two cents. The evidence for a European-Clovis connection you'd have to send in a U.P.S. box, at least."

Robson Bonnichsen, the director of the Center for the Study of the First Americans, at Oregon State University, is another archeologist whose research is challenging the established theories. "There is a presumption, written into almost every textbook on prehistory, that Paleo-Americans such as Clovis are the direct ancestors of today's Native Americans," he told me. "But now we have a very limited number of skeletons from that early time, and it's not clear that that's true. We're getting some hints from people working with genetic data

that these earliest populations might have some shared genetic characteristics with latter-day European populations. A lot more research is needed to sort all this out. Now, for the first time, we have the technology to do this research, especially in molecular biology. Which is why we *must* study Kennewick."

This summer, Bonnichsen hopes to go to France and recover human hair from Solutrean and other Upper Paleolithic sites. He will compare DNA from that hair with DNA taken from naturally shed Paleo-American hair recovered from the United States to see if there is a genetic link. Human hair can survive thousands of years in the ground, and, using new techniques, Bonnichsen and his research team have been finding hair in the places where people worked and camped.

BONNICHSEN and most other archeologists tend to favor the view that if the ancestors of Clovis once lived in Europe they came to America via Asia—the Beringian Walk theory with somewhat different people doing the walking. C. Vance Haynes, Jr., the country's top Paleo-Indian geochronologist, who is a professor of anthropology at the University of Arizona, and is a plaintiff in the lawsuit, said to me, "When I look at Clovis and ask myself where in the world the culture was derived from, I would say Europe." In an article on the origins of Clovis, Haynes noted that there were extraordinary resemblances between New World Clovis and groups that lived in Czechoslovakia and Ukraine twenty thousand years ago. He noted at least nine "common traits" shared by Clovis and certain Eastern European cultures: large blades, end scrapers, burins, shaft wrenches, cylindrical bone points, knapped bone, unifacial flake tools, red ochre, and circumferentially chopped mammoth tusks. He also pointed out that an eighteen-thousand-year-old burial site of two children near Lake Baikal, in Central Asia, exhibits remarkable similarities to what appears to be a Clovis burial site of two cremated children in Montana. The similarities extend beyond tools and points buried with the remains: red ochre, a kind of iron oxide, was placed in both graves. This suggests a migratory group carrying its technology from Europe across Asia. "If you want to speculate, I see a band moving eastward from Europe through Siberia, and meeting people there, and having cultural differences," Haynes said to me. "Any time there's conflict, it drives people, and maybe it just drove them right across the Bering land bridge. And exploration could have been as powerful a driving force thirteen thousand years ago as it was in 1492."

ONCE the Clovis people or their predecessors reached the New World, what happened to them? This is the second—and equally controversial—half of the theory: that the Clovis people or their immediate successors, the Folsom people, may have been supplanted by the ancestors of today's Native Americans. In this scenario, Kennewick Man may have been part of a remnant Caucasoid population related to Clovis and Folsom. Dennis Stanford, of the Smithsonian, said to me, "For a long time, I've held the theory that the Clovis and the Folsom were overwhelmed by a migration of Asians over the Bering land bridge. It may not just have been a genetic swamping or a pushing aside. The north Asians may have been carrying diseases that the Folsom and the Clovis had no resistance to"—just as European diseases wiped out a large percentage of the Native American population after the arrival of Columbus. Stanford explained that at several sites the Paleo-Indian tradition of Clovis and Folsom was abruptly replaced by Archaic Indian traditions, which had advanced but very different lithic technologies. The abruptness of the transition and the sharp change in technology, Stanford feels, suggest a rapid replacement of Folsom by the Archaic cultural complex, rather than an evolution from one into the other. The Archaic spear point embedded in Kennewick Man's hip could even be evidence of an ancient conflict: the Archaic Indian tradition was just beginning to appear in the Pacific Northwest at the time of Kennewick Man's death.

OWSLEY and other physical anthropologists who have studied the skulls of the earliest Americans say that the living population they most closely match is the mysterious Ainu, the aboriginal inhabitants of the Japanese islands. Called the Hairy People by the Japanese, the Ainu are considered by some researchers to be a Causasoid group who, before mixing with the Japanese, in the late nineteenth and twentieth centuries, had European faces, wavy hair, thick beards, and a European-type distribution of body hair. Early travellers reported that some also had blue eyes. Linguists have not been able to connect the Ainu language with any other on earth. The American Museum of Natural History, in New York, has a collection of nineteenth-century photographs of pure-blood Ainu, which I have examined: they stare from the glass plates like fierce, black-bearded Norwegians.

Historically, the Ainu have been the "Indians" of Japan. After the ancestors of the Japanese migrated from the mainland a couple of thousand years ago, they fought the Ainu and pushed them into the northernmost islands of the Japanese archipelago. The Japanese later discriminated against the Ainu, forcing their children to attend Japanese schools, and suppressing their religion and their language. Today, most Ainu have lost their language and many of their distinct physical characteristics, although there has recently been a movement among them to recapture their traditions, their religion, their language, and their songs. Like the American Indians, the Ainu suffer high rates of alcoholism. The final irony is that the Ainu, like the American Indians for Americans, have become a popular Japanese tourist attraction. Many Ainu now make a living doing traditional dances and selling handicrafts to Japanese tourists. The Japanese are as fascinated by the Ainu as we are by the Indians. The stories are mirror images of each other, with only the races changed.

If the Ainu are a remnant population of those people who crossed into America thirteen or more millennia ago, then they are right where one might expect to find them, in the extreme eastern part

of Asia. Stanford says, "That racial type goes all the way to Europe, and I suspect that originally they were the same racial group at both ends. At that point in time, this racial diversification hadn't developed. They could have come into the New World from two directions at once, east and west."

THERE is a suspicion among anthropologists that some of the people behind the effort to rebury Kennewick and other ancient skeletons are afraid that the bones could show that the earliest Americans were Caucasoid. I asked Armand Minthorn about this. "We're not afraid of the truth," he said calmly. "We already know our truth. We're not telling the scientists what *their* truth is."

The Umatillas were infuriated by the research that Chatters did on the skeleton before it was seized by the Corps. "Scientists have dug up and studied Native Americans for decades," Minthorn wrote. "We view this practice as desecration of the body and a violation of our most deeply-held religious beliefs." Chatters told me that he had received "vitriolic" and "abusive" telephone calls from tribe members, accusing him of illegalities and racism. (The latter was an odd charge, since Chatters's wife is of Native American descent.) A client of his, he said, received an unsigned letter from one of the tribes, telling the client not to work with Chatters anymore. "They're going to ruin my livelihood," the forensic consultant said.

In a larger sense, the anger of the Umatillas and other Native American tribes is understandable, and even justified. If you look into the acquisition records of most large, old natural-history museums, you will see a history of unethical, and even grisly, collecting practices. Fresh graves were dug up and looted, sometimes in the dead of night. "It is most unpleasant to steal bones from a grave," the eminent anthropologist Franz Boas wrote in his diary just around the turn of the century, "but what is the use, someone has to do it." Indian skulls were bought and sold among collectors like arrowheads and pots. Skeletons were exhibited with no regard for

ANTHROPOLOGY

Anthropologists 1, Army Corps 0

American anthropologists won a round in an important legal battle last week, when a U.S. District Court in Portland, Oregon, ordered the Army Corps of Engineers to reopen the case of 9300-year-old human skeleton found on federal land in Washington state. The corps, which has jurisdiction over the skeleton—one of the oldest in the Americas—had planned to turn it over to American Indian tribes for reburial last October under the 1990 Native American Graves Protection and Repatriation Act (NAGPRA). But scientists sued to prevent the handover until they had studied the bones.

Now, although the court stopped short of allowing study of the skeleton, it opened the door for future research by recognizing that the scientists have a legitimate claim that must be considered. "It's a landmark ruling," exulted Alan Schneider, the Portland attorney for the scientists. "This is the first case where a court has held that a third party like a scientist has standing to challenge a government agency's overenforcement of NAGPRA."

The male skeleton, known as "Kennewick Man" for the town where it was discovered in the bank of the Columbia River last summer, is a rare representative of the earliest people to inhabit the Americas. It also had a projectile point embedded in its pelvis and facial features that may be Caucasoid-like, offering clues about early Americans (*Science*, 11 October 1996, p. 172).

As soon as the skeleton was discovered, Smithsonian Institution skeletal biologist Douglas Owsley sought permission to study it, and a team at the University of California, Davis, extracted ancient DNA from a bit of its finger bone for analysis. Owsley and others say that without scientific study it's impossible to know whether the skeleton has a biological or cultural tie to any living people—knowledge needed to determine which tribe, if any, should receive it.

But Owsley says he was told by the corps "flat-out that there would be no scientific study." The corps ordered a halt to the DNA work as soon as it learned of it in October, and announced that it would hand over the skeleton to the Confederated Tribes of the Umatilla Indian Reservation under the auspices of NAGPRA, which requires remains or cultural objects to be given to a culturally affiliated tribe. It later rescinded that order but has kept the skeleton locked away and unavailable for study. Court documents suggest that corps officials were concerned about alienating the Indians. In an 18 September 1996 e-mail message, a corps official wrote: "All risk to us seems to be associated with not repatriating the remains."

In the new decision, U.S. District Court Magistrate John Jelderks invalidated all the corps' orders in the case and criticized the "flawed" procedures used by the agency, which he said "acted before it had all the evidence or fully appreciated the scope of the problem." Jelderks also asked the corps to report back to him with its decision on the case and to answer several questions, including whether repatriation under NAGPRA required a biological or cultural link between bones and living tribes, and how such a link would be determined.

Corps attorney Daria Zane declined to comment for the record. As for the scientists, although they still can't study the skeleton, they're pleased. "Hopefully, the Army Corps will come to its senses," says Owsley. "They are using government lawyers and taxpayers' dollars to argue against academic freedom."

—**Ann Gibbons**

From *Science*, Vol. 277, July 11, 1997, p. 173. © 1997 by The American Association for the Advancement of Science. Reprinted by permission

tribal sensitivities. During the Indian wars, warriors who had been killed on the battlefield were sometimes decapitated by Army doctors so that scientists back in the East could study their heads. The American Museum of Natural History "collected" six live Inuits in Greenland and brought them to New York to study; four of them died of respiratory diseases, whereupon the museum macerated their corpses and installed the bones in its collection. When I worked at the museum, in the nineteen-eighties, entire hallways were lined with glass cases containing Indian bones and mummified body parts—a small fraction of the museum's collection, which includes an estimated twenty thousand or more human remains, of all races. Before NAGPRA, the Smithsonian had some thirty-five thousand sets of human remains in storage; around eighteen thousand of them were Native American.

Now angry Native Americans, armed with NAGPRA and various state reburial laws, are emptying such museums of bones and grave goods. Although most anthropologists agree that burials identified with particular tribes should be returned, many have been horrified to discover that some tribes are trying to get everything—even skeletons and priceless funerary objects that are thousands of years old.

An amendment that was introduced in Congress last January would tighten NAGPRA further. The amended law could have the effect of hindering much archeology in the United States involving human remains, and add to the cost of construction projects that inadvertently uncover human bones. (Or perhaps the law would merely guarantee that such remains would be quietly destroyed.)

Native Americans have already claimed and reburied two of the earliest skeletons, the Buhl Burial and the Hourglass Cave skeleton, both of which apparently had some Caucasoid characteristics. The loss of the Buhl Burial was particularly significant to anthropologists, because it was more than ten thousand years old—a thousand years older than Kennewick Man—and had been found buried with its grave goods. The skeleton, of a woman between eighteen and twenty years old, received, in the opinion of some anthropologists, inadequate study before it was turned over to the Shoshone-Bannock tribe. The Northern

Paiute have asked that the Spirit Cave mummy be reburied. If these early skeletons are all put back in the ground, anthropologists say, much of the history of the peopling of the Americas will be lost.

WHEN Darwin proposed his theory of natural selection, it was seized upon and distorted by economists, social engineers, and politicians, particularly in England: they used it to justify all sorts of vicious social and economic policies. The scientific argument about the original peopling of the Americas threatens to be distorted in a similar way. Some tabloids and radio talk shows have referred to Kennewick as a "white man" and have suggested that his discovery changes everything with respect to the rights of Native Americans in this country. James

Chatters said to me, "There are some less racially enlightened folks in the neighborhood who are saying, 'Hey, our ancestors were here first, so we don't owe the Indians anything.' "

This is clearly racist nonsense: these new theories cannot erase or negate the existing history of genocide, broken treaties, and repression. But it does raise an interesting question: If the original inhabitants of the New World were Europeans who were pushed out by Indians, would it change the Indians' position in the great moral landscape?

"No," Stanford said in reply to this question. "Whose ancestors are the people who were pushed out? And who did the pushing? The answer is that we're all the descendants of those folks. If you go back far enough, eventually we all have a common ancestor—*we're all the*

same. When the story is finally written, the peopling of the Americas will turn out to be far more complicated than anyone imagined. There have been a lot of people who came here, at many different times. Some stayed and some left, some made it and some didn't, some got pushed out and some did the pushing. It's the history of humankind: the tough guy gets the ground."

Chatters put it another way. "We didn't go digging for this man. He fell out—he was actually a volunteer. I think it would be wrong to stick him back in the ground without waiting to hear the story he has to tell. We need to look at things as human beings, not as one race or another. The message this man brings to us is one of unification: there may be some commonality in our past that will bring us together."

Black, White, Other

Racial categories are cultural constructs masquerading as biology

Jonathan Marks

New York-born Jonathan Marks earned an undergraduate degree in natural science at Johns Hopkins. After getting his Ph.D. in anthropology, Marks did a post-doc in genetics at the University of California at Davis and is now an associate professor of anthropology at Yale University. He is the coauthor, with Edward Staski, of the introductory textbook Evolutionary Anthropology *(San Diego: Harcourt, Brace Jovanovich, 1992). His new book,* Human Biodiversity: Genes, Race, and History *is published (1995) by Aldine de Gruyter.*

While reading the Sunday edition of the *New York Times* one morning last February, my attention was drawn by an editorial inconsistency. The article I was reading was written by attorney Lani Guinier. (Guinier, you may remember, had been President Clinton's nominee to head the civil rights division at the Department of Justice in 1993. Her name was hastily withdrawn amid a blast of criticism over her views on political representation of minorities.) What had distracted me from the main point of the story was a photo caption that described Guinier as being "half-black." In the text of the article, Guinier had described herself simply as "black."

How can a person be black and half black at the same time? In algebraic terms, this would seem to describe a situation where $x = \frac{1}{2}x$, to which the only solution is $x = 0$.

The inconsistency in the *Times* was trivial, but revealing. It encapsulated a longstanding problem in our use of racial categories—namely, a confusion between biological and cultural heredity. When Guinier is described as "half-black," that is a statement of biological ancestry, for one of her two parents is black. And when Guinier describes herself as black, she is using a cultural category, according to which one can either be black or white, but not both.

Race—as the term is commonly used—is inherited, although not in a strictly biological fashion. It is passed down according to a system of folk heredity, an all-or-nothing system that is different from the quantifiable heredity of biology. But the incompatibility of the two notions of race is sometimes starkly evident—as when the state decides that racial differences are so important that interracial marriages must be regulated or outlawed entirely. Miscegenation laws in this country (which stayed on the books in many states through the 1960s) obliged the legal system to define who belonged in what category. The resulting formula stated that anyone with one-eighth or more black ancestry was a "negro." (A similar formula, defining Jews, was promulgated by the Germans in the Nuremberg Laws of the 1930s.)

Applying such formulas led to the biological absurdity that having one black great-grandparent was sufficient to define a person as black, but having seven white great grandparents was insufficient to define a person as white. Here, race and biology are demonstrably at odds. And the problem is not semantic but conceptual, for race is presented as a category of nature.

Human beings come in a wide variety of sizes, shapes, colors, and forms—or, because we are visually oriented primates, it certainly seems that way. We also come in larger packages called populations; and we are said to belong to even larger and more confusing units, which have long been known as races. The history of the study of human variation is to a large extent the pursuit of those human races—the attempt to identify the small number of fundamentally distinct kinds of people on earth.

This scientific goal stretches back two centuries, to Linnaeus, the father of biological systematics, who radically established *Homo sapiens* as one species within a group of animals he called Primates. Linnaeus's system of naming groups within groups logically implied further breakdown. He consequently sought to establish a number of subspecies within *Homo sapiens*. He identified five: four geographical species (from Europe, Asia, Africa, and America) and one grab-bag subspecies called *monstrosus*. This category was dropped by subsequent researchers (as was Linnaeus's use of criteria such as personality and dress to define his subspecies).

While Linnaeus was not the first to divide humans on the basis of the continents on which they lived, he had given the division a scientific stamp. But in attempting to determine the proper number of subspecies, the heirs of Linnaeus always seemed to find different answers, depending upon the criteria they applied. By the mid-twentieth century, scores of anthropologists—led by Harvard's Earnest Hooton—had expended enormous energy on the problem. But these scholars could not convince one another about the precise nature of the fundamental divisions of our species.

Part of the problem—as with the *Times's* identification of Lani Guinier—was that we humans have two constantly intersecting ways of thinking about the divisions among us. On the one hand, we like to think of "race"—as Linnaeus did—as an objective, biological category. In this sense, being a member of a race is supposed to be the equivalent of being a member of a species or of a phylum—except that race, on the analogy of subspecies, is an even narrower (and presumably more exclusive and precise) biological category.

The other kind of category into which we humans allocate ourselves—when we say "Serb" or "Hutu" or "Jew" or "Chicano" or "Republican" or "Red Sox fan"—is cultural. The label refers to little or nothing in the natural attributes of its members. These members may not live in the same region and may not even know many others like themselves. What they share is neither strictly nature nor strictly community. The groupings are constructions of human social history.

Membership in these *un*biological groupings may mean the difference between life and death, for they are the categories that allow us to be identified (and accepted or vilified) socially. While membership in (or allegiance to) these categories may be assigned or adopted from birth, the differentia that mark members from nonmembers are symbolic and abstract; they serve to distinguish people who cannot be readily distinguished by nature. So important are these symbolic distinctions that some of the strongest animosities are often expressed between very similar-looking peoples. Obvious examples are Bosnian Serbs and Muslims, Irish and English, Huron and Iroquois.

Obvious natural variation is rarely so important as cultural difference. One simply does not hear of a slaughter of the short people at the hands of the tall, the glabrous at the hands of the hairy, the red-haired at the hands of the brown-haired. When we do encounter genocidal violence between different looking peoples, the two groups are invariably socially or culturally distinct as well. Indeed, the tragic frequency of hatred and genocidal violence between biologically indistinguishable peoples implies that biological differences such as skin color are not motivations but, rather, excuses. They allow nature to be invoked to reinforce group identities and antagonisms that would exist without these physical distinctions. But are there any truly "racial" biological distinctions to be found in our species?

Obviously, if you compare two people from different parts of the world (or whose ancestors came from different parts of the world), they will differ physically, but one cannot therefore define three or four or five basically different kinds of people, as a biological notion of race would imply. The anatomical properties that distinguish people—such as pigmentation, eye form, body build—are not clumped in discrete groups, but distributed along geographical gradients, as are nearly all the genetically determined variants detectable in the human gene pool.

These gradients are produced by three forces. Natural selection adapts populations to local circumstances (like climate) and thereby differentiates them from other populations. Genetic drift (random fluctuations in a gene pool) also differentiates populations from one another, but in non-adaptive ways. And gene flow (via intermarriage and other child-producing unions) acts to homogenize neighboring populations.

In practice, the operations of these forces are difficult to discern. A few features, such as body build and the graduated distribution of the sickle cell anemia gene in populations from western Africa, southern Asia, and the Mediterranean can be plausibly related to the effects of selection. Others, such as the graduated distribution of a small deletion in the mitochondrial DNA of some

East Asian, Oceanic, and Native American peoples, or the degree of flatness of the face, seem unlikely to be the result of selection and are probably the results of random biohistorical factors. The cause of the distribution of most features, from nose breadth to blood group, is simply unclear.

The overall result of these forces is evident, however. As Johann Friedrich Blumenbach noted in 1775, "you see that all do so run into one another, and that one variety of mankind does so sensibly pass into the other, that you cannot mark out the limits between them." (Posturing as an heir to Linnaeus, he nonetheless attempted to do so.) But from humanity's gradations in appearance, no defined groupings resembling races readily emerge. The racial categories with which we have become so familiar are the result of our imposing arbitrary cultural boundaries in order to partition gradual biological variation.

Unlike graduated biological distinctions, culturally constructed categories are ultrasharp. One can be French or German, but not both; Tutsi or Hutu, but not both; Jew or Catholic, but not both; Bosnian Muslim or Serb, but not both; black or white, but not both. Traditionally, people of "mixed race" have been obliged to choose one and thereby identify themselves unambiguously to census takers and administrative bookkeepers—a practice that is now being widely called into question.

A scientific definition of race would require considerable homogeneity within each group, and reasonably discrete differences between groups, but three kinds of data militate against this view: First, the groups traditionally described as races are not at all homogeneous. Africans and Europeans, for instance, are each a collection of biologically diverse populations. Anthropologists of the 1920s widely recognized *three* European races: Nordic, Alpine, and Mediterranean. This implied that races could exist within races. American anthropologist Carleton Coon identified *ten* European races in 1939. With such protean use, the term race came to have little value in describing actual biological entities within *Homo sapiens*. The scholars were not only grappling with a broad north-south gradient in human appearance across Europe, they were trying to bring the data into line with their belief in profound and fundamental constitutional differences between groups of people.

But there simply isn't one European race to contrast with an African race, nor three, nor ten: the question (as scientists long posed it) fails to recognize the actual patterning of diversity in the human species. Fieldwork revealed, and genetics later quantified, the existence of far more biological diversity within any group than between groups. Fatter and thinner people exist everywhere, as do people with type O and type A blood. What generally varies from one population to the next is the *proportion* of people in these groups expressing the trait or gene. Hair color varies strikingly among Europeans and native Australians, but little among other peoples. To focus on discovering differences between presumptive races, when the vast majority of detectable variants do not help differentiate them, was thus to define a very narrow—if not largely illusory—problem in human biology. (The fact that Africans are biologically more diverse than Europeans, but have rarely been split into so many races, attests to the cultural basis of these categorizations.)

Second, differences between human groups are only evident when contrasting geographical extremes. Noting these extremes, biologists of an earlier era sought to identify representatives of "pure," primordial races presumably located in Norway, Senegal, and Thailand. At no time, however, was our species composed of a few populations within which everyone looked pretty much the same. Ever since some of our ancestors left Africa to spread out through the Old World, we humans have always lived in the "in-between" places. And

human populations have also always been in genetic contact with one another. Indeed, for tens of thousands of years, humans have had trade networks; and where goods flow, so do genes. Consequently, we have no basis for considering *extreme* human forms the most pure, or most representative, of some ancient primordial populations. Instead, they represent populations adapted to the most disparate environments.

And third, between each presumptive "major" race are unclassifiable populations and people. Some populations of India, for example, are darkly pigmented (or "black"), have Europeanlike ("Caucasoid") facial features, but inhabit the continent of Asia (which should make them "Asian"). Americans might tend to ignore these "exceptions" to the racial categories, since immigrants to the United States from West Africa, Southeast Asia, and northwest Europe far outnumber those from India. The very existence of unclassifiable peoples undermines the idea that there are just three human biological groups in the Old World. Yet acknowledging the biological distinctiveness of such groups leads to a rapid proliferation of categories. What about Australians? Polynesians? The Ainu of Japan?

Categorizing people is important to any society. It is, at some basic psychological level, probably necessary to have group identity about who and what you are, in contrast to who and what you are not. The concept of race, however, specifically involves the recruitment of biology to validate those categories of self-identity.

Mice don't have to worry about that the way humans do. Consequently, classifying them into subspecies entails less of a responsibility for a scientist than classifying humans into sub-species

does. And by the 1960s, most anthropologists realized they could not defend any classification of *Homo sapiens* into biological subspecies or races that could be considered reasonably objective. They therefore stopped doing it, and stopped identifying the endeavor as a central goal of the field. It was a biologically intractable problem—the old square-peg-in-a-round-hole enterprise; and people's lives, or welfares, could well depend on the ostensibly scientific pronouncement. Reflecting on the social history of the twentieth century, that was a burden anthropologists would no longer bear.

This conceptual divorce in anthropology—of cultural from biological phenomena was one of the most fundamental scientific revolutions of our time. And since it affected assumptions so rooted in our everyday experience, and resulted in conclusions so counterintuitive—like the idea that the earth goes around the sun, and not vice-versa—it has been widely underappreciated.

Kurt Vonnegut, in *Slaughterhouse Five*, describes what he remembered being taught about human variation: "At that time, they were teaching that there was absolutely no difference between anybody. They may be teaching that still." Of course there are biological differences between people, and between populations. The question is: How are those differences patterned? And the answer seems to be: Not racially. Populations are the only readily identifiable units of humans, and even they are fairly fluid, biologically similar to populations nearby, and biologically different from populations far away.

In other words, the message of contemporary anthropology is: You may group humans into a small number of races if you want to, but you are denied biology as a support for it.

Racial Odyssey

Boyce Rensberger

The human species comes in an artist's palette of colors: sandy yellows, reddish tans, deep browns, light tans, creamy whites, pale pinks. It is a rare person who is not curious about the skin colors, hair textures, bodily structures and facial features associated with racial background. Why do some Africans have dark brown skin, while that of most Europeans is pale pink? Why do the eyes of most "white" people and "black" people look pretty much alike but differ so from the eyes of Orientals? Did one race evolve before the others? If so, is it more primitive or more advanced as a result? Can it be possible, as modern research suggests, that there is no such thing as a pure race? These are all honest, scientifically worthy questions. And they are central to current research on the evolution of our species on the planet Earth.

Broadly speaking, research on racial differences has led most scientists to three major conclusions. The first is that there are many more differences among people than skin color, hair texture and facial features. Dozens of other variations have been found, ranging from the shapes of bones to the consistency of ear wax to subtle variations in body chemistry.

The second conclusions is that the overwhelming evolutionary success of the human species is largely due to its great genetic variability. When migrating bands of our early ancestors reached a new environment, at least a few already had physical traits that gave them an edge in surviving there. If the coming centuries bring significant environmental changes, as many believe they will, our chances of surviving them will be immeasurably enhanced by our diversity as a species.

There is a third conclusion about race that is often misunderstood. Despite our wealth of variation and despite our constant, everyday references to race, no one has ever discovered a reliable way of distinguishing one race from another. While it is possible to classify a great many people on the basis of certain physical features, there are no known feature or groups·of features that will do the job in all cases.

Skin color won't work. Yes, most Africans from south of the Sahara and their descendants around the world have skin that is darker than that of most Europeans. But there are millions of people in India, classified by some anthropologists as members of the Caucasoid, or "white," race who have darker skins than most Americans who call themselves black. And there are many Africans living in sub-Sahara Africa today whose skins are no darker than the skins of many Spaniards, Italians, Greeks or Lebanese.

What about stature as a racial trait? Because they are quite short, on the average, African Pygmies have been considered racially distinct from other dark-skinned Africans. If stature, then, is a racial criterion, would one include in the same race the tall African Watusi and the Scandinavians of similar stature?

The little web of skin that distinguishes Oriental eyes is said to be a particular feature of the Mongoloid race. How, then, can it be argued that the American Indian, who lacks this epicanthic fold, is Mongoloid?

Even more hopeless as racial markers are hair color, eye color, hair form, the shapes of noses and lips or any of the other traits put forth as typical of one race or another.

NO NORMS

Among the tall people of the world there are many black, many white and many in between. Among black people of the world there are many with kinky hair, many with straight or wavy hair, and many in between. Among the broad-nosed, full-lipped people of the world there are many with dark skins, many with light skins and many in between.

How did our modern perceptions of race arise? One of the first to attempt a scientific classification of peoples was Carl von Linné, better known as Linnaeus. In 1735, he published a classification that remains the standard today. As Linnaeus saw it there were four races, classifiable geographically and by skin color. The names Linnaeus gave them were *Homo sapiens Africanus nigrus* (black African human being), *H. sapiens Americanus rubescens* (red American human being), *H. sapiens Asiaticus fuscusens* (brownish Asian human being), and *H. sapiens Europaeus albescens* (white European human being). All, Linnaeus recognized, were members of a single human species.

A species includes all individuals that are biologically capable of interbreeding and producing fertile offspring. Most matings between species are fruitless, and even when they succeed, as when a horse and a donkey interbreed and produce a mule, the progeny are sterile. When a poodle mates with a collie, however, the offspring are fertile, showing that both dogs are members of the same species.

Even though Linnaeus's system of nomenclature survives, his classifications were discarded, especially after voyages of discovery revealed that there were many more kinds of people than could be pigeonholed into four categories. All over the world there are small populations that don't fit. Among the better known are:

- The so-called Bushmen of southern Africa, who look as much Mongoloid as Negroid.
- The Negritos of the South Pacific, who do look Negroid but are very far from Africa and have no known links to that continent.
- The Ainu of Japan, a hairy aboriginal people who look more Caucasoid than anything else.
- The Lapps of Scandinavia, who look as much like Eskimos as like Europeans.

- The aborigines of Australia, who often look Negroid but many of whom have straight or wavy hair and are often blond as children.
- The Polynesians, who seem to be a blend of many races, the proportions differing from island to island.

To accommodate such diversity, many different systems of classification have been proposed. Some set up two or three dozen races. None has ever satisfied all experts.

CLASSIFICATION SYSTEM

Perhaps the most sweeping effort to impose a classification upon all the peoples of the world was made by the American anthropologist Carleton Coon. He concluded there are five basic races, two of which have major subdivisions: Caucasoids; Mongoloids; full-size Australoids (Australian aborigines); dwarf Australoids (Negritos—Andaman Islanders and similar peoples); full-size Congoids (African Negroids); dwarf Congoids (African Pygmies); and Capoids (the so-called Bushmen and Hottentots).

In his 1965 classic, *The Living Races of Man,* Coon hypothesized that before A.D. 1500 there were five pure races—five centers of human population that were so isolated that there was almost no mixing.

Each of these races evolved independently, Coon believed, diverging from a pre-*Homo sapiens* stock that was essentially the same everywhere. He speculated that the common ancestor evolved into *Homo sapiens* in five separate regions at five different times, beginning about 35,000 years ago. The populations that have been *Homo sapiens* for the shortest periods of time, Coon said, are the world's "less civilized" races.

The five pure races remained distinct until A.D. 1500; then Europeans started sailing the world, leaving their genes—as sailors always have—in every port and planting distant colonies. At about the same time, thousands of Africans were captured and forcibly settled in many parts of the New World.

That meant the end of the five pure races. But Coon and other experts held that this did not necessarily rule out the idea of distinct races. In this view, there *are* such things as races; people just don't fit into them very well anymore.

The truth is that there is really no hard evidence to suggest that five or any particular number of races evolved independently. The preponderance of evidence today suggests that as traits typical of fully modern people arose in any one place, they spread quickly to all human populations. Advances in intelligence were almost certainly the fastest to spread. Most anthropologists and geneticists now believe that human beings have always been subject to migrating and mixing. In other words, there probably never were any such things as pure races.

Race mixing has not only been a fact of human history but is, in this day of unprecedented global mobility, taking place at a more rapid rate than ever. It is not farfetched to envision the day when, generations hence, the entire "complexion" of major population centers will be different. Meanwhile, we can see such changes taking place before our eyes, for they are a part of everyday reality.

HYBRID VIGOR

Oddly, those who assert scientific validity for their notions of pure and distinct races seem oblivious of a basic genetic principle that plant and animal breeders know well: too much inbreeding can lead to proliferation of inferior traits. Cross-breeding with different strains often produces superior combinations and "hybrid vigor."

The striking differences among people may very well be a result of constant genetic mixing. And as geneticists and ecologists know, in diversity lies strength and resilience.

To understand the origin and proliferation of human differences, one must first know how Darwinian evolution works.

Evolution is a two-step process. Step one is mutation: somehow a gene in the ovary or testes of an individual is altered, changing the molecular configu-

DISEASE ORIGINS

The gene for sickle cell anemia, a disease found primarily among black people, appears to have evolved because its presence can render its bearer resistant to malaria. Such a trait would have obvious value in tropical Africa.

A person who has sickle cell anemia must have inherited genes for the disease from both parents. If a child inherits only one sickle cell gene, he or she will be resistant to malaria but will not have the anemia. Paradoxically, inheriting genes from both parents does not seem to affect resistance to malaria.

In the United States, where malaria is practically nonexistent, the sickle cell gene confers no survival advantage and is disappearing. Today only about 1 out of every 10 American blacks carries the gene.

Many other inherited diseases are found only in people from a particular area. Tay-Sachs disease, which often kills before the age of two, is almost entirely confined to Jews from parts of Eastern Europe and their descendants elsewhere. Paget's disease, a bone disorder, is found most often among those of English descent. Impacted wisdom teeth are a common problem among Asians and Europeans but not among Africans. Children of all races are able to digest milk because their bodies make lactase, the enzyme that breaks down lactose, or milk sugar. But the ability to digest lactose in adulthood is a racially distributed trait.

About 90 percent of Orientals and blacks lose this ability by the time they reach adulthood and become quite sick when they drink milk.

Even African and Asian herders who keep cattle or goats rarely drink fresh milk. Instead, they first treat the milk with fermentation bacteria that break down lactose, in a sense predigesting it. They can then ingest the milk in the form of yogurt or cheese without any problem.

About 90 percent of Europeans and their American descendants, on the other hand, continue to produce the enzyme throughout their lives and can drink milk with no ill effects.

---

ration that stores instructions for forming a new individual. The children who inherit that gene will be different in some way from their ancestors.

Step two is selection: for a racial difference, or any other evolutionary change to arise, it must survive and be passed through several generations. If the mutation confers some disadvantage, the individual dies, often during embryonic development. But if the change is beneficial in some way, the individual should have a better chance of thriving than relatives lacking the advantage.

NATURAL SELECTION

If a new trait is beneficial, it will bring reproductive success to its bearer. After several generations of multiplication, bearers of the new trait may begin to outnumber nonbearers. Darwin called this natural selection to distinguish it from the artificial selection exercised by animal breeders.

Skin color is the human racial trait most generally thought to confer an evolutionary advantage of this sort. It has long been obvious in the Old World that the farther south one goes, the darker the skin color. Southern Europeans are usually somewhat darker than northern Europeans. In North Africa, skin colors are darker still, and, as one travels south, coloration reaches its maximum at the Equator. The same progressions holds in Asia, with the lightest skins to the north. Again, as one moves south, skin color darkens, reaching in southern India a "blackness" equal to that of equatorial Africans.

This north-south spectrum of skin color derives from varying intensities of the same dark brown pigment called melanin. Skin cells simply have more or less melanin granules to be seen against a background that is pinkish because of the underlying blood vessels. All races can increase their melanin concentration by exposure to the sun.

What is it about northerly latitudes in the Northern Hemisphere that favors less pigmentation and about southerly latitudes that favors more? Exposure to intense sunlight is not the only reason why people living in southerly latitudes are dark. A person's susceptibility to rickets and skin cancer, his ability to withstand cold and to see in the dark may also be related to skin color.

The best-known explanation says the body can tolerate only a narrow range of intensities of sunlight. Too much causes sunburn and cancer, while too little deprives the body of vitamin D, which is synthesized in the skin under the influence of sunlight. A dark complexion protects the skin from the harmful effects of intense sunlight. Thus, albinos born in equatorial regions have a high rate of skin cancer. On the other hand, dark skin in northerly latitudes screens out sunlight needed for the synthesis of vitamin D. Thus, dark-skinned children living in northern latitudes had high rates of rickets—a bone-deforming disease caused by a lack of vitamin D—before their milk was routinely fortified. In the sunny tropics, dark skin admits enough light to produce the vitamin.

Recently, there has been some evidence that skin colors are linked to differences in the ability to avoid injury from the cold. Army researchers found that during the Korean War blacks were more susceptible to frostbite than were whites. Even among Norwegian soldiers in World War II, brunettes had a slightly higher incidence of frostbite than did blonds.

EYE PIGMENTATION

A third link between color and latitude involves the sensitivity of the eye to various wavelengths of light. It is known that dark-skinned people have more pigmentation in the iris of the eye and at the back of the eye where the image falls. It has been found that the less pigmented the eye, the more sensitive it is to colors at the red end of the spectrum. In situations illuminated with reddish light, the northern European can see more than a dark African sees.

It has been suggested that Europeans developed lighter eyes to adapt to the longer twilights of the North and their greater reliance on firelight to illuminate caves.

Although the skin cancer-vitamin D hypothesis enjoys wide acceptance, it may well be that resistance to cold, possession of good night vision and other yet unknown factors all played roles in the evolution of skin colors.

Most anthropologists agree that the original human skin color was dark brown, since it is fairly well established that human beings evolved in the tropics of Africa. This does not, however, mean that the first people were Negroids, whose descendants, as they moved north, evolved into light-skinned Caucasoids. It is more likely that the skin color of various populations changed several times from dark to light and back as people moved from one region to another.

Consider, for example, that long before modern people evolved, *Homo erectus* had spread throughout Africa, Europe and Asia. The immediate ancestor of *Homo sapiens, Homo erectus,* was living in Africa 1.5 million years ago and in Eurasia 750,000 years ago. The earliest known forms of *Homo sapiens* do not make their appearance until somewhere between 250,000 and 500,000 years ago. Although there is no evidence of the skin color of any hominid fossil, it is probable that the *Homo erectus* population in Africa had dark skin. As subgroups spread into northern latitudes, mutations that reduced pigmentation conferred survival advantages on them and lighter skins came to predominate. In other words, there were probably black *Homo erectus* peoples in Africa and white ones in Europe and Asia.

Did the black *Homo erectus* populations evolve into today's Negroids and the white ones in Europe into today's Caucasoids? By all the best evidence, nothing like this happened. More likely, wherever *Homo sapiens* arose it proved so superior to the *Homo erectus* populations that it eventually replaced them everywhere.

If the first *Homo sapiens* evolved in Africa, they were probably dark-skinned; those who migrated northward into Eurasia lost their pigmentation. But it is just as possible that the first *Homo sapiens* appeared in northern climes, descendants of white-skinned *Homo erectus.* These could have migrated southward toward Africa, evolving darker skins. All modern races, incidentally, arose long after the brain had reached its present size in all parts of the world.

North-south variations in pigmentation are quite common among mammals and birds. The tropical races tend to be darker in fur and feather, the desert races tend to be brown, and those near the Arctic Circle are lighter colored.

There are exceptions among humans. The Indians of the Americas, from the Arctic to the southern regions of South America, do not conform to the north-south scheme of coloration. Though most think of Indians as being reddish-brown, most Indians tend to be relatively light skinned, much like their presumed Mongoloid ancestors in Asia. The ruddy complexion that lives in so many stereotypes of Indians is merely what years of heavy tanning can produce in almost any light-skinned person. Anthropologists explain the color consistency as a consequence of the relatively recent entry of people into the Americas—probably between 12,000 and 35,000 years ago. Perhaps they have not yet had time to change.

Only a few external physical differences other than color appear to have adaptive significance. The strongest cases can be made for nose shape and stature.

WHAT'S IN A NOSE

People native to colder or drier climates tend to have longer, more beak-shaped noses than those living in hot and humid regions. The nose's job is to warm and humidify air before it reaches sensitive lung tissues. The colder or drier the air is, the more surface area is needed inside the nose to get it to the right temperature or humidity. Whites tend to have longer and beakier noses than blacks or Orientals. Nevertheless, there is great variation within races. Africans in the highlands of East Africa have longer noses than Africans from the hot, humid lowlands, for example.

Stature differences are reflected in the tendency for most northern peoples to have shorter arms, legs and torsos and to be stockier than people from the tropics. Again, this is an adaptation to heat or cold. One way of reducing heat loss is to have less body surface, in relation to weight or volume, from which heat can escape. To avoid overheating, the most desirable body is long limbed and lean. As a result, most Africans tend to be lankier than northern Europeans. Arctic peoples are the shortest limbed of all.

Hair forms may also have a practical role to play, but the evidence is weak. It has been suggested that the more tightly curled hair of Africans insulates the top of the head better than does straight or wavy hair. Contrary to expectation, black hair serves better in this role than white hair. Sunlight is absorbed and converted to heat at the outer surface of the hair blanket; it radiates directly into the air. White fur, common on Arctic animals that need to absorb solar heat, is actually transparent and transmits light into the hair blanket, allowing the heat to form within the insulating layer, where it is retained for warmth.

Aside from these examples, there is little evidence that any of the other visible differences among the world's people provide any advantage. Nobody knows, for example, why Orientals have epicanthic eye folds or flatter facial profiles. The thin lips of Caucasoids and most Mongoloids have no known advantages over the Negroid's full lips. Why should middle-aged and older Caucasoid men go bald so much more frequently than the men of other races? Why does the skin of Bushmen wrinkle so heavily in the middle and later years? Or why does the skin of Negroids resist wrinkling so well? Why do the Indian men in one part of South America have blue penises? Why do Hottentot women have such unusually large buttocks?

There are possible evolutionary explanations for why such apparently useless differences arise.

One is a phenomenon known as sexual selection. Environmentally adaptive traits arise, Darwin thought, through natural selection—the environment itself chooses who will thrive or decline. In sexual selection, which Darwin also suggested, the choice belongs to the prospective mate.

In simple terms, ugly individuals will be less likely to find mates and reproduce their genes than beautiful specimens will. Take the blue penis as an example. Women might find it unusually attractive or perhaps believe it to be endowed with special powers. If so, a man born with a blue penis will find many more opportunities to reproduce his genes than his ordinary brothers.

Sexual selection can also operate when males compete for females. The moose with the larger antlers or the lion with the more imposing mane will stand a better chance of discouraging less well-endowed males and gaining access to females. It is possible that such a process operated among Caucasoid males, causing them to become markedly hairy, especially around the face.

ATTRACTIVE TRAITS

Anthropologists consider it probable that traits such as the epicanthic fold or the many regional differences in facial features were selected this way.

Yet another method by which a trait can establish itself involves accidental selection. It results from what biologists call genetic drift.

Suppose that in a small nomadic band a person is born with perfectly parallel fingerprints instead of the usual loops, whorls or arches. That person's children would inherit parallel fingerprints, but they would confer no survival advantages. But if our family decides to strike out on its own, it will become the founder of a new band consisting of its own descendants, all with parallel fingerprints.

Events such as this, geneticists and anthropologists believe, must have occurred many times in the past to produce

the great variety within the human species. Among the apparently neutral traits that differ among populations are:

Ear Wax

There are two types of ear wax. One is dry and crumbly and the other is wet and sticky. Both types can be found in every major population, but the frequencies differ. Among northern Chinese, for example, 98 percent have dry ear wax. Among American whites, only 16 percent have dry ear wax. Among American blacks the figure is 7 percent.

Scent Glands

As any bloodhound knows, every person has his or her own distinctive scent. People vary in the mixture of odoriferous compounds exuded through the skin—most of it coming from specialized glands called apocrine glands. Among whites, these are concentrated in the armpits and near the genitals and anus. Among blacks, they may also be found on the chest and abdomen. Orientals have hardly any apocrine glands at all. In the words of the Oxford biologist John R. Baker, "The Europids and Negrids are smelly, the Mongoloids scarcely or not at all." Smelliest of all are northern European, or so-called Nordic, whites. Body odor is rare in Japan. It was once thought to indicate a European in the ancestry and to be a disease requiring hospitalization.

Blood Groups

Some populations have a high percentage of members with a particular blood group. American Indians are overwhelmingly group O—100 percent in some regions. Group A is most common among Australian aborigines and the Indians in western Canada. Group B is frequent in northern India, other parts of Asia and western Africa.

Advocates of the pure-race theory once seized upon blood groups as possibly unique to the original pure races. The proportions of groups found today, they thought, would indicate the degree of mixing. It was subsequently found that chimpanzees, our closest living relatives, have the same blood groups as humans.

Taste

PTC (phenylthiocarbamide) is a synthetic compound that some people can taste and others cannot. The ability to taste it has no known survival value, but it is clearly an inherited trait. The proportion of persons who can taste PTC varies in different populations: 50 to 70 percent of Australian aborigines can taste it, as can 60 to 80 percent of all Europeans. Among East Asians, the percentage is 83 to 100 percent, and among Africans, 90 to 97 percent.

Urine

Another indicator of differences in body chemistry is the excretion of a compound known as BAIB (beta-amino-isobutyric acid) in urine. Europeans seldom excrete large quantities, but high levels of excretion are common among Asians and American Indians. It had been shown that the differences are not due to diet.

No major population has remained isolated long enough to prevent any unique genes from eventually mixing with those of neighboring groups. Indeed, a map showing the distribution of so-called traits would have no sharp boundaries, except for coastlines. The intensity of a trait such as skin color, which is controlled by six pairs of genes and can therefore exist in many shades, varies gradually from one population to another. With only a few exceptions, every known genetic possibility possessed by the species can be found to some degree in every sizable population.

EVER-CHANGING SPECIES

One can establish a system of racial classification simply by listing the features of populations at any given moment. Such a concept of race is, however, inappropriate to a highly mobile and ever-changing species such as *Homo sapiens*. In the short view, races may seem distinguishable, but in biology's long haul, races come and go. New ones arise and blend into neighboring groups to create new and racially stable populations. In time, genes from these groups flow into other neighbors, continuing the production of new permutations.

Some anthropologists contend that at the moment American blacks should be considered a race distinct from African blacks. They argue that American blacks are a hybrid of African blacks and European whites. Indeed, the degree of mixture can be calculated on the basis of a blood component known as the Duffy factor.

In West Africa, where most of the New World's slaves came from, the Duffy factor is virtually absent. It is present in 43 percent of American whites. From the number of American blacks who are now "Duffy positive" it can be calculated that whites contributed 21 percent of the genes in the American black population. The figure is higher for blacks in northern and western states and lower in the South. By the same token, there are whites who have black ancestors. The number is smaller because of the tendency to identify a person as black even if only a minor fraction of his ancestors were originally from Africa.

The unwieldiness of race designations is also evident in places such as Mexico where most of the people are, in effect, hybrids of Indians (Mongoloid by some classifications) and Spaniards (Caucasoid). Many South American populations are tri-hybrids—mixtures of Mongoloid, Caucasoid and Negroid. Brazil is a country where the mixture has been around long enough to constitute a racially stable population. Thus, in one sense, new races have been created in the United States, Mexico and Brazil. But in the long run, those races will again change.

Sherwood Washburn, a noted anthropologist, questions the usefulness of racial classification: "Since races are open systems which are intergrading, the number of races will depend on the purpose of the classification. I think we should require people who propose a classification of races to state in the first place why they wish to divide the human species."

The very notion of a pure race, then, makes no sense. But, as evolutionists know full well, a rich genetic diversity within the human species most assuredly *does*.

Culture, Not Race, Explains Human Diversity

By Mark Nathan Cohen

WE ARE NOT LIKELY to convince students that racist views are wrong if we teach them only about biology and ignore culture. The basic facts about the broad patterns of human biological variation and "race" are fairly clear and well established: Individual human beings undeniably differ in myriad ways, and each specific difference may be important. Variation in skin color, for example, affects susceptibility to sunburn, skin cancer, and rickets. Variations in body build affect the ability to keep warm or cool; variations in the shape of the nose affect its ability in warm and moisten the air inhaled. Unseen genetic variations protect some people from (and predispose other people to) diseases ranging from malaria and smallpox to diabetes and cancer.

However, "races" as imagined by the public do not actually exist. Any definition of "race" that we attempt produces more exceptions than sound classifications. No matter what system we use, most people don't fit.

As almost every introductory textbook in physical anthropology explains, the distinctions among human populations are generally graded, not abrupt. In other words, skin color comes in a spectrum from dark to light, not just in black or white; noses come in a range of shapes, not just broad or narrow. Furthermore, the various physical traits such as skin color and nose shape (plus the enormous number of invisible traits) come in an infinite number of combinations; one cannot predict other traits by knowing one trait that a person possesses. A person with dark skin can have any blood type and can have a broad nose (a combination common in West Africa), a narrow nose (as many East Africans do), or even blond hair (a combination seen in Australia and New Guinea).

Of the 50,000 to 100,000 pairs of genes needed to make a human being, perhaps 35,000 to 75,000 are the same in all people, and 15,000 to 25,000 may take different forms in different people, thus accounting for human variation. But only a tiny number of these genes

---

We ... have to stop teaching or accepting the idea that humans are divided into three races—Caucasian, Negroid, and Mongoloid— an idea that is at least 50 years out of date.

From the *Chronicle of Higher Education*, April 17, 1998, pp. B4-B5. © 1998 by Mark Nathan Cohen. Reprinted by permission.

Many scholars of colonialism have described how members of oppressed minorities create a sense of community by choosing to ignore their colonizers' culture.

affect what many people might consider to be racial traits. For example, geneticists believe that skin color is based on no more than 4 to 10 pairs of genes. The genes of black and white Americans probably are 99.9 per cent alike.

In addition, studies of the human family tree based on detailed genetic analysis suggest that traits such as skin color are not even good indicators of who is related to whom, because the traits occur independently in several branches of the human family. When we consider the pairs of genes that may differ among humans, we see that, beyond the genes determining skin color, black people from Africa, Australia, and the south of India are not particularly closely related to each other genetically.

All of this means that variations among "races" cannot possibly explain the differences in behavior or intelligence that people think they see. Although black Americans on average receive lower scores on standardized tests than do white Americans, neither "race" is actually a biological group. Skin color alone cannot account for the differences in group averages. Although we know a great deal about the role of many of our genes—for instance, which ones cause sickle-cell anemia—no genes are known to control differences in specific behavior or in intelligence among human groups. Even if someone discovered such genes, we have no reason to assume that they would correlate with skin color any more than most other genes do.

Anthropologists and other academics must do a better job of communicating these facts to our students and to the public at large. But even if we make sure that everyone understands these facts, racism will persist—unless we convince people that a different explanation for variations in behavior makes more sense. The alternative that we need to emphasize is the concept of culture—but a concept of culture far deeper and more sophisticated than that taught by many multiculturalists.

The anthropological concept of culture can be explained best by an analogy with language. Just as language is more than vocabulary, culture is more than, say, art and music. Language has rules of grammer and sound that limit and give structure to communication, usually without conscious thought. In English, for example, we use word order to convey the relationship among the words in a sentence. Latin uses suffixes to show relationships; Swahili uses prefixes. Even languages as similar as French, Spanish, and Italian use different subsets of the many sounds the human mouth can make.

Equally, culture structures our behavior, thoughts, perceptions, values, goals, morals, and cognitive processes—also usually without conscious thought. Just as each language is a set of arbitrary conventions shared by those who speak the language, so each culture is made up of its own arbitrary conventions. Many languages work perfectly well; many cultures do, too.

Just as familiarity with the language of one's childhood makes it harder to learn the sounds and grammer of another language, so one's culture tends to blind one to alternatives. All of us—Americans as well as members of remote Amazonian tribes—are governed by culture. Our choices in life are circumscribed largely by arbitrary rules, and we have a hard time seeing the value of other people's choices and the shortcomings of our own. For example, many societies exchange goods not for profit, but to foster social relationships, as most Americans do within their own families. Some peoples prefer to maintain their surrounding environment rather than to seek "progress." Other groups have a very different sense than we do of the balance between individual freedom and responsibility to one's community.

Besides teaching our students about the importance of culture, we need to revive the anthropological concept of cultural relativism—perhaps the most important concept that liberal education can teach. The enemies of relativism have claimed that the concept means that everything is equal—that no moral judgments are possible, that Americans must accept whatever other people do. But this is not what relativism means.

What it does mean is that we must look carefully at what other people are doing and try to understand their behavior in context before we judge it. It means that other people may not share our desires or our perceptions. It also means that we have to recognize the arbitrary nature of our own choices and be willing to reexamine them by learning about the choices that other people have made. In medicine, for example, many cultures have long tried to treat the whole patient, mind and body together, which some of our doctors are just beginning to focus on, having been trained instead to concentrate on treating a particular disease.

THE KEY POINT is that what we see as "racial" differences in behavior may reflect the fact that people have different values, make different choices, operate within different cultural "grammars," and categorize things (and therefore think) in different ways. Over the years, we have learned that culture shapes many things we once thought were determined by biology, including sexuality, aggression, perception, and susceptibility to disease. But many people still confuse the effects of biology and those of culture.

For example, we still use analogy problems to test students' skill in logic, which we then define as innate, genetically driven intelligence, which, some argue, differs by "race." But solving the problems depends on putting items into categories, and the categories we use are cultural, not universal. Giving the correct answer depends less on inherent intelligence than it does on knowing the classification rules used by those who created the test. Most U.S. students would flunk analogy problems put to them by Mesoamerican peasants, who divide things into "hot" and "cold"—

In addition, psychologists, in particular, must stop assuming that only one pattern exists for human cognition, perception, formation of categories, and so forth. All too many psychologists believe that standardized I.Q. tests are equally valid for assessing individuals from different cultural backgrounds. Operating from that mistaken assumption, they teach a narrow view of "correct" human behavior, which promotes racism.

We must realize that students from some minority groups are likely to do badly on I.Q. tests for a variety of reasons beyond the poor health, nutrition, and education that many of them have

even noticing what someone from another group would consider crucial. A student who did not grow up in the middle class might well believe that to appear smart, one should give a slow, thoughtful answer—not a snap answer or sound bite. Such a student will probably be put off by the idea of competitiveness and individual ranking, and thus by the whole experience of the I.Q. test.

ANTHROPOLOGISTS must do a better job of communicating these and other important facts about human biological and cultural variation to their students and to the public at

Cultural relativism is the only road to tolerance and real freedom of thought, because it lets us get outside the blinders imposed by our own culture.

categories that, in their usage, go far beyond physical temperature.

To communicate the importance of culture to students, we have to make some adjustments in our teaching. First, we have to stop teaching world and American history in ways that deny the contributions of others and prevent thoughtful analysis of our own actions. Teaching along these lines has improved, but it can be better yet. For instance, how many students are taught that George Washington and Benjamin Franklin were speculators who wanted to enrich themselves with land on the Western frontier—land the British had intended to reserve for Native American use?

We also have to stop teaching or accepting the idea that humans are divided into three races—Caucasian, Negroid, and Mongoloid—an idea that is at least 50 years out of date. I constantly face students who have been taught the concept of three races in the 1990s by high-school and college teachers in other social sciences.

experienced because of poverty. The content of the tests is biased toward students in the mainstream, both in terms of the subject matter of the questions and in more-subtle ways, such as expectations about what is important in a problem.

And some minority students see no reason to try to do well on I.Q. tests. They may not expect to go to college. Their sense of self-worth may depend on their lack of interest in the mainstream culture, which they feel has rejected them. Many scholars of colonialism have described how members of oppressed minorities create a sense of community by choosing to ignore their colonizers' culture. John Ogbu, an anthropologist at the University of California at Berkeley, has shown that the same phenomenon occurs in U.S. schools.

Thus, scores on I.Q. tests depend not only on how "smart" one is, but also on familiarity with middleclass, white American culture. Different cultural groups are likely to pay more attention to different parts of test questions, not

large. Rather than revel in the details of obscure populations, our courses in cultural anthropology should focus on interpretations of significant contemporary events. We have to demonstrate to students that not all events outside—and even within—the United States can be understood from the single cultural perspective that other social sciences tend to teach.

We can be more outspoken in the media about alternative perspectives on international events, seeking opportunities to explain cultural practices and conventions in other countries that may lead to their different decisions or priorities. Indeed, anthropologists have to be more confident about the significance of their discipline and more willing to assert the importance of their knowledge. We need to find ways to work both with faculty members in other disciplines and with policy makers to convey our awareness that, in virtually every contemporary problem, different participants not only speak differently but also think differently.

For example, we could attempt to understand the behavior of Iranians in the United States' 20-year-old confrontation with Iran by looking objectively at their culture and history, and at ours. One of the most obvious points is that traditional Iranian leadership involves a blend of religious and secular power that is foreign to Americans. The result is that leaders of the two countries have very different agendas. And many Iranians have a different sense of the proper balance among business, profit, family, community, and spirituality.

We must show our students that they, just as much as our adversaries or "primitive peoples" in isolated cultures, are bound by the arbitrary rules of their own culture. If we don't teach these points, we are failing to show that racist assumptions about why people behave as they do are not legitimate.

We also have to stop focusing on details of the human fossil record in our introductory physical-anthropology courses. Basic courses should deal instead with human variation in a variety of populations, by looking at such issues as fertility, mortality, growth, nutrition, adaptation to the environment, and disease. It would be productive to teach, for example, that hypertension and diabetes are not simply natural results of individual genetic endowment or the aging process but, in fact, rarely occur among people whose food does not go through the commercial processing that is customary in the West.

Finally, administrators and professors across the disciplines need to recognize that it is crucial for all students to understand the concepts of culture and cultural relativism—not the inaccurate caricatures of these concepts that have been bandied about in the culture wars, but the sophisticated, carefully defined and nuanced versions that anthropologists have evolved over decades of work. Cultural relativism is the only road to tolerance and real freedom of thought, because it lets us get outside the blinders imposed by our own culture. It must be built into courses taught in core curricula—not taught only as an elective frill if a student happens to sign up for an anthropology course.

Mark Nathan Cohen is a professor of anthropology at the State University of New York College at Plattsburgh and the author of Culture of Intolerance: Chauvinism, Class, and Racism in the United States *(Yale University Press, 1998).*

COMMENTARY

The Tall and the Short of It

BY BARRY BOGIN

BARRY BOGIN is a professor of anthropology at the University of Michigan in Dearborn and the author of Patterns of Human Growth.

BAFFLED BY YOUR FUTURE PROSPECTS? As a biological anthropologist, I have just one word of advice for you: plasticity. *Plasticity* refers to the ability of many organisms, including humans, to alter themselves—their behavior or even their biology—in response to changes in the environment. We tend to think that our bodies get locked into their final form by our genes, but in fact we alter our bodies as the conditions surrounding us shift, particularly as we grow during childhood. Plasticity is as much a product of evolution's fine-tuning as any particular gene, and it makes just as much evolutionary good sense. Rather than being able to adapt to a single environment, we can, thanks to plasticity, change our bodies to cope with a wide range of environments. Combined with the genes we inherit from our parents, plasticity accounts for what we are and what we can become.

Anthropologists began to think about human plasticity around the turn of the century, but the concept was first clearly defined in 1969 by Gabriel Lasker, a biological anthropologist at Wayne State University in Detroit. At that time scientists tended to consider only those adaptations that were built into the genetic makeup of a person and passed on automatically to the next generation. A classic example of this is the ability of adults in some human societies to drink milk. As children, we all produce an enzyme called lactase, which we need to break down the sugar lactose in our mother's milk. In many of us, however, the lactase gene slows down dramatically as we approach adolescence—probably as the result of another gene that regulates its activity. When that regulating gene turns down the production of lactase, we can no longer digest milk.

Lactose intolerance—which causes intestinal gas and diarrhea—affects between 70 and 90 percent of African Americans, Native Americans, Asians, and people who come from around the Mediterranean. But others, such as people of central and western European descent and the Fulani of West Africa, typically have no problem drinking milk as adults. That's because they are descended from societies with long histories of raising goats and cattle. Among these people there was a clear benefit to being able to drink milk, so natural se-lection gradually changed the regulation of their lactase gene, keeping it functioning throughout life.

That kind of adaptation takes many centuries to become established, but Lasker pointed out that there are two other kinds of adaptation in humans that need far less time to kick in. If people have to face a cold winter with little or no heat, for example, their metabolic rates rise over the course of a few weeks and they produce more body heat. When summer returns, the rates sink again.

Lasker's other mode of adaptation concerned the irreversible, lifelong modification of people as they develop—that is, their plasticity. Because we humans take so many years to grow to adulthood, and because we live in so many different environments, from forests to cities and from deserts to the Arctic, we are among the world's most variable species in our physical form and behavior. Indeed, we are one of the most plastic of all species.

One of the most obvious manifestations of human malleability is our great range of height, and it is a subject I've made a special study of for the last 25 years. Consider these statistics: in 1850 Americans were the tallest people in the world, with American men averaging

5'6". Almost 150 years later, American men now average 5'8", but we have fallen in the standings and are now only the third tallest people in the world. In first place are the Dutch. Back in 1850 they averaged only 5'4"—the shortest men in Europe—but today they are a towering 5'10". (In these two groups, and just about everywhere else, women average about five inches less than men at all times.)

In an age when DNA is king, it's worth considering why Americans are no longer the world's tallest people, and some Guatemalans no longer pygmies.

So what happened? Did all the short Dutch sail over to the United States? Did the Dutch back in Europe get an infusion of "tall genes"? Neither. In both America and the Netherlands life got better, but more so for the Dutch, and height increased as a result. We know this is true thanks in part to studies on how height is determined. It's the product of plasticity in our childhood and in our mothers' childhood as well. If a girl is undernourished and suffers poor health, the growth of her body, including her reproductive system, is usually reduced. With a shortage of raw materials, she can't build more cells to construct a bigger body; at the same time, she has to invest what materials she can get into repairing already existing cells and tissues from the damage caused by disease. Her shorter stature as an adult is the result of a compromise her body makes while growing up.

Such a woman can pass on her short stature to her child, but genes have nothing to do with it for either of them. If she becomes pregnant, her small reproductive system probably won't be able to supply a normal level of nutrients and oxygen to her fetus. This harsh environment reprograms the fetus to grow more slowly than it would if the woman was healthier, so she is more likely to give birth to a smaller baby. Low-birth-weight babies (weighing less than 5.5 pounds) tend to continue their prenatal program of slow growth through childhood. By the time they are teenagers, they are usually significantly shorter than people of normal birth weight. Some particularly striking evidence of this reprogramming comes from studies on monozygotic twins, which develop from a single fertilized egg cell and are therefore identical genetically. But in certain cases, monozygotic twins end up being nourished by unequal portions of the placenta. The twin with the smaller fraction of the placenta is often born with low birth weight, while the other one is normal. Follow-up studies show that this difference between the twins can last throughout their lives.

As such research suggests, we can use the average height of any group of people as a barometer of the health of their society. After the turn of the century both the United States and the Netherlands began to protect the health of their citizens by purifying drinking water, installing sewer systems, regulating the safety of food, and, most important, providing better health care and diets to children. The children responded to their changed environment by growing taller. But the differences in Dutch and American societies determined their differing heights today. The Dutch decided to provide public health benefits to all the public, including the poor. In the United States, meanwhile, improved health is enjoyed most by those who can afford it. The poor often lack adequate housing, sanitation, and health care. The difference in our two societies can be seen at birth: in 1990 only 4 percent of Dutch babies were born at low birth weight, compared with 7 percent in the United States. For white Americans the rate was 5.7 percent, and for black Americans the rate was a whopping 13.3 percent. The disparity between rich and poor in the United States carries through to adulthood: poor Americans are shorter than the better-off by about one inch. Thus, despite great affluence in the United States, our average height has fallen to third place.

People are often surprised when I tell them the Dutch are the tallest people in the world. Aren't they shrimps compared with the famously tall Tutsi (or "Watusi," as you probably first encountered them) of Central Africa? Actually, the supposed great height of the Tutsi is one of the most durable myths from the age of European exploration. Careful investigation reveals that today's Tutsi men average 5'7" and that they have maintained that average for more than 100 years. That means that back in the 1800s, when puny European men first met the Tutsi, the Europeans suffered strained necks from looking up all the time. The two-to-three-inch difference in average height back then could easily have turned into fantastic stories of African giants by European adventures and writers.

The Tutsi could be as tall or taller than the Dutch if equally good health care and diets were available in Rwanda and Burundi, where the Tutsi live. But poverty rules the lives of most African people, punctuated by warfare, which makes the conditions for growth during childhood even worse. And indeed, it turns out that the Tutsi and other Africans who migrate to Western Europe or North America at young ages end up taller than Africans remaining in Africa.

At the other end of the height spectrum, Pygmies tell a similar story. The shortest people in the world today are the Mbuti, the Efe, and other Pygmy peoples of Central Africa. Their average stature is almost 4'9" for adult men and 4'6" for women. Part of the reason Pygmies are short is indeed genetic: some evidently lack the genes for producing the growth-promoting hormones that course through other people's bodies, while others are genetically incapable of using these hormones to trigger the cascade of reactions that lead to growth. But another important reason for their small size is environmental. Pygmies living as hunter-gatherers in the forests of Central African countries appear to be undernourished, which further limits their growth. Pygmies who live on farms and ranches outside the forest are better fed than their hunter-gatherer relatives and are taller as well. Both genes and nutrition thus account for the size of Pygmies.

Peoples in other parts of the world have also been labeled pygmies, such as some groups in Southeast Asia and the Maya of Guatemala. Well-meaning explorers and scientists have often claimed that they are genetically short, but here we encounter another myth of height. A group of extremely short people in New Guinea, for example, turned out to eat a diet deficient in iodine and other essential nutrients. When they were supplied with cheap mineral and vitamin supplements, their supposedly genetic short stature vanished in their children, who grew to a more normal height.

ANOTHER WAY FOR THESE SO-CALLED pygmies to stop being pygmies is to immigrate to the United States. In my own research, I study the growth of two groups of Mayan children. One group lives in their homeland of Guatemala, and the other is a group of refugees living in the United States. The Maya in Guatemala live in the village of San Pedro, which has no safe source of drinking water. Most of the water is contaminated with fertilizers and pesticides used on nearby agricultural fields. Until recently, when a deep well was dug, the townspeople depended on an unreliable supply of water from rain-swollen streams. Most homes still lack running water and have only pit toilets. The parents of the Mayan children work mostly at clothing factories and are paid only a few dollars a day.

I began working with the schoolchildren in this village in 1979, and my research shows that most of them eat only 80 percent of the food they need. Other research shows that almost 30 percent of the girls and 20 percent of the boys are deficient in iodine, that most of the children suffer from intestinal parasites, and that many have persistent ear and eye infections. As a consequence, their health is poor and their height reflects it: they average about three inches shorter than better-fed Guatemalan children.

The Mayan refugees I work with in the United States live in Los Angeles and in the rural agricultural community of Indiantown in central Florida. Although the adults work mostly in minimum-wage jobs, the children in these communities are generally better off than their counterparts in Guatemala. Most Maya arrived in the 1980s as refugees escaping a civil war as well as a political system that threatened them and their children. In the United States they found security and started new lives, and before long their children began growing faster and bigger. My data show that the average increase in height among the first generation of these immigrants was 2.2 inches, which means that these so-called pygmies have undergone one of the largest single-generation increases in height ever recorded. When people such as my own grandparents migrated from the poverty of rural life in Eastern Europe to the cities of the United States just after World War I, the increase in height of the next generation was only about one inch.

One reason for the rapid increase in stature is that in the United States the Maya have access to treated drinking water and to a reliable supply of food. Especially critical are school breakfast and lunch programs for children from low-income families, as well as public assistance programs such as the federal Woman, Infants, and Children (WIC) program and food stamps. That these programs improve health and growth is no secret. What is surprising is how fast they work. Mayan mothers in the United States tell me that even their babies are bigger and healthier than the babies they raised in Guatemala, and hospital statistics bear them out. These women must be enjoying a level of health so improved from that of their lives in Guatemala that their babies are growing faster in the womb. Of course, plasticity means that such changes are dependent on external conditions, and unfortunately the rising height—and health—of the Maya is in danger from political forces that are attempting to cut funding for food stamps and the WIC program. If that funding is cut, the negative impact on the lives of poor Americans, including the Mayan

refugees, will be as dramatic as were the former positive effects.

Height is only the most obvious example of plasticity's power; there are others to be found everywhere you look. The Andes-dwelling Quechua people of Peru are well-adapted to their high-altitude homes. Their large, barrel-shaped chests house big lungs that inspire huge amounts of air with each breath, and they manage to survive on the lower pressure of oxygen they breathe with an unusually high level of red blood cells. Yet these secrets of mountain living are

One way for the so-called pygmies of Guatemala to stop being pygmies is to immigrate to the United States.

not hereditary. Instead the bodies of young Quechua adapt as they grow in their particular environment, just as those of European children do when they live at high altitudes.

Plasticity may also have a hand in determining our risks for developing a number of diseases. For example, scientists have long been searching for a cause for Parkinson's disease. Because Parkinson's tends to run in families, it is natural to think there is a genetic cause. But while a genetic mutation linked to some types of Parkinson's disease was reported in mid-1997, the gene accounts for only a fraction of people with the disease. Many more people with Parkinson's do not have the gene, and not all people with the mutated gene develop the disease.

Ralph Garruto, a medical researcher and biological anthropologist at the National Institutes of Health, is investigating the role of the environment and human plasticity not only in Parkinson's but in Lou Gehrig's disease as well. Garruto and his team traveled to the islands of Guam and New Guinea, where rates of both diseases are 50 to 100 times higher than in the United States. Among the native Chamorro people of Guam these diseases kill one person out of every five over the age of 25. The scientists found that both diseases are linked to a shortage of calcium in the diet. This shortage sets off a cascade of events that result in the digestive system's absorbing too much of the aluminum present in the diet. The aluminum wreaks havoc on various parts of the body, including the brain, where it destroys neurons and eventually causes paralysis and death.

The most amazing discovery made by Garruto's team is that up to 70 percent of the people they studied in Guam had some brain damage, but only 20 percent progressed all the way to Parkinson's or Lou Gehrig's disease. Genes and plasticity seem to be working hand in hand to produce these lower-than-expected rates of disease. There is a certain amount of genetic variation in the ability that all people have in coping with calcium shortages—some can function better than others. But thanks to plasticity, it's also possible for people's bodies to gradually develop ways to protect themselves against aluminum poisoning. Some people develop biochemical barriers to the aluminum they eat, while others develop ways to prevent the aluminum from reaching the brain.

An appreciation of plasticity may temper some of our fears about these diseases and even offer some hope. For if Parkinson's and Lou Gehrig's diseases can be prevented among the Chamorro by plasticity, then maybe medical researchers can figure out a way to produce the same sort of plastic changes in you and me. Maybe Lou Gehrig's disease and Parkinson's disease—as well as many other, including some cancers—aren't our genetic doom but a product of our development, just like variations in human height. And maybe their danger will in time prove as illusory as the notion that the Tutsi are giants, or the Maya pygmies—or Americans still the tallest of the tall.

Unit Selections

Key Points to Consider

❖ What is "forensic anthropology"? How can it be applied to modern life?

❖ Will the human demographic explosion amount to a death sentence for primates? Defend your answer.

❖ Is there any way to prevent epidemics in the human species? How?

❖ What do you think should be done to alleviate the worldwide AIDS epidemic?

❖ Are humans inherently violent?

❖ Should we attempt to take control of our genetic future? Why or why not?

❖ What social policy issues are involved in the nature versus nurture debate?

❖ What relevance does the concept of natural selection have to the treatment of disease?

❖ Is technology evolving too fast? Why or why not?

 **Links** | **www.dushkin.com/online/**

These sites are annotated on pages 4 and 5.

Anthropology continues to evolve as a discipline, not only in the tools and techniques of the trade, but also in the application of whatever knowledge we stand to gain about ourselves. It is in this context that Patrick Huyghe, in "Profile of an Anthropologist: No Bone Unturned," describes "forensic anthropology," a whole new field involving the use of physical similarities and differences between people in order to identify human remains. Sometimes an awareness of our biological and behavioral past may even make the difference in bodily health. Lori Oliwenstein (in "Dr. Darwin") talks about how the symptoms of disease must first be interpreted as to whether they represent part of the aggressive strategy of microbes or the defensive mechanisms of the patient before treatment can be applied.

One theme that ties the articles of this section together, then, is that they have as much to do with the present as they have to do with the past.

As we reflect upon where we have been and how we came to be as we are in the evolutionary sense, the inevitable question arises as to what will happen next. (See "Wonders: How Fast Is Technology Evolving?" by W. Brian Arthur.) This is the most difficult issue of all, since our biological future depends so much on long-range ecological trends that no one seems to be able to predict. There is no better example of this problem than the recent explosion of new diseases, as described in "The Viral Superhighway" by George Armelagos and in "HIV 1998: The Global Picture" by Jonathan Mann and Daniel Tarantola. Some wonder if we will even survive long enough as a species to experience any significant biological changes. Perhaps our capacity for knowledge is outstripping the wisdom to use it wisely, and the consequent destruction of our earthly environments and wildlife (as recounted in "Gorilla Warfare" by Craig Stanford) is placing us in ever greater danger of creating the circumstances of our own extinction.

Counterbalancing this pessimism is the view, implied by Robert Sussman, in "Exploring Our Basic Human Nature: Are Humans Inherently Violent?" that because it has been our conscious decision making (and not the genetically predetermined behavior that characterizes some species) that has gotten us into this mess, then it will be the conscious will of our generation and future generations that will get us out. But, can we wait much longer for humanity to collectively come to its senses? Or is it already too late?

Profile of an Anthropologist

No Bone Unturned

Patrick Huyghe

The research of some physical anthropologists and archaeologists involves the discovery and analysis of old bones (as well as artifacts and other remains). Most often these bones represent only part of a skeleton or maybe the mixture of parts of several skeletons. Often these remains are smashed, burned, or partially destroyed. Over the years, physical anthropologists have developed a remarkable repertoire of skills and techniques for teasing the greatest possible amount of information out of sparse material remains.

Although originally developed for basic research, the methods of physical anthropology can be directly applied to contemporary human problems. . . . In this profile, we look briefly at the career of Clyde C. Snow, a physical anthropologist who has put these skills to work in a number of different settings. . . .

As you read this selection, ask yourself the following questions:

• Given what you know of physical anthropology, what sort of work would a physical anthropologist do for the Federal Aviation Administration?
• What is anthropometry? *How might anthropometric surveys of pilots and passengers help in the design of aircraft equipment?*
• What is forensic anthropology? *How can a biological anthropologist be an expert witness in legal proceedings?*

Clyde Snow is never in a hurry. He knows he's late. He's always late. For Snow, being late is part of the job. In fact, he doesn't usually begin to work until death has stripped some poor individual to the bone, and no one—neither the local homicide detectives nor the pathologists—can figure out who once gave identity to the skeletonized remains. No one, that is, except a shrewd, laconic, 60-year-old forensic anthropologist.

Snow strolls into the Cook County Medical Examiner's Office in Chicago on this brisk October morning wearing a pair of Lucchese cowboy boots and a three-piece pin-striped suit. Waiting for him in autopsy room 160 are a bunch of naked skeletons found in Illinois, Wisconsin, and Minnesota since his last visit. Snow, a native Texan who now lives in rural Oklahoma, makes the trip up to Chicago some six times a year. The first case on his agenda is a pale brown skull found in the garbage of an abandoned building once occupied by a Chicago cosmetics company.

Snow turns the skull over slowly in his hands, a cigarette dangling from his fingers. One often does. Snow does not seem overly concerned about mortality, though its tragedy surrounds him daily.

"There's some trauma here," he says, examining a rough edge at the lower back of the skull. He points out the area to Jim Elliott, a homicide detective with the Chicago police. "This looks like a chopping blow by a heavy bladed instrument. Almost like a decapitation." In a place where the whining of bone saws drifts through hallways and the sweet-sour smell of death hangs in the air, the word surprises no one.

Snow begins thinking aloud. "I think what we're looking at here is a female, or maybe a small male, about thirty to forty years old. Probably Asian." He turns the skull upside down, pointing out the degree of wear on the teeth. "This was somebody who lived on a really rough diet. We don't normally find this kind of dental wear in a modern Western population."

"How long has it been around?" Elliott asks.

Snow raises the skull up to his nose. "It doesn't have any decompositional odors," he says. He pokes a finger in the skull's nooks and crannies. "There's no soft tissue left. It's good and dry. And it doesn't show signs of having been buried. I would say that this has been lying around in an attic or a box for years. It feels like a souvenir skull," says Snow.

Souvenir skulls, usually those of Japanese soldiers, were popular with U.S. troops serving in the Pacific during World War II; there was also a trade in skulls during the Vietnam War years. On closer inspection, though, Snow begins to wonder about the skull's Asian origins—the broad nasal aperture and the jutting forth of the upper-tooth-bearing part of the face suggest Melanesian features. Sifting through the objects found in the abandoned building with the skull, he finds several loose-leaf albums of 35-millimeter transparencies documenting life among the highland tribes of New Guinea. The slides, shot by an anthropologist, include graphic scenes of ritual warfare. The skull, Snow concludes, is more likely to be a trophy from one of these tribal battles than the result of a local Chicago homicide.

"So you'd treat it like found property?" Elliott asks finally. "Like somebody's garage-sale property?"

"Exactly," says Snow.

Clyde Snow is perhaps the world's most sought-after forensic anthropologist. People have been calling upon him to identify skeletons for more than a quarter of a century. Every year he's involved in some 75 cases of identification, most of them without fanfare. "He's an old scudder who doesn't have to blow his own whistle," says Walter Birkby, a forensic anthropologist at the University of Arizona. "He know's he's good."

Yet over the years Snow's work has turned him into something of an unlikely celebrity. He has been called upon to identify the remains of the Nazi war criminal Josef Mengele, reconstruct the face of the Egyptian boy-king Tutankhamen, confirm the authenticity of the body autopsied as that of President John F. Kennedy, and examine the skeletal remains of General Custer's men at the battlefield of the Little Bighorn. He has also been involved in the grim task of identifying the bodies in some of the United States' worst airline accidents.

Such is his legend that cases are sometimes attributed to him in which he played no part. He did not, as the *New York Times* reported, identify the remains of the crew of the *Challenger* disaster. But the man is often the equal of his myth. For the past four years, setting his personal safety aside, Snow has spent much of his time in Argentina, searching for the graves and identities of some of the thousands who "disappeared" between 1976 and 1983, during Argentina's military regime.

Snow did not set out to rescue the dead from oblivion. For almost two decades, until 1979, he was a physical anthropologist at the Civil Aeromedical Institute, part of the Federal Aviation Administration in Oklahoma City. Snow's job was to help engineers improve aircraft design and safety features by providing them with data on the human frame.

One study, he recalls, was initiated in response to complaints from a flight attendants' organization. An analysis of accident patterns had revealed that inadequate restraints on flight attendants' jump seats were leading to deaths and injuries and that aircraft doors weighing several hundred pounds were impeding evacuation efforts. Snow points out that ensuring the survival of passengers in emergencies is largely the flight attendants' responsibility. "If they are injured or killed in a crash, you're going to find a lot of dead passengers."

Reasoning that equipment might be improved if engineers had more data on the size and strength of those who use it, Snow undertook a study that required meticulous measurement. When his report was issued in 1975, Senator William Proxmire was outraged that $57,800 of the taxpayers' money had been spent to caliper 423 airline strewardesses from head to toe. Yet the study, which received one of the senator's dubious Golden Fleece Awards, was firmly supported by both the FAA and the Association of Flight Attendants. "I can't imagine," says Snow with obvious delight, "how much coffee Proxmire got spilled on him in the next few months."

It was during his tenure at the FAA that he developed an interest in forensic work. Over the years the Oklahoma police frequently consulted the physical anthropologist for help in identifying crime victims. "The FAA figured it was a kind of community service to let me work on these cases," he says.

The experience also helped to prepare him for the grim task of identifying the victims of air disasters. In December 1972, when a United Airlines plane crashed outside Chicago, killing 43 of the 61 people aboard (including the wife of Watergate conspirator Howard Hunt, who was found with $10,000 in her purse), Snow was brought in to help examine the bodies. That same year, with Snow's help, forensic anthropology was recognized as a specialty by the American Academy of Forensic Sciences. "It got a lot of anthropologists interested in forensics," he says, "and it made a lot of pathologists out there aware that there were anthropologists who could help them."

Each nameless skeleton poses a unique mystery for Snow. But some, like the second case awaiting him back in the autopsy room at the Cook County morgue, are more challenging than others. This one is a real chiller. In a large cardboard box lies a jumble of bones along with a tattered leg from a pair of blue jeans, a sock shrunk tightly around the bones of a foot, a pair of Nike running shoes without shoelaces, and, inside the hood of a blue windbreaker, a mass of stringy, blood-caked hair. The remains were discovered frozen in ice about 20 miles outside Milwaukee. A rusted bicycle was found lying close by. Paul Hibbard, chief deputy medical examiner for Waukesha County, who brought the skeleton to Chicago, says no one has been reported missing.

Snow lifts the bones out of the box and begins reconstructing the skeleton on an autopsy table. "There are two hundred six bones and thirty-two teeth in the human body," he says, "and each has a story to tell." Because bone is dynamic, living tissue, many of life's significant events—injuries, illness, childbearing—leave their mark on the body's internal framework. Put together the stories told by these bones, he says, and what you have is a person's "osteobiography."

Snow begins by determining the sex of the skeleton, which is not always obvious. He tells the story of a skeleton that was brought to his FAA office in the late 1970s. It had been found along with some women's clothes and a purse in a local back lot, and the police had assumed that it was female. But when Snow examined the bones, he realized that "at six foot three, she would have probably have been the tallest female in Oklahoma."

Then Snow recalled that six months earlier the custodian in his building had suddenly not shown up for work. The man's supervisor later mentioned to Snow, "You know, one of these days when they find Ronnie, he's going to be dressed as a woman." Ronnie, it turned out, was a weekend transvestite. A copy of his dental records later confirmed that the skeleton in women's clothing was indeed Snow's janitor.

The Wisconsin bike rider is also male. Snow picks out two large bones that look something like twisted oysters—the innominates, or hipbones, which along with the sacrum, or lower backbone, form the pelvis. This pelvis is narrow and steep-walled like a male's, not broad and shallow like a female's. And the sciatic notch (the V-shaped space where the sciatic nerve passes through the hipbone) is narrow, as is normal in a male. Snow can also determine a skeleton's sex by checking the size of the mastoid processes (the bony knobs at the base of the skull) and the prominence of the brow ridge, or by measuring the head of an available limb bone, which is typically broader in males.

From an examination of the skull he concludes that the bike rider is "predominantly Caucasoid." A score of bony traits help the forensic anthropologist assign a skeleton to one of the three major racial groups: Negroid, Caucasoid, or Mongoloid. Snow notes that the ridge of the boy's nose is high and salient, as it is in whites. In Negroids and Mongoloids (which include American Indians as well as most Asians) the nose tends to be broad in relation to its height. However, the boy's nasal margins are somewhat smoothed down, usually a Mongoloid feature. "Possibly a bit of American Indian admixture," says Snow. "Do you have Indians in your area?" Hibbard nods.

Age is next. Snow takes the skull and turns it upside down, pointing out the basilar joint, the junction between the two major bones that form the underside of the skull. In a child the joint would still be open to allow room for growth, but here the joint has fused—something that usually happens in the late teen years. On the other hand, he says, pointing to the zigzagging lines on the dome of the skull, the cranial sutures are open. The cranial sutures, which join the bones of the braincase, begin to fuse and disappear in the mid-twenties.

Next Snow picks up a femur and looks for signs of growth at the point where the shaft meets the knobbed end. The thin plates of cartilage—areas of incomplete calcification—that are visible at this point suggest that the boy hadn't yet attained his full height. Snow double-checks with an examination of the pubic symphysis, the joint where the

two hipbones meet. The ridges in this area, which fill in and smooth over in adulthood, are still clearly marked. He concludes that the skeleton is that of a boy between 15 and 20 years old.

"One of the things you learn is to be pretty conservative," says Snow. "It's very impressive when you tell the police, 'This person is eighteen years old,' and he turns out to be eighteen. The problem is, if the person is fifteen you've blown it—you probably won't find him. Looking for a missing person is like trying to catch fish. Better get a big net and do your own sorting."

Snow then picks up a leg bone, measures it with a set of calipers, and enters the data into a portable computer. Using the known correlation between the height and length of the long limb bones, he quickly estimates the boy's height. "He's five foot six and a half to five foot eleven," says Snow. "Medium build, not excessively muscular, judging from the muscle attachments that we see." He points to the grainy ridges that appear where muscle attaches itself to the bone. The most prominent attachments show up on the teenager's right arm bone, indicating right-handedness.

Then Snow examines the ribs one by one for signs of injury. He finds no stab wounds, cuts, or bullet holes, here or elsewhere on the skeleton. He picks up the hyoid bone from the boy's throat and looks for the tell-tale fracture signs that would suggest the boy was strangled. But, to Snow's frustration, he can find no obvious cause of death. In hopes of identifying the missing teenager, he suggests sending the skull, hair, and boy's description to Betty Pat Gatliff, a medical illustrator and sculptor in Oklahoma who does facial reconstructions.

Six weeks later photographs of the boy's likeness appear in the *Milwaukee Sentinel*. "If you persist long enough," says Snow, "eighty-five to ninety percent of the cases eventually get positively identified, but it can take anywhere from a few weeks to a few years."

Snow and Gatliff have collaborated many times, but never with more glitz than in 1983, when Snow was commissioned by Patrick Barry, a Miami orthopedic surgeon and amateur Egyptologist,

to reconstruct the face of the Egyptian boy-king Tutankhamen. Normally a facial reconstruction begins with a skull, but since Tutankhamen's 3,000-year-old remains were in Egypt, Snow had to make do with the skull measurements from a 1925 postmortem and X-rays taken in 1975. A plaster model of the skull was made, and on the basis on Snow's report—"his skull is Caucasoid with some Negroid admixtures"—Gatliff put a face on it. What did Tutankhamen look like? Very much like the gold mask on his sarcophagus, says Snow, confirming that it was, indeed, his portrait.

Many cite Snow's use of facial reconstructions as one of his most important contributions to the field. Snow, typically self-effacing, says that Gatliff "does all the work." The identification of skeletal remains, he stresses, is often a collaboration between pathologists, odontologists, radiologists, and medical artists using a variety of forensic techniques.

One of Snow's last tasks at the FAA was to help identify the dead from the worst airline accident in U.S. history. On May 25, 1979, a DC-10 crashed shortly after takeoff from Chicago's O'Hare Airport, killing 273 people. The task facing Snow and more than a dozen forensic specialists was horrific. "No one ever sat down and counted," says Snow, "but we estimated ten thousand to twelve thousand pieces or parts of bodies." Nearly 80 percent of the victims were identified on the basis of dental evidence and fingerprints. Snow and forensic radiologist John Fitzpatrick later managed to identify two dozen others by comparing postmortem X-rays with X-rays taken during the victim's lifetime.

Next to dental records, such X-ray comparisons are the most common way of obtaining positive identifications. In 1978, when a congressional committee reviewed the evidence on John F. Kennedy's assassination, Snow used X-rays to show that the body autopsied at Bethesda Naval Hospital was indeed that of the late president and had not—as some conspiracy theorists believed—been switched.

The issue was resolved on the evidence of Kennedy's "sinus print," the scalloplike pattern on the upper margins of the sinuses that is visible in X-rays of the forehead. So characteristic is a person's sinus print that courts throughout the world accept the matching of antemortem and postmortem X-rays of the sinuses as positive identification.

Yet another technique in the forensic specialist's repertoire is photo superposition. Snow used it in 1977 to help identify the mummy of a famous Oklahoma outlaw named Elmer J. McCurdy, who was killed by a posse after holding up a train in 1911. For years the mummy had been exhibited as a "dummy" in a California funhouse—until it was found to have a real human skeleton inside it. Ownership of the mummy was eventually traced back to a funeral parlor in Oklahoma, where McCurdy had been embalmed and exhibited as "the bandit who wouldn't give up."

Using two video cameras and an image processor, Snow superposed the mummy's profile on a photograph of McCurdy that was taken shortly after his death. When displayed on a single monitor, the two coincided to a remarkable degree. Convinced by the evidence, Thomas Noguchi, then Los Angeles County corner, signed McCurdy's death certificate ("Last known occupation: Train robber") and allowed the outlaw's bones to be returned to Oklahoma for a decent burial.

It was this technique that also allowed forensic scientists to identify the remains of the Nazi "Angel of Death," Josef Mengele, in the summer of 1985. A team of investigators, including Snow and West German forensic anthropologist Richard Helmer, flew to Brazil after an Austrian couple claimed that Mengele lay buried in a grave on a São Paulo hillside. Tests revealed that the stature, age, and hair color of the unearthed skeleton were consistent with information in Mengele's SS files; yet without X-rays or dental records, the scientists still lacked conclusive evidence. When an image of the reconstructed skull was superposed on 1930s photographs of Mengele, however, the match was eerily compelling. All doubts were removed a few months later when Mengele's dental X-rays were tracked down.

In 1979 Snow retired from the FAA to the rolling hills of Norman, Oklahoma, where he and his wife, Jerry, live in a sprawling, early-1960s ranch house. Unlike his 50 or so fellow forensic anthropologists, most of whom are tied to academic positions, Snow is free to pursue his consultancy work full-time. Judging from the number of miles that he logs in the average month, Snow is clearly not ready to retire for good.

His recent projects include a reexamination of the skeletal remains found at the site of the Battle of the Little Bighorn, where more than a century ago Custer and his 210 men were killed by Sioux and Cheyenne warriors. Although most of the enlisted men's remains were moved to a mass grave in 1881, an excavation of the battlefield in the past few years uncovered an additional 375 bones and 36 teeth. Snow, teaming up again with Fitzpatrick, determined that these remains belonged to 34 individuals.

The historical accounts of Custer's desperate last stand are vividly confirmed by their findings. Snow identified one skeleton as that of a soldier between the ages of 19 and 23 who weighed around 150 pounds and stood about five foot eight. He'd sustained gunshot wounds to his chest and left forearm. Heavy blows to his head had fractured his skull and sheared off his teeth. Gashed thigh bones indicated that his body was later dismembered with an ax or hatchet.

Given the condition and number of the bodies, Snow seriously questions the accuracy of the identifications made by the original nineteenth-century burial crews. He doubts, for example, that the skeleton buried at West Point is General Custer's.

For the last four years Snow has devoted much of his time to helping two countries come to terms with the horrors of a much more recent past. As part of a group sponsored by the American Association for the Advancement of Science, he has been helping the Argentinian National Commission on Disappeared Persons to determine the fate of some of those who vanished during their country's harsh military rule: between 1976 and 1983 at least 10,000 people were systematically swept off the streets by roving

death squads to be tortured, killed, and buried in unmarked graves. In December 1986, at the invitation of the Aquino government's Human Rights Commission, Snow also spent several weeks training Philippine scientists to investigate the disappearances that occurred under the Marcos regime.

But it is in Argentina where Snow has done the bulk of his human-rights work. He has spent more than 27 months in and around Buenos Aires, first training a small group of local medical and anthropology students in the techniques of forensic investigation, and later helping them carefully exhume and examine scores of the *desaparecidos,* or disappeared ones.

Only 25 victims have so far been positively identified. But the evidence has helped conflict seven junta members and other high-ranking military and police officers. The idea is not necessarily to identify all 10,000 of the missing, says Snow. "If you have a colonel who ran a detention center where maybe five hundred people were killed, you don't have to nail them with five hundred deaths. Just one or two should be sufficient to get him convicted." Forensic evidence from Snow's team may be used to prosecute several other military officers, including General Suarez Mason. Mason is the former commander of the I Army Corps in Buenos Aires and is believed to be responsible for thousands of disappearances. He was recently extradited from San Francisco back to Argentina, where he is expected to stand trial this winter [1988].

The investigations have been hampered by a frustrating lack of ante mortem information. In 1984, when commission lawyers took depositions from relatives and friends of the disappeared, they often failed to obtain such basic information as the victim's height, weight, or hair color. Nor did they ask for the missing person's X-rays (which in Argentina are given to the patient) or the address of the victim's dentist. The problem was compounded by the inexperience of those who carried out the first mass exhumations prior to Snow's arrival. Many of the skeletons were inadvertently destroyed by bulldozers as they were brought up.

Every unearthed skeleton that shows signs of gunfire, however, helps to erode the claim once made by many in the Argentinian military that most of the *desaparecidos* are alive and well and living in Mexico City, Madrid, or Paris. Snow recalls the case of a 17-year-old boy named Gabriel Dunayavich, who disappeared in the summer of 1976. He was walking home from a movie with his girlfriend when a Ford Falcon with no license plates snatched him off the street. The police later found his body and that of another boy and girl dumped by the roadside on the outskirts of Buenos Aires. The police went through the motions of an investigation, taking photographs and doing an autopsy, then buried the three teenagers in an unmarked grave.

A decade later Snow, with the help of the boy's family, traced the autopsy reports, the police photographs, and the grave of the three youngsters. Each of them had four or five closely spaced bullet wounds in the upper chest—the signature, says Snow, of an automatic weapon. Two also had wounds on their arms from bullets that had entered behind the elbow and exited from the forearm.

"That means they were conscious when they were shot," says Snow. "When a gun was pointed at them, they naturally raised their arm." It's details like these that help to authenticate the last moments of the victims and bring a dimension of reality to the judges and jury.

Each time Snow returns from Argentina he says that this will be the last time. A few months later he is back in Buenos Aires. "There's always more work to do," he says. It is, he admits quietly, "terrible work."

"These were such brutal, cold-blooded crimes," he says. "The people who committed them not only murdered; they had a system to eliminate all trace that their victims even existed."

Snow will not let them obliterate their crimes so conveniently. "There are human-rights violations going on all around the world," he says. "But to me murder is murder, regardless of the motive. I hope that we are sending a message to governments who murder in the name of politics that they can be held to account."

Gorilla Warfare

*The fate of Africa's rarest apes hinges on
battles over their territory as well as their taxonomy*

By Craig B. Stanford

HIGH AMONG THE VIRUNGA VOLCA-noes, along the eastern edge of the Democratic Republic of Congo (DRC), there lives a group of gorillas with little interest in international politics. Day by day and week by week they wander through meadows of bracken fern, eating bamboo and nettles, mating in polygynous groups and fastidiously grooming one another. Although there are only around 600 mountain gorillas left in the world—half of them here and half in Uganda's aptly named Bwindi-Impenetrable National Park—the gorillas themselves seem unconcerned about that fact. Their most aggressive, most territorial act toward people is to bite a farmer on the behind now and again.

But if the forest, to a gorilla's eye, seems peaceful and unbroken, from a human perspective it is riven by disputes, crosshatched by historical, political and biological borders. The volcanoes themselves may be dormant, but they straddle three of the most incendiary places on earth: the southwestern tip of Uganda, the easternmost edge of Congo (formerly Zaïre and officially known as the DRC) and the northernmost lip of Rwanda. The first has a history of violent dictatorship; the other two are still shuddering with brutal conflicts. On any given day, therefore, the visitor cannot be sure whether he will run into gorillas or guerrillas on the trail.

The gorillas don't care: every year they walk across the spine of the volcanoes, spend several months in Uganda, then walk back home to Congo. But the camera-laden Westerners who have hiked for hours to watch them have no such lux-ury. Political instability put an end to gorilla viewing in Congo as well as in Rwanda. Tourists are still free to watch the Bwindi gorillas, but when the Virunga gorillas reach the Congo border the tourists have to stop and watch them disappear into the trees. Then the tourists turn and head back to camp.

This past March 1st, however, that peaceful pattern was suddenly threatened. That morning a group of guerrillas plunged across the Congo border and into Uganda, in violation of international law, in search of unsuspecting ecotourists. Known as the Interahamwe, the guerrillas were Hutus from Rwanda who for years had fought the Tutsi minority that ruled their country. When war erupted in Rwanda in 1994, after a plane carrying the Rwandan president was shot down, the Interahamwe began a brutal campaign of genocide against the Tutsi. In the ensuing slaughter, carried out by both sides, often with machetes, between 10,000 and 20,000 Interahamwe were chased into eastern Congo, where dense forest and political chaos afforded them the best possible cover. There they regrouped and licked their wounds, stealing radios and other supplies to maintain their ragtag revolutionary army.

With Ugandan soldiers patrolling nearby, the Ugandan army did not suspect that the rebels would risk an attack, especially on Western tourists. But the Interahamwe were furious not only with the Rwandan government, they were also incensed with the Ugandan and Western governments, which had supported Rwanda. And so, on that March dawn, more than a hundred rebels charged into Bwindi-Impenetra-ble park, armed with assault rifles, and headed toward the ecotourism center at Buhoma. Soon a warden was dead and fourteen Western tourists had been seized and taken hostage in the forest. Among them was Mitchell A. Keiver, a Canadian research assistant who worked with me at a nearby camp where I study gorillas and chimpanzees.

UNTIL RECENTLY, MOUNTAIN GORILLAS had good reason not to fear the region's bloody politics: they were worth far too much alive. Before neighboring Rwanda slid into genocidal chaos, gorilla tourism among the Virunga volcanoes was that nation's second-largest source of revenue, after

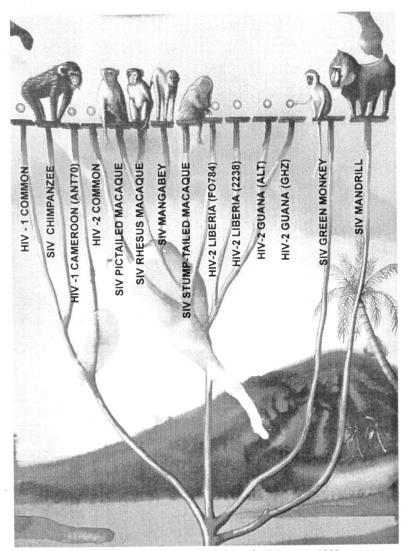

Alexis Rockman, HIV's Extended Family (Primates), *1993*

The mountain gorillas of Bwindi and the Virungas, officially known as *Gorilla gorilla beringei*, are a relict population that long ago migrated from the west and evolved in isolation. Much farther west are the western lowland gorillas, which make up a second subspecies, and, compared with their montane cousins, there are lots of them. The western lowland gorilla was the first to be named and therefore carries the doubly redundant moniker *Gorilla gorilla gorilla*. Estimates of its population vary greatly, but a conservative count would put it between 100,000 and 150,000.

The rest of African gorilladom is represented by *Gorilla gorilla graueri*, the eastern lowland gorilla, with traits somewhere between those of the lowlanders to the west and the highlanders to the east. According to a recent survey carried out by the Wildlife Conservation Society, about 7,000 eastern lowland gorillas are left, all in war-torn Congo.

L OWLAND GORILLAS ARE SMALLER, grayer and shorter-haired than gorillas in the mountains. Unlike their more sedentary cousins, lowland gorillas may travel miles each day in search of fruit. The lowland social system, in turn, seems to reflect that mobility: it would be unusual indeed to find mountain gorillas wandering far from the core of their group, whereas lowlanders disperse on a daily basis.

All three gorilla subspecies are in trouble, but G. g. beringei is clearly in the direst straits. In the past several hundred years, hunters, farmers and loggers have painted the gorillas into a mountaintop corner. The villages and farms that surround them are among the most densely populated on the continent, leaving few resources for the gorillas; logging has reduced their habitat to a few small patches of forest, and poachers have sold the babies to zoos and chopped up the adults to sell as tourist trinkets. The 600 that remain would be a paltry number for any species, but for gorillas, which reproduce once every four years, it could be disastrously small.

Logging and hunting are closely linked in western and central Africa. Every year, nearly 16,000 square miles of forest are felled on the continent. And

coffee. As many as a million Rwandans were killed during the massacres, and two million more were forced to flee, but only five gorillas were killed by the military. No matter who ended up controlling Rwanda, both sides must have reasoned, the gorillas had to remain safe and sound.

Mountain gorillas owe much of their fame, and thus their value as tourist attractions, to the pioneering field research of the primatologist Dian Fossey and the dramatization of her life in the 1988 film *Gorillas in the Mist*. Fossey's work reversed the popular image of gorillas, born of countless Tarzan and King Kong movies. Rather than bloodthirsty, chest-thumping behemoths—testosterone in black leather—gorillas were depicted as the easygoing animals they are, threatened by people rather than threat-

ening to them. They are herbivores, after all, and exceedingly shy ones at that. A 400-pound male, unaccustomed to tourists, will bolt into the forest, trailing a stream of diarrhea, at the mere sight of a person.

Fossey's work, and the movie about her, sparked an international effort to protect the gorillas. But that effort, in some ways, depends on borders and classifications as arbitrary and disputed as the invisible line that separates Uganda from Congo. Like most animal and plant species, gorillas have traditionally been broken into subspecies, or races, by taxonomists. According to the current classification, there are three kinds of gorilla in Africa (though that number may soon rise), each with a different look, lifestyle and habitat, and a dramatically different population status.

as logging roads reach into the great forest tracts, they give hunters with rifles access to apes that once had only tribes with spears to fear. The logging companies themselves encourage hunting, which feeds their workers in forest camps, and gorillas and chimpanzees are sold as table items to wealthy and middle-class Africans, who see bush meat as a status symbol. "People don't want beef or chicken anymore—they call it 'white man's meat,'" one conservationist told *The New York Times*. As a result, the bush-meat trade kills some 3,000 lowland gorillas every year. Even western lowland gorillas, with their population of 100,000, will not withstand such pressures for long.

The one point of light in this bleak vista is ecotourism. In recent years Uganda has recovered from its years of political chaos to become something of a model African country for wildlife conservation. The Ugandan Wildlife Authority has built a thriving industry around the viewing of mountain gorillas. In Bwindi the country boasts a UNESCO World Heritage site, as well as one of the most beautiful forest landscapes on the African continent. In Buhoma, the site of the now-infamous kidnappings, several camps were set up for gorilla-watching. The fanciest of them, run by the world-renowned safari organization Abercrombie and Kent, took visitors back in time to the Africa of Ernest Hemingway. Great green tents sheltered tourists from the frequent rain, local cooks prepared exotic feasts, and waiters in white shirts and bow ties served mixed drinks on hot

afternoons. The only concession to modernity was that no animal was killed there. Tourists, as they say, shot only with their cameras, not with guns.

I T IS A SCENE TO PROVOKE A SNEER, PERhaps, from some conservationists. But the fact is, in Bwindi, eco-tourism *works*. Poaching is under control and the remaining gorilla habitat is protected in national parks and reserves. So comfortable have the gorillas grown around people in recent years that they have taken to foraging in farmers' fields (hence the occasional butt bite), and they allow people to sit by them. That approachability, combined with the animals' rarity, their fame and their relatively sedentary ways (compared to the lifestyle of their wide-ranging lowland cousins), have made the mountain gorillas into ideal tourist attractions.

Indeed, gorilla-watching has become the centerpiece of the modern tourist's African safari. People pay as much as $3,130 for package tours in which they slog through rugged, wet terrain in order to spend an hour a day sitting with a group of habituated gorillas. Multiply that by several groups of gorillas, each visited by six high-rolling tourists a day, and the tourist income adds up quickly. Every year the farmers who live around Bwindi and the Virungas, scratching out a living from the steep volcanic slopes, get 20 percent of the revenue collected from gorilla-tracking permits. And gorilla tourism heavily subsidizes the Ugandan Wildlife Authority's budget as well. The fate of the mountain gorillas,

in other words, is inseparable from the fate of the people who live around them.

Thanks to ecotourism, mountain gorillas are among the best-protected primates in the world. But a double-edged sword hangs over the area, one edge political, the other taxonomic. Deprive the gorillas of their safeguarded wildlife preserves, and loggers and poachers will soon drive them to extinction. Change their biological classification, and their conservation status could be put in jeopardy. Both lines of protection, unfortunately, are in danger of being cut.

I N 1996 THE PRIMATOLOGISTS ESTEBAN E. Sarmiento of the American Museum of Natural History in New York City, and Thomas M. Butynski and Jan Kalina of Zoo Atlanta proposed a new classification scheme for mountain gorillas. The Bwindi gorillas, they argued, should be recognized as a new and separate subspecies from the Virunga gorillas. The suggestion made a certain sense. The Bwindi and Virunga gorillas already live in distinct groups, about twenty-five miles apart. Five hundred years ago, before subsistence farmers finished clearing the land between the two mountain ranges, the two gorilla habitats were one. Since the separation, however, the populations no longer mingle or interbreed, and their differing habitats—or, perhaps, the random accumulation of genetic mutations in both groups—have, over the centuries, wrought differences in the two groups' looks and behavior.

Because my research project began just after Sarmiento, Butynski and Kalina published their paper, I was in a

FAMILY PORTRAIT

G ENES DO NOT ALWAYS A SPECIES MAKE. ACCORDING TO A PAPER published in the April issue of the *Proceedings of the National Academy of Sciences* (PNAS), the breadth of genetic variation across the various primate species is anything but consistent. Indeed, a single band of fifty-five West African chimpanzees has more genetic variation than all of humanity.

Far to one side of the phylogenetic tree, like a branch badly in need of pruning, sit the orangutans. Although Sumatran and Bornean orangutans ostensibly belong to one species, DNA from the two groups is as distinct as chimpanzee DNA is from bonobo DNA. Gorillas are almost as motley a group. Although mountain and eastern lowland gorillas share much of the same DNA, western lowland gorillas have genes so varied that the animals could easily be split into three or more subspecies. Over on the other half of

the primate tree, the genes within any given chimpanzee subspecies are relatively uniform compared to those within gorilla and orangutan subspecies. Yet compared to those of their close cousins, human beings, they are a veritable genetic menagerie. The reason for the shrunken human gene pool, the authors of the PNAS paper suggest, is that our ancestors once nearly went extinct—though exactly when and how is still unclear.

If genes alone determined species, in other words, taxonomists would have their work cut out for them. In addition to coming up with a new name for one group of orangutans and a few new names for gorillas, they would have to fend off Christian fundamentalists: The fact that humanity is just a hairless species of chimpanzee would be a hard sell in any era.

—BURKHARD BILGER

perfect position to observe those differences. From our camp— a collection of mud-and-thatch huts clinging to the steep side of a mountain—gorillas and chimpanzees can be heard in the forest below around the clock. And we spend most of our waking hours collecting information on the animals' habits. The Bwindi gorillas, we have found, tend to eat more fruit than the Virunga gorillas, and so they travel farther each day. Whereas the Virunga gorillas almost always sleep on the ground, the Bwindi gorillas sometimes sleep in trees. Most obviously, the Virunga gorillas have shaggier coats and a slightly larger build.

Are such differences enough to redefine the two populations as separate subspecies? That depends on your definition of the term. In 1942 the biologist Ernst Mayr defined a subspecies as a population of animals that has adapted genetically to a unique environmental niche and that reproduces in isolation from others of the species. Two populations of squirrels that live on opposite sides of the Grand Canyon might fit that definition, and so would the Bwindi and Virunga gorillas. But over the years the concept of subspecies has undergone an evolution of its own. Some biologists now define it genetically; others focus on social structure or the mechanisms of mate recognition; still others proceed mathematically, by grouping physical traits through cladistic analysis. In general, the various factions can be grouped into two larger camps—the "lumpers" and the "splitters"—but even members of the same camp agree on few specifics. Mayr himself, in 1982, recanted his earlier view, describing subspecies as merely convenient pigeonholing devices for taxonomists.

To most laypeople, genetic differences seem the most plausible way to divide up species. The longer two populations live and breed apart, the more distinctive mutations accumulate in their genes. What could be clearer and simpler? Yet speciation and evolutionary distance are not intimately connected. Two animal populations can look quite different from each other—Great Danes and Chihuahuas,

for instance—yet remain closely related genetically. Or they can look very much alike—the orangutans from Borneo and those from Sumatra are a good example—even though they diverged from the same ancestral stocks many thousands or even millions of years ago [see "Family Portrait"].

FACED WITH SO MANY VARIABLES, AND so little agreement about how to weigh them, biologists have little choice but to use a scorecard approach. And that substantially complicates the classification question for mountain gorillas. In 1996 Karen J. Garner and Oliver A. Ryder, two geneticists at the Zoological Society of San Diego in California, compared the mitochondrial DNA of all the gorilla populations. Whereas mountain-gorilla DNA is quite different from lowland-gorilla DNA, they found, the DNA of the Bwindi and Virunga populations is virtually indistinguishable. How, then, should those groups be classified? Should they be put in separate subspecies, on the basis of their looks, behavior and biogeography, or should they be kept together, on the basis of their genes?

In this case, I believe one must side with the lumpers. Genetic evidence may be no trump card in matters of speciation, but among mountain gorillas the physical and behavioral differences are even less definitive. Gorillas in the Virungas live more than 10,000 feet above sea level, in cold, misty meadows rich in herbs and ground cover but poor in fruit. Bwindi gorillas live in lower and warmer mountains, and so they have access to more fruit trees. Hence the physical circumstances alone can explain why Bwindi gorillas eat more fruit and travel farther to find it, whereas Virunga gorillas make do with foliage. Gorillas are smart, flexible primates, and they have simply adapted their behavior to local conditions.

The physical differences between the groups are just as easy to explain. When my cat and I moved from Michigan to southern California some years ago, she stopped putting on a layer of fat and thick fur every winter. That is because cats, like many animals, have an inherent ability to respond to low nighttime

temperatures, decreasing daylight hours and other signs of approaching cold. Gorillas can no doubt respond to their environment in a similar way.

As abstract and academic as the speciation debate seems, it could become a matter of life or death for mountain gorillas. If the Bwindi gorillas are biologically split from the Virunga gorillas, the world has effectively "lost" half its mountain gorillas, and gained another critically endangered gorilla population. Some would argue that the new status might draw even more attention to the gorillas' plight, enhancing their protection. But what if one subspecies thrives while the other declines? If the bloodlines must be kept separate, gorillas from one group could not be recruited into the other, and the weak group might well become extinct. Taxonomists, it seems, could end up killing the mountain gorilla as effectively as poachers do.

At the same time, political instability is a rising threat to the tourist business. For years, Rwandan rebels have occasionally shot at gorilla-watchers in Congo, and in 1994, when genocidal war broke out in Rwanda, most gorilla research and ecotourism in that country and Congo were shut down. Tourists and investigators still felt relatively insulated from the trouble. But as early as last August there were clear signs that they were not. That month a band of Interahamwe rebels kidnapped a group of eight Western tourists who were following gorillas in Congo. Four of those hostages—two New Zealanders and two Swedes—were never heard from again.

Only six months later, the incident at Buhoma brought the danger across the border to Uganda.

WHEN I FIRST HEARD ABOUT THE rebel attack I was in Pasadena, California, with my wife and three children, only three days after returning from Africa. I had originally planned to bring the family to my field station, as I had done when working in Tanzania. But the logistical hassles finally convinced my wife and me against it.

In the middle of the night, therefore, when an E-mail message arrived from a colleague in Rwanda, saying "ATTACK ON BUHOMA. MORE DETAILS LATER," I felt a faint trace of relief mixed with my concern. Unfortunately, the details that followed soon dashed that small comfort.

FOURTEEN HOSTAGES were rounded up, ordered to take off their shoes and pushed into the forest barefoot. Later eight of them were murdered.

Although the park rangers managed to wound at least four of the attackers, they were quickly overwhelmed. In the next hour the tourism center and several vehicles were set on fire, and the deputy warden, Paul Ross Wagaba, was killed. When the fourteen hostages, nearly all of them tourists, had been rounded up, they were divided into two groups, ordered to take off their shoes, and pushed into the forest barefoot on a forced march back to Congo.

At first I didn't believe that Mitch could be among them. He and I had parted only a few days before in Kampala, the Ugandan capital, and the itinerary he described to me then would not have put him in Buhoma on the morning of the attack. When I called to reassure his parents, however, at their farm in Three Hills, Alberta, Mitch's mother told me that he had called her after I left Uganda. He had changed his plans slightly, she said, and had been in the ecotourism center on that morning after all. There were thirteen tourists in the group, a Canadian official had told her, and one investigator.

A hike through the Ugandan forest can be exhausting even for someone in great shape; for those hostages it must have been excruciating. Mitch's group of six hostages climbed for several hours, up steep forest trails. When they reached the border between Uganda and Congo, they convinced the rebels to express their grievances in a two-page note, composed in French on the spot. "This is a punishment for the Anglo-Saxons who sold us out," the note read, in part. "You are protecting the minority and oppressing the majority." Then Mitch and the others were released. They later learned that the eight hostages in the second group had been brutally murdered.

A WEEK AFTER HIS ORDEAL MITCH flew home to Three Hills. But he is already back in Uganda, working for the International Gorilla Conservation Programme in Kabale. Although I canceled my next research trip, my Ugandan field assistants continue to collect data every day at my camp, and the national park was reopened to tourism only a month after the attack had shut it down. Three tourists a day, on average, have visited the gorillas since, but that is down from twelve or more before the attack. If poaching patrols and basic operations are to continue, the shortfall in income will have to be made up by nongovernmental organizations; regardless, gorilla conservation efforts are likely to suffer.

There is still cause for hope. Two years ago representatives from Congo, Rwanda and Uganda met for the first time to define shared goals for gorilla conservation. Although their political troubles run deep, their economic incentives for protecting the gorillas run even deeper. Nevertheless, as effective as ecotourism can be in Bwindi, the attack was a lesson in its limitations. Tourist dollars may give local villagers a reason to share in the ethic of protecting gorillas. But money alone can't put out the political fires that so recently spread to Buhoma.

In the past two centuries the people of Africa have suffered under the same forces that have decimated gorilla populations and destroyed much of the animals' habitat. Myriad governments have been born and died, their countries carved and recarved at the whims of colonial powers and dictators, their people pushed together or shoved apart with no regard for their own ancient ethnic rivalries. Uganda was a British colony; Congo was a Belgian colony; Rwanda was part of German East Africa and then forced into the Belgian League of Nations. All three countries are inherently flammable, and Western aid and Western tourists, in some ways, only add fuel to the blaze.

WHEREAS MY COLLEAGUES AND I were understandably preoccupied with the Western victims, many Ugandans reacted differently. The deaths of a few tourists at the rebels' hands, they pointed out, provoked nearly as much sympathy and outrage from Westerners as the deaths of a million Rwandans in 1994. Once again Westerners were demonstrating their neocolonial myopia, showing more concern for white people who happen to be living or traveling in Africa than for the Africans themselves.

The mountain gorillas, for their part, know too little to care: their travels mock the very idea of political nation states. As long as they remain pawns in the battle zone that is east-central Africa, however, they are in mortal danger, and even the most dedicated conservationists are helpless to protect them. Through science we may try to fix what people have botched through politics and economics. We may declare some species endangered and others healthy, drawing new borders between species or erasing them altogether. But in Africa, the divisions that matter the most are the ones between people, not between animals, and they cut deeper than any border on a map. Only by healing those rifts can we ever hope to save the mountain gorillas.

CRAIG B. STANFORD is an associate professor of anthropology at the University of Southern California in Los Angeles. He is the author of CHIMPANZEE AND RED BOLOBUS, *published last year by Harvard University Press, and* THE HUNTING APES, *published this year by Princeton University Press.*

HIV 1998:
The Global Picture

Worldwide, the populations most affected by the AIDS virus are often the least empowered to confront it effectively

by Jonathan M. Mann and Daniel J. M. Tarantola

In 1996, after more than a decade of relentless rises, deaths from AIDS finally declined in the U.S. The drop appears to have stemmed mainly from the introduction of powerful therapies able to retard the activity of HIV, the virus responsible for AIDS. Other economically advantaged nations, including France and Britain, have documented declines as well. But the trend in industrial countries is not representative of the world as a whole.

Further, the international pandemic of HIV infection and AIDS—composed of thousands of separate epidemics in communities around the globe—is expanding rapidly, particularly in the developing nations, where the vast majority of people reside. Since the early 1980s more than 40 million individuals have contracted HIV, and almost 12 million have died (leaving at least eight million orphans), according to UNAIDS, a program sponsored by the United Nations. In 1997 alone, nearly six million people—close to

16,000 a day—acquired HIV, and some 2.3 million perished from it, including 460,000 children.

This grim picture reflects some other unpalatable facts. Over the years, resources devoted to battling the pandemic have been apportioned along societal lines. Although more than 90 percent of HIV-infected people live in developing nations, well over 90 percent of the money for care and prevention is spent in industrial countries. This disparity explains why the new HIV-taming therapies, costing annually upward of $10,000 per person, have had no impact

in the developing nations; by and large, these countries lack the infrastructure and funds to provide the medicines. In a few locales in the developing world, notably in parts of Uganda and Thailand, public health campaigns seem to be slowing the rate of infection. Yet those are the exceptions: in most other places, the situation is worsening.

HIV is spreading especially quickly in sub-Saharan Africa and in Southeast Asia. The region below the Sahara now houses two thirds of the globe's HIV-infected population and about 90 percent of all infected children. In areas of Bot-

HIV'S GLOBAL EFFECT is devastating. Sub-Saharan Africa and South and Southeast Asia are home to the greatest number of HIV-infected people (map). The disease is also spreading fastest in those areas (graph). In the U.S., as in most nations, populations that encounter the most discrimination often face the highest risk of acquiring AIDS (bar chart). For instance, in 1996 black men and Hispanic men were, respectively, 51 and 25 times more likely than white women to be diagnosed.

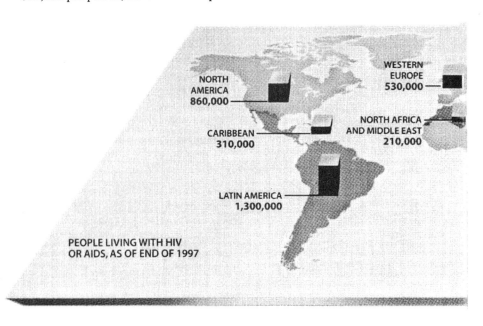

PEOPLE LIVING WITH HIV
OR AIDS, AS OF END OF 1997

NORTH AMERICA 860,000

WESTERN EUROPE 530,000

CARIBBEAN 310,000

NORTH AFRICA AND MIDDLE EAST 210,000

LATIN AMERICA 1,300,000

swana, Swaziland and several provinces of South Africa, one in four adults is afflicted; in many African countries, life expectancy, which had been rising since the 1950s, is falling. Unprotected heterosexual sex accounts for most of HIV's spread in sub-Saharan Africa, but the problem is compounded by contamination of the blood supply. At least a quarter of the 2.5 million units of blood administered in Africa (mostly to women and children) is not screened for the AIDS virus.

In Southeast Asia the epidemic is dominated by India (with three to five million HIV-infected individuals) and Thailand. It is now also raging in Burma and is expanding further into Vietnam and China.

Mirroring the dichotomy between the developing and the industrial nations, certain populations within nations are suffering a disproportionate number of infections. Epidemiologists have been dismayed to uncover a societal-level factor influencing the distribution: groups whose human rights are least respected are most affected. As epidemics mature within communities and countries, the brunt of the epidemic often shifts from the primary population in which HIV first appeared to those who

were socially marginalized or discriminated against before the epidemic began.

Those who are discriminated against—whether because of their gender, race or economic status or because of cultural, religious or political affiliations—may have limited or no access to preventive information and to health and social services and may be particularly vulnerable to sexual and other forms of exploitation. And later, if they become infected, they may similarly be denied needed care and social support. Stigmatization therefore pursues its course unabated, deepening individual susceptibility and, as a result, collective vulnerability to the spread of HIV and to its effects.

As a case in point, 10 years ago in the U.S., whites accounted for 60 percent of AIDS cases and blacks and Hispanics for 39 percent. By 1996, 38 percent of new cases were diagnosed in whites and 61 percent in blacks and Hispanics. Further, between 1995 and 1996, the incidence of AIDS declined by 13 percent in whites but not at all in blacks.

Social marginalization, manifested in lack of educational and economic opportunity, also magnifies risk in the developing world. In Brazil, for example, the bulk of AIDS cases once affected people who had at least a secondary school education; now more than half the cases arise in people who have attended only primary school, if that. Moreover, in much of the world, women (who account for more than 40 percent of all HIV infections) have low social status and lack the power to insist on condom use or other safe-sex practices; they will be unable to protect themselves until their social status improves. Recognizing

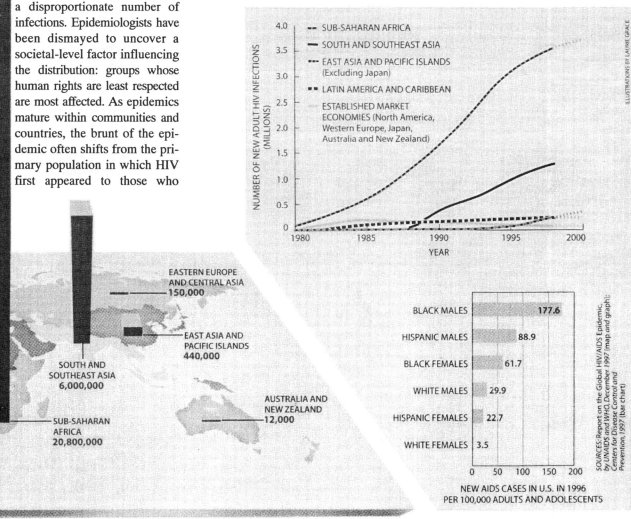

the societal-level roots of vulnerability to HIV and AIDS, UNAIDS has incorporated advancement of human rights in its global-prevention strategy.

What will the future bring? In the short run, it seems likely that the international epidemic will become even more concentrated in the developing countries (that is, mainly in the Southern Hemisphere), where it will expand. Explosive new epidemics (as in southern Africa and Cambodia) will coexist with areas of slower HIV spread, and HIV will enter areas where infection has not yet been detected. And the already overwhelming burden of care will increase enormously.

In the industrial nations the epidemic will slow, at least for some populations, but it will take a higher toll on socially marginalized groups. The cost of care will rise substantially, as more and more infected individuals receive aggressive therapy. In both the Southern and Northern hemispheres, efforts to address societal forces that enhance vulnerability to HIV infection and AIDS will most likely proceed slowly, countered by significant resistance from social elites.

Control of the pandemic will require the extensive broadening of prevention programs. But given the large numbers of people at risk for infection and the difficulty of effecting behavioral change, the expansion of prevention programs will have to be coupled with greater efforts to develop HIV vaccines. The research establishment—both governmental and private—and international organizations must give highest priority to finding a vaccine and making it available to those who need it most: the marginalized populations who are bearing the brunt of the global HIV and AIDS pandemic.

The Authors

JONATHAN M. MANN and DANIEL J. M. TARANTOLA are co-editors of the 1996 report AIDS in the World II and, until recently, were colleagues at the François-Xavier Bagnoud Center for Health and Human Rights at the Harvard School of Public Health. In January, Mann, formerly director of the center and founding director of the WHO Global Program on AIDS, became dean of the School of Public Health at Allegheny University of the Health Sciences in Philadelphia. Tarantola is director of the International AIDS Program at the center and founding co-chair of the international MAP (Monitoring the AIDS Pandemic) network.

The Viral Superhighway

*Environmental disruptions and international travel have brought on
a new era in human illness, one marked by diabolical new diseases*

By George J. Armelagos

*So the Lord sent a pestilence upon Is-
rael from the morning until the ap-
pointed time; and there died of the
people from Dan to Beer-sheba seventy thousand men.*
—2 Sam. 24:15

SWARMS OF CROP-DESTROYING LOCUSTS, rivers fouled with blood, lion-headed horses breathing fire and sulfur: the Bible presents a lurid assortment of plagues, described as acts of retribution by a vengeful God. Indeed, real-life epidemics—such as the influenza outbreak of 1918, which killed 21 million people in a matter of months—can be so sudden and deadly that it is easy, even for nonbelievers, to view them as angry messages from the beyond.

How reassuring it was, then, when the march of technology began to give people some control over the scourges of the past. In the 1950s the Salk vaccine, and later, the Sabin vaccine, dramatically reduced the incidence of polio. And by 1980 a determined effort by health workers worldwide eradicated smallpox, a disease that had afflicted humankind since earliest times with blindness, disfigurement and death, killing nearly 300 million people in the twentieth century alone.

But those optimistic years in the second half of our century now seem, with hindsight, to have been an era of inflated expectations, even arrogance. In 1967 the surgeon general of the United States, William H. Stewart, announced that vic-

tory over infectious diseases was imminent—a victory that would close the book on modern plagues. Sadly, we now know differently. Not only have deadly and previously unimagined new illnesses such as AIDS and Legionnaires' disease emerged in recent years, but historical diseases that just a few decades ago seemed to have been tamed are returning in virulent, drug-resistant varieties. Tuberculosis, the ancient lung disease that haunted nineteenth-century Europe, afflicting, among others, Chopin, Dostoyevski and Keats, is aggressively mutating into strains that defy the standard medicines; as a result, modern TB victims must undergo a daily drug regimen so elaborate that health-department workers often have to personally monitor patients to make sure they comply [see "A Plague Returns," by Mark Earnest and John A. Sbarbaro, September/October 1993]. Meanwhile, bacteria and viruses in foods from chicken to strawberries to alfalfa sprouts are sickening as many as 80 million Americans each year.

And those are only symptoms of a much more general threat. Deaths from infectious diseases in the United States rose 58 percent between 1980 and 1992. Twenty-nine new diseases have been reported in the past twenty-five years, a few of them so bloodcurdling and bizarre that descriptions of them bring to mind tacky horror movies. Ebola virus, for instance, can in just a few days reduce a healthy person to a bag of teem-

ing flesh spilling blood and organ parts from every orifice. Creutzfeldt-Jakob disease, which killed the choreographer George Balanchine in 1983, eats away at its victims' brains until they resemble wet sponges. Never slow to fan mass hysteria, Hollywood has capitalized on the phenomenon with films such as *Outbreak*, in which a monkey carrying a deadly new virus from central Africa infects unwitting Californians and starts an epidemic that threatens to annihilate the human race.

The reality about infectious disease is less sensational but alarming nonetheless. Gruesome new pathogens such as Ebola are unlikely to cause a widespread epidemic because they sicken and kill so quickly that victims can be easily identified and isolated; on the other hand, the seemingly innocuous practice of overprescribing antibiotics for bad colds could ultimately lead to untold deaths, as familiar germs evolve to become untreatable. We are living in the twilight of the antibiotic era: within our lifetimes, scraped knees and cut fingers may return to the realm of fatal conditions.

Through international travel, global commerce and the accelerating destruction of ecosystems worldwide, people are inadvertently exposing themselves to a Pandora's box of emerging microbial threats. And the recent rumblings of biological terrorism from Iraq highlight the appalling potential of disease organisms for being manipulated to vile ends. But although it may appear that the apocalypse has arrived, the truth is that people today are not facing a unique predicament. Emerging diseases have long loomed like a shadow over the human race.

PEOPLE AND PATHOGENS HAVE A LONG history together. Infections have been detected in the bones of human ancestors more than a million years old, and evidence from the mummy of the Egyptian pharaoh Ramses V suggests that he may have died from smallpox more than 3,000 years ago. Widespread outbreaks of disease are also well documented. Between 1347 and 1351 roughly a third of the population of medieval Europe was wiped out by bubonic plague, which is carried by fleas that live on rodents. In 1793, 10 percent of the population of Philadelphia succumbed to yellow fever, which is spread by mosquitoes. And in 1875 the son of a Fiji chief came down with measles after a ceremonial trip to Australia. Within four months more than 20,000 Fijians were dead from the imported disease, which spreads through the air when its victims cough or sneeze.

According to conventional wisdom in biology, people and invading microorganisms evolve together: people gradually become more resistant, and the microorganisms become less virulent. The result is either mutualism, in which the relation benefits both species, or commensalism, in which one species benefits without harming the other. Chicken pox and measles, once fatal afflictions, now exist in more benign forms. Logic would suggest, after all, that the best interests of an organism are not served if it kills its host; doing so would be like picking a fight with the person who signs your paycheck.

But recently it has become clear to epidemiologists that the reverse of that cooperative paradigm of illness can also be true: microorganisms and their hosts sometimes exhaust their energies devising increasingly powerful weaponry and defenses. For example, several variants of human immunodeficiency virus (HIV) may compete for dominance within a person's body, placing the immune system under ever-greater siege. As long as a virus has an effective mechanism for jumping from one person to another, it can afford to kill its victims [see "The Deadliest Virus," by Cynthia Mills, January/February 1997].

If the competition were merely a question of size, humans would surely win: the average person is 10^{17} times the size of the average bacterium. But human beings, after all, constitute only one species, which must compete with 5,000 kinds of viruses and more than 300,000 species of bacteria. Moreover, in the twenty years it takes humans to produce a new generation, bacteria can reproduce a half-million times. That disparity enables pathogens to evolve ever more virulent adaptations that quickly outstrip human responses to them. The scenario is governed by what the English zoologist Richard Dawkins of the University of Oxford and a colleague have called the "Red Queen Principle." In Lewis Carroll's *Through the Looking Glass* the Red Queen tells Alice she will need to run faster and faster just to stay in the same place. Staving off illness can be equally elusive.

THE CENTERS FOR DISEASE CONTROL and Prevention (CDC) in Atlanta, Georgia, has compiled a list of the most recent emerging pathogens. They include:
- *Campylobacter,* a bacterium widely found in chickens because of the commercial practice of raising them in cramped, unhealthy conditions. It causes between two million and eight million cases of food poisoning a year in the United States and between 200 and 800 deaths.
- *Escherichia coli* 0157:H7, a dangerously mutated version of an often harmless bacterium. Hamburger meat from Jack in the Box fast-food restaurants that was contaminated with this bug led to the deaths of at least four people in 1993.
- Hantaviruses, a genus of fast-acting, lethal viruses, often carried by rodents, that kill by causing the capillaries to leak blood. A new hantavirus known as *sin nombre* (Spanish for "nameless") surfaced in 1993 in the southwestern United States, causing the sudden and mysterious deaths of thirty-two people.
- HIV, the deadly virus that causes AIDS (acquired immunodeficiency syndrome). Although it was first observed in people as recently as 1981, it has spread like wildfire and is now a global scourge, affecting more than 30 million people worldwide.
- The strange new infectious agent that causes bovine spongiform encephalopathy, or mad cow disease, which recently threw the British meat industry and consumers into a panic. This bizarre agent, known as a prion, or "proteinaceous infectious particle," is also responsible for Creutzfeldt-Jakob disease, the brain-eater I mentioned earlier. A Nobel Prize was awarded last year to the biochemist Stanley B. Prusiner of the University of California, San Francisco, for his discovery of the prion.
- *Legionella pneumophila,* the bacterium that causes Legionnaires' disease. The microorganism thrives in wet environments; when it lodges in air-conditioning systems or the mist machines in supermarket produce sections, it can be expelled into the air, reaching people's lungs. In 1976 thirty-four participants at an American Legion convention in Philadelphia died—the incident that led to the discovery and naming of the disease.
- *Borrelia burgdorferi,* the bacterium that causes Lyme disease. It is carried by ticks that live on deer and white-footed mice. Left untreated, it can cause crippling, chronic problems in the nerves, joints and internal organs.

HOW IRONIC, GIVEN SUCH A ROGUES' gallery of nasty characters, that just a quarter-century ago the Egyptian demographer Abdel R. Omran could observe that in many modern industrial nations the major killers were no longer infectious diseases. Death, he noted, now came not from outside but rather from within the body, the result of gradual deterioration. Omran traced the change to the middle of the nineteenth century, when the industrial revolution took hold in the United States and parts of Europe. Thanks to better nutrition, improved public-health measures and medical advances such as mass immunization and the introduction of antibiotics, microorganisms were brought under control. As people began living longer, their aging bodies succumbed to "diseases of civilization": cancer, clogged arteries, diabetes, obesity and osteoporosis. Omran was the first to formally recognize that shift in the disease environment. He called it an "epidemiological transition."

Like other anthropologists of my generation, I learned of Omran's theory early in my career, and it soon became a basic tenet—a comforting one, too, implying as it did an end to the supremacy of microorganisms. Then, three years ago, I began working with the anthropologist Kathleen C. Barnes of Johns Hopkins University in Baltimore, Maryland, to formulate an expansion of Omran's ideas. It occurred to us that his epidemiological transition had not been a unique event. Throughout history human populations have undergone shifts in their relations with disease—shifts, we noted, that are always linked to major changes in the way people interact with the environment. Barnes and I, along with James Lin, a master's student at Johns Hopkins University School of Hygiene and Public Health, have since developed a new theory: that there have been not one but three major epidemiological transitions; that each one has been sparked by human activities; and that we are living through the third one right now.

The first epidemiological transition took place some 10,000 years ago, when people abandoned their nomadic existence and began farming. That profoundly new way of life disrupted ecosystems and created denser living conditions that led, as I will soon detail, to new diseases. The second epidemiological transition was the salutary one Omran singled out in 1971, when the war against infectious diseases seemed to have been won. And in the past two decades the emergence of illnesses such as hepatitis C, cat scratch disease (caused by the bacterium *Bartonella henselae*), Ebola and others on CDC's list has created a third epidemiological transition, a disheartening set of changes that in many ways have reversed the effects of the second transition and coincide with the shift to globalism. Burgeoning population growth and urbanization, widespread environmental degradation, including global warming and tropical deforestation, and radically improved methods of transportation have given rise to new ways of contracting and spreading disease.

We are, quite literally, making ourselves sick.

WHEN EARLY HUMAN ANCESTORS moved from African forests onto the savanna millions of years ago, a few diseases came along for the ride. Those "heirloom" species—thus designated by the Australian parasitologist J. F. A. Sprent because they had afflicted earlier primates—included head and body lice; parasitic worms such as pinworms, tapeworms and liver flukes; and possibly herpes virus and malaria.

For 99.8 percent of the five million years of human existence, hunting and gathering was the primary mode of subsistence. Our ancestors lived in small groups and relied on wild animals and plants for their survival. In their foraging rounds, early humans would occasionally have contracted new kinds of illnesses through insect bites or by butchering and eating disease-ridden animals. Such events would not have led to widespread epidemics, however, because groups of people were so sparse and widely dispersed.

Global Warming could allow the mosquitoes that carry dengue fever to survive as far north as New York City.

About 10,000 years ago, at the end of the last ice age, many groups began to abandon their nomadic lifestyles for a more efficient and secure way of life. The agricultural revolution first appeared in the Middle East; later, farming centers developed independently in China and Central America. Permanent villages grew up, and people turned their attention to crafts such as toolmaking and pottery. Thus when people took to cultivating wheat and barley, they planted the seeds of civilization as well.

With the new ways, however, came certain costs. As wild habitats were transformed into urban settings, the farmers who brought in the harvest with their flint-bladed sickles were assailed by grim new ailments. Among the most common was scrub typhus, which is carried by mites that live in tall grasses, and causes a potentially lethal fever. Clearing vegetation to create arable fields brought farmers frequently into mite-infested terrain.

Irrigation brought further hazards. Standing thigh-deep in watery canals, farm workers were prey to the worms that cause schistosomiasis. After living within aquatic snails during their larval stage, those worms emerge in a free-swimming form that can penetrate human skin, lodge in the intestine or urinary tract, and cause bloody urine and other serious maladies. Schistosomiasis was well known in ancient Egypt, where outlying fields were irrigated with water from the Nile River; descriptions of its symptoms and remedies are preserved in contemporary medical papyruses.

The domestication of sheep, goats and other animals cleared another pathway for microorganisms. With pigs in their yards and chickens roaming the streets, people in agricultural societies were constantly vulnerable to pathogens that could cross interspecies barriers. Many such organisms had long since reached commensalism with their animal hosts, but they were highly dangerous to humans. Milk from infected cattle could transmit tuberculosis, a slow killer that eats away at the lungs and causes its victims to cough blood and pus. Wool and skins were loaded with anthrax, which can be fatal when inhaled and, in modern times, has been developed by several nations as a potential agent of biological warfare. Blood from infected cattle, injected into people by biting insects such as the tsetse fly, spread sleeping sickness, an often-fatal disease marked by tremors and protracted lethargy.

A SECOND MAJOR EFFECT OF AGRICULture was to spur population growth and, perhaps more important, density. Cities with populations as high as 50,000 had developed in the Near East by 3000 B.C. Scavenger species such as rats, mice and sparrows, which congregate wherever large groups of people live, exposed city dwellers to bubonic plague, typhus and rabies. And now that people were crowded together, a new pathogen could quickly start an epidemic. Larger populations also enabled diseases such as measles, mumps, chicken pox and smallpox to persist in an endemic form—always present, afflicting part of the population while sparing those with acquired immunity.

Thus the birth of agriculture launched humanity on a trajectory that has again and again brought people into contact with new pathogens. Tilling soil and raising livestock led to more energy-intensive ways of extracting resources from the earth—to lumbering, coal mining, oil drilling. New resources led to increasingly complex social organization, and to new and more frequent contacts between various societies. Loggers today who venture into the rain forest disturb previously untouched creatures and give them, for the first time, the chance to attack humans. But there is nothing new about this drama; only the players have changed. Some 2,000 years ago the introduction of iron tools to sub-Saharan Africa led to a slash-and-burn style of agriculture that brought people into contact with *Anopheles gambiae,* a mosquito that transmits malaria.

Improved transportation methods also help diseases extend their reach: microorganisms cannot travel far on their own, but they are expert hitchhikers. When the Spanish invaded Mexico in the early 1500s, for instance, they brought with them diseases that quickly raged through Tenochtitlán, the stately, temple-filled capital of the Aztec Empire. Smallpox, measles and influenza wiped out millions of Central America's original inhabitants, becoming the invisible weapon in the European conquest.

IN THE PAST THREE DECADES PEOPLE AND their inventions have drilled, polluted, engineered, paved, planted and deforested at soaring rates, changing the biosphere faster than ever before. The combined effects can, without hyperbole, be called a global revolution. After all, many of them have worldwide repercussions: the widespread chemical contamination of waterways, the thin-

ning of the ozone layer, the loss of species diversity. And such global human actions have put people at risk for infectious diseases in newly complex and devastating ways. Global warming, for instance, could expose millions of people for the first time to malaria, sleeping sickness and other insect-borne illnesses; in the United States, a slight overall temperature increase would allow the mosquitoes that carry dengue fever to survive as far north as New York City.

Major changes to the landscape that have become possible in the past quarter-century have also triggered new diseases. After the construction of the Aswan Dam in 1970, for instance, Rift Valley fever infected 200,000 people in Egypt, killing 600. The disease had been known to affect livestock, but it was not a major problem in people until the vast quantities of dammed water became a breeding ground for mosquitoes. The insects bit both cattle and humans, helping the virus jump the interspecies barrier.

In the eastern United States, suburbanization, another relatively recent phenomenon, is a dominant factor in the emergence of Lyme disease—10,000 cases of which are reported annually. Thanks to modern earth-moving equipment, a soaring economy and population pressures, many Americans have built homes in formerly remote, wooded areas. Nourished by lawns and gardens and unchecked by wolves, which were exterminated by settlers long ago, the deer population has exploded, exposing people to the ticks that carry Lyme disease.

Meanwhile, widespread pollution has made the oceans a breeding ground for microorganisms. Epidemiologists have suggested that toxic algal blooms—fed by the sewage, fertilizers and other contaminants that wash into the oceans—harbor countless viruses and bacteria. Thrown together into what amounts to a dirty genetic soup, those pathogens can undergo gene-swapping and mutations, engendering newly antibiotic-resistant strains. Nautical traffic can carry ocean pathogens far and wide: a devastating outbreak of cholera hit Latin America in 1991 after a ship from Asia unloaded its contaminated ballast water into the harbor of Callao, Peru. Cholera causes diarrhea so severe its victims can

die in a few days from dehydration; in that outbreak more than 300,000 people became ill, and more than 3,000 died.

The modern world is becoming—to paraphrase the words of the microbiologist Stephen S. Morse of Columbia University—a viral superhighway. Everyone is at risk.

Our newly global society is characterized by huge increases in population, international travel and international trade—factors that enable diseases to spread much more readily than ever before from person to person and from continent to continent. By 2020 the world population will have surpassed seven billion, and half those people will be living in urban centers. Beleaguered third-world nations are already hard-pressed to provide sewers, plumbing and other infrastructure; in the future, clean water and adequate sanitation could become increasingly rare. Meanwhile, political upheavals regularly cause millions of people to flee their homelands and gather in refugee camps, which become petri dishes for germs.

More than 500 million people cross international borders each year on commercial flights. Not only does that traffic volume dramatically increase the chance a sick person will infect the inhabitants of a distant area when she reaches her destination; it also exposes the sick person's fellow passengers to the disease, because of poor air circulation on planes. Many of those passengers can, in turn, pass the disease on to others when they disembark.

THE GLOBAL ECONOMY THAT HAS arisen in the past two decades has established a myriad of connections between far-flung places. Not too long ago bananas and oranges were rare treats in northern climes. Now you can walk into your neighborhood market and find food that has been flown and trucked in from all over the world: oranges from Israel, apples from New Zealand, avocados from California. Consumers in affluent nations expect to be able to buy whatever they want whenever they want it. What people do not generally realize, however, is that this global network of food production and delivery provides countless pathways for pathogens. Raspberries from Guatemala, carrots from Peru and coconut milk from Thailand have been responsible for recent outbreaks of food poisoning in the United States. And the problem cuts both ways: contaminated radish seeds and frozen beef from the United States have ended up in Japan and South Korea.

Finally, the widespread and often indiscriminate use of antibiotics has played a key role in spurring disease. Forty million pounds of antibiotics are manufactured annually in the United States, an eightyfold increase since 1954. Dangerous microorganisms have evolved accordingly, often developing antibiotic-resistant strains. Physicians are now faced with penicillin-resistant gonorrhea, multiple-drug-resistant tuberculosis and E. coli variants such as 0157:H7. And frighteningly, some enterococcus bacteria have become resistant to all known antibiotics. Enterococcus infections are rare, but staphylococcus infections are not, and many strains of staph bacteria now respond to just one antibiotic, vancomycin. How long will it be before run-of-the-mill staph infections—in a boil, for instance, or in a surgical incision—become untreatable?

ALTHOUGH CIVILIZATION CAN EXPOSE people to new pathogens, cultural progress also has an obvious countervailing effect: it can provide tools—medicines, sensible city planning, educational campaigns about sexually transmitted diseases—to fight the encroachments of disease. Moreover, since biology seems to side with microorganisms anyway, people have little choice but to depend on protective cultural practices to keep pace: vaccinations, for instance, to confer immunity, combined with practices such as handwashing by physicians between patient visits, to limit contact between people and pathogens.

All too often, though, obvious protective measures such as using only clean hypodermic needles or treating urban drinking water with chlorine are neglected, whether out of ignorance or a wrongheaded emphasis on the short-term financial costs. The worldwide disparity in wealth is also to blame: not surprisingly, the advances made during the second epidemiological transition were limited largely to the affluent of the industrial world.

Such lapses are now beginning to teach the bitter lesson that the delicate balance between humans and invasive microorganisms can tip the other way again. Overconfidence—the legacy of the second epidemiological transition—has made us especially vulnerable to emerging and reemerging diseases. Evolutionary principles can provide this useful corrective: in spite of all our medical and technological hubris, there is no quick fix. If human beings are to overcome the current crisis, it will be through sensible changes in behavior, such as increased condom use and improved sanitation, combined with a commitment to stop disturbing the ecological balance of the planet.

The Bible, in short, was not far from wrong: We do bring plagues upon ourselves—not by sinning, but by refusing to heed our own alarms, our own best judgment. The price of peace—or at least peaceful coexistence—with the microorganisms on this planet is eternal vigilance.

George J. Armelagos is a professor of anthropology at Emory University in Atlanta, Georgia. He has coedited two books on the evolution of human disease: PALEOPATHOLOGY AT THE ORIGINS OF AGRICULTURE, *which deals with prehistoric populations, and* DISEASE IN POPULATIONS IN TRANSITION, *which focuses on contemporary societies.*

Exploring Our Basic Human Nature

Are Humans Inherently Violent?

by Robert W. Sussman

Robert Sussman is professor of anthropology at Washington University at St. Louis and editor of the American Anthropologist, the journal of the American Anthropological Association.

Are human beings forever doomed to be violent? Is aggression fixed within our genetic code, an inborn action pattern that threatens to destroy us? Or, as asked by Richard Wrangham and Dale Peterson in their recent book, *Demonic Males: Apes and the Origins of Human Violence,* can we get beyond our genes, beyond our essential "human nature"?

Wrangham and Peterson's belief in the importance of violence in the evolution and nature of humans is based on new primate research that they assert demonstrates the continuity of aggression from our great ape ancestors. The authors argue that 20–25 years ago most scholars believed human aggression was unique. Research at that time had shown great apes to be basically non-aggressive gentle creatures. Furthermore, the separation of humans from our ape ancestors was thought to have occurred 15–20 million years ago (Mya). Although Raymond Dart, Sherwood Washburn, Robert Ardrey, E. O. Wilson and others had argued through much of the 20th century that hunting, killing, and extreme aggressive behaviors were biological traits inherited from our earliest hominid

hunting ancestors, many anthropologists still believed that patterns of aggression were environmentally determined and culturally learned behaviors, not inherited characteristics.

Demonic Males discusses new evidence that killer instincts are not unique to humans, but rather shared with our nearest relative, the common chimpanzee. The authors argue that it is this inherited propensity for killing that allows hominids and chimps to be such good hunters.

According to Wrangham and Peterson, the split between humans and the common chimpanzee was only 6–8 Mya. Furthermore, humans may have split from the chimpanzee-bonobo line after gorillas, with bonobos (pygmy chimps) separating from chimps only 2.5 Mya. Because chimpanzees may be the modern ancestor of all these forms, and because the earliest australopithecines were quite chimpanzee-like, Wrangham speculates (in a separate article) that "chimpanzees are a conservative species and an amazingly good model for the ancestor of hominids" (1995, reprinted in Sussman 1997:106). If modern chimpanzees and modern humans share certain behavioral traits, these traits have "long evolutionary roots" and are likely to be fixed, biologically inherited parts of our basic human nature and not culturally determined.

Wrangham argues that chimpanzees are almost on the brink of humanness:

> Nut-smashing, root-eating, savannah-using chimpanzees, resembling our ancestors, and capable by the way of extensive bipedalism. Using ant-wands, and sandals, and bowls, meat-sharing, hunting cooperatively. Strange paradox ... a species trembling on the verge of hominization, but so conservative that it has stayed on that edge. ... (1997:107).

Wrangham and Peterson (1996:24) claim that only two animal species, chimpanzees and humans, live in patrilineal, male-bonded communities "with intense, male initiated territorial aggression, including lethal raiding into neighboring communities in search of vulnerable enemies to attack and kill." Wrangham asks:

> Does this mean chimpanzees are naturally violent? Ten years ago it wasn't clear. ... In this cultural species, it may turn out that one of the least variable of all chimpanzee behaviors is the intense competition between males, the violent aggression they use against strangers, and their willingness to maim and kill those that frustrate their goals. ... As the picture of chimpanzee society settles into focus, it now includes infanticide, rape and regular battering of females by males (1997:108).

214   Reprinted with permission from *Anthro Notes,* National Museum of Natural History, Publication for Educators, Fall 1997, pp. 1-6, 17-19.

Since humans and chimpanzees share these violent urges, the implication is that human violence has long evolutionary roots. "We are apes of nature, cursed over six million years or more with a rare inheritance, a Dostoyevskyan demon. . . . The coincidence of demonic aggression in ourselves and our closest kin bespeaks its antiquity" (1997:108–109).

Intellectual Antecedents

From the beginning of Western thought, the theme of human depravity runs deep, related to the idea of humankind's fall from grace and the emergence of original sin. This view continues to pervade modern "scientific" interpretations of the evolution of human behavior. Recognition of the close evolutionary relationship between humans and apes, from the time of Darwin's *Descent of Man* (1874) on, has encouraged theories that look to modern apes for evidence of parallel behaviors reflecting this relationship.

By the early 1950s, large numbers of australopithecine fossils and the discovery that the large-brained "fossil" ancestor from Piltdown, in England, was a fraud, led to the realization that our earliest ancestors were more like apes than like modern humans. Accordingly, our earliest ancestors must have behaved much like other non-human primates. This, in turn, led to a great interest in using primate behavior to understand human evolution and the evolutionary basis of human nature. The subdiscipline of primatology was born.

Raymond Dart, discoverer of the first australopithecine fossil some thirty years earlier, was also developing a different view of our earliest ancestors. At first Dart believed that australopithecines were scavengers barely eking out an existence in the harsh savanna environment. But from the fragmented and damaged bones found with the australopithecines, together with dents and holes in these early hominid skulls, Dart eventually concluded that this species had used bone, tooth and antler tools to kill, butcher and eat their prey, as well as to kill one another. This hunting hypothesis (Cartmill 1997:511) "was linked from

the beginning with a bleak, pessimistic view of human beings and their ancestors as instinctively bloodthirsty and savage." To Dart, the australopithecines were:

> confirmed killers: carnivorous creatures that seized living quarries by violence, battered them to death, tore apart their broken bodies, dismembered them limb from limb, slaking their ravenous thirst with the hot blood of victims and greedily devouring livid writhing flesh (1953:209).

Cartmill, in a recent book (1993), shows that this interpretation of early human morality is reminiscent of earlier Greek and Christian views. Dart's (1953) own treatise begins with a 17th century quote from the Calvinist R. Baxter: "of all the beasts, the man-beast is the worst/to others and himself the cruellest foe."

Between 1961–1976, Dart's view was picked up and extensively popularized by the playwright Robert Ardrey (*The Territorial Imperative, African Genesis*). Ardrey believed it was the human competitive and killer instinct, acted out in warfare, that made humans what they are today. "It is war and the instinct for territory that has led to the great accomplishments of Western Man. Dreams may have inspired our love of freedom, but only war and weapons have made it ours" (1961:324).

Man the Hunter

In the 1968 volume *Man the Hunter*, Sherwood Washburn and Chet Lancaster presented a theory of "The evolution of hunting," emphasizing that it is this behavior that shaped human nature and separated early humans from their primate relatives.

> To assert the biological unity of mankind is to affirm the importance of the hunting way of life. . . . However much conditions and customs may have varied locally, the main selection pressures that forged the species were the same. The biology, psychology and customs that separate us from the apes . . . we owe to the hunters of time past . . . for those who would understand the origins and nature of human behavior there is no choice but to

try to understand "Man the Hunter" (1968:303).

Rather than amassing evidence from modern hunters and gatherers to prove their theory, Washburn and Lancaster (1968:299) use the 19th-century concept of cultural "survivals": behaviors that persist as evidence of an earlier time but are no longer useful in society.

> Men enjoy hunting and killing, and these activities are continued in sports even when they are no longer economically necessary. If a behavior is important to the survival of a species . . . then it must be both easily learned and pleasurable (Washburn & Lancaster, p. 299).

Man the Dancer

Using a similar logic for the survival of ancient "learned and pleasurable" behaviors, perhaps it could easily have been our propensity for dancing rather than our desire to hunt that can explain much of human behavior. After all, men and women love to dance; it is a behavior found in all cultures but has even less obvious function today than hunting. Our love of movement and dance might explain, for example, our propensity for face-to-face sex, and even the evolution of bipedalism and the movement of humans out of trees and onto the ground.

Could the first tool have been a stick to beat a dance drum, and the ancient Laetoli footprints evidence of two individuals going out to dance the "Afarensis shuffle"? Although it takes only two to tango, a variety of social interactions and systems might have been encouraged by the complex social dances known in human societies around the globe.

Sociobiology and E. O. Wilson

In the mid-1970s, E. O. Wilson and others described a number of traits as genetically based and therefore human universals, including territoriality, male-female bonds, male dominance over females, and extended maternal care leading to matrilineality. Wilson argued that the genetic basis of these traits was indicated by their relative constancy among our primate relatives and by their persistence throughout human evolution and in human societies. Elsewhere, I have shown that these characteristics are neither general primate traits nor human universals (Sussman 1995). Wilson, however, argued that these were a product of evolutionary hunting past.

For at least a million years—probably more—Man engaged in a hunting way of life, giving up the practice a mere 10,000 years ago. . . . Our innate social responses have been fashioned through this life style. With caution, we can compare the most widespread hunter-gatherer qualities with similar behavior displayed by some of the non-human primates that are closely related to Man. Where the same pattern of traits occurs in . . . most or all of those primates—we can conclude that it has been subject to little evolution (Wilson 1976, in Sussman 1997:65–66).

Wilson's theory of sociobiology, the evolution of social behavior, argued that:

(1) the goal of living organisms is to pass on one's genes at the expense of all others;

(2) an organism should only cooperate with others if:

(a) they carry some of his/her own genes (kin selection) or

(b) if at some later date the others might aid you (reciprocal altruism).

To sociobiologists, evolutionary morality is based on an unconscious need to multiply our own genes, to build group cohesion in order to win wars. We should not look down on our warlike, cruel nature but rather understand its success when coupled with "making nice" with **some** other individuals or groups. The genetically driven "making nice" is the basis of human ethics and morality.

Throughout recorded history the conduct of war has been common . . . some of the noblest traits of mankind, including team play, altruism, patriotism, bravery . . . and so forth are the genetic product of warfare (Wilson 1975:572–3).

The evidence for any of these universals or for the tenets of sociobiology is as weak as was the evidence for Dart's, Ardrey's and Washburn and Lancaster's theories of innate aggression. Not only are modern gatherer-hunters and most apes remarkably non-aggressive, but in the 1970s and 1980s studies of fossil bones and artifacts have shown that early humans were not hunters, and that weapons were a later addition to the human repertoire. In fact, C. K. Brain (1981) showed that the holes and dents in Dart's australopithecine skulls matched perfectly with fangs of leopards or with impressions of rocks pressing against the buried fossils. Australopithecines apparently were the hunted, not the hunters (Cartmill 1993, 1997).

Beyond Our Genes

Wrangham and Peterson's book goes beyond the assertion of human inborn aggression and propensity towards violence. The authors ask the critical question: Are we doomed to be violent forever because this pattern is fixed within our genetic code or can we go beyond our past?—get out of our genes, so to speak.

The authors believe that we can look to the bonobo or pygmy chimpanzee as one potential savior, metaphorically speaking.

Bonobos, although even more closely related to the common chimpanzee than humans, have become a peace-loving, love-making alternative to chimpanzee-human violence. How did this happen? In chimpanzees and humans, females of the species select partners that are violent . . . "while men have evolved to be demonic males, it seems likely that women have evolved to prefer demonic males . . . as long as demonic males are the most successful reproducers, any female who mates with them is provided with sons who themselves will likely be good reproducers" (Wrangham and Peterson 1996:239). However, among pygmy chimpanzees females form alliances and have chosen to mate with less aggressive males. So, after all, it is not violent males that have caused humans and chimpanzees to be their inborn, immoral, dehumanized selves, it is rather, poor choices by human and chimpanzee females.

Like Dart, Washburn, and Wilson before them, Wrangham and Peterson believe that killing and violence is inherited from our ancient relatives of the past. However, unlike these earlier theorists, Wrangham and Peterson argue that is not a trait unique to hominids, nor is it a by-product of hunting. In fact, it is just this violent nature and a natural "blood lust" that makes both humans and chimpanzees such good hunters. It is the bonobos that help the authors come to this conclusion. Because bonobos have lost the desire to kill, they also have lost the desire to hunt.

. . . do bonobos tell us that the suppression of personal violence carried with it the suppression of predatory aggression? The strongest hypothesis at the moment is that bonobos came from a chimpanzee-like ancestor that hunted monkeys and hunted one another. As they evolved into bonobos, males lost their demonism, becoming less aggressive to each other. In so doing they lost their lust for hunting monkeys, too. . . . Murder and hunting may be more closely tied together than we are used to thinking (Wrangham and Peterson 1996:219).

The Selfish Gene Theory

Like Ardrey, Wrangham and Peterson believe that blood lust ties killing and hunting tightly together but it is the killing that drives hunting in the latter's argument. This lust to kill is based upon the sociobiological tenet of the selfish gene. "The general principle that behavior evolves to serve selfish ends has been widely accepted; and the idea that humans might have been favored by natural selection to hate and to kill their enemies has become entirely, if tragically, reasonable" (Wrangham and Peterson 1996:23).

As with many of the new sociobiological or evolutionary anthropology theories, I find problems with both the theory itself and with the evidence used to support it. Two arguments that humans and chimpanzees share biologically fixed behaviors are: (1) they are more closely related to each other than chimpanzees are to gorillas; (2) chimpanzees are a good model for our earliest ancestor and retain conservative traits that should be shared by both.

The first of these statements is still hotly debated and, using various genetic evidence, the chimp-gorilla-human triage is so close that it is difficult to tell exact divergence time or pattern among the three. The second statement is just not true. Chimpanzees have been evolving for as long as humans and gorillas, and there is no reason to believe ancestral chimps were similar to present-day chimps. The fossil evidence for the last 5–8 million years is extremely sparse, and it is likely that many forms of apes have become extinct just as have many hominids.

Furthermore, even if the chimpanzee were a good model for the ancestral hominid, and was a conservative representative of this phylogenetic group, this would not mean that humans would necessarily share specific behavioral traits. As even Wrangham and Peterson emphasize, chimps, gorillas, and bonobos all behave very differently from one another in their social behavior and in their willingness to kill conspecifics.

Evidence Against "Demonic Males"

The proof of the "Demonic Male" theory does not rest on any theoretical grounds but must rest solely on the evidence that violence and killing in chimpanzees and in humans are behaviors that are similar in pattern; have ancient, shared evolutionary roots; and are inherited. Besides killing of conspecifics, Wrangham "includes infanticide, rape, and regular battering of females by males" as a part of this inherited legacy of violent behaviors shared by humans and chimpanzees (1997:108).

Wrangham and Peterson state: "That chimpanzees and humans kill members of neighboring groups of their own species is . . . a startling exception to the normal rule for animals" (1996:63). "Fighting adults of almost all species normally stop at winning: They don't go on to kill" (1996:155). However, as Wrangham points out there are exceptions, such as lions, wolves, spotted hyenas, and I would add a number of other

predators. In fact, most species do not have the weapons to kill one another as adults.

Just how common is conspecific killing in chimpanzees? This is where the real controversy may lie. Jane Goodall described the chimpanzee as a peaceful, non-aggressive species during the first 24 years of study at Gombe (1950–1974). During one year of concentrated study, Goodall observed 284 agonistic encounters: of these 66% were due to competition for introduced bananas, and only 34% "could be regarded as attacks occurring in 'normal' aggressive contexts" (1968:278). Only 10 percent of the 284 attacks were classified as 'violent,' and "even attacks that appeared punishing to me often resulted in no discernable injury. . . . Other attacks consisted merely of brief pounding, hitting or rolling of the individual, after which the aggressor often touched or embraced the other immediately (1968:277).

Chimpanzee aggression before 1974 was considered no different from patterns of aggression seen in many other primate species. In fact, Goodall explains in her 1986 monograph, *The Chimpanzees of Gombe,* that she uses data mainly from after 1975 because the earlier years present a "very different picture of the Gombe chimpanzees" as being "far more peaceable than humans" (1986:3). Other early naturalists' descriptions of chimpanzee behavior were consistent with those of Goodall and confirmed her observations. Even different communities were observed to come together with peaceful, ritualized displays of greeting (Reynolds and Reynolds 1965; Suguyama 1972; Goodall 1968).

Then, between 1974 and 1977, five adult males from one subgroup were attacked and disappeared from the area, presumably dead. Why after 24 years did the patterns of aggression change? Was it because the stronger group saw the weakness of the other and decided to improve their genetic fitness? But surely there were stronger and weaker animals and subgroups before this time. Perhaps we can look to Goodall's own perturbations for an answer. In 1965, Goodall began to provide "restrictive human-controlled feeding." A few years later she realized that

the constant feeding was having a marked effect on the behavior of the chimps. They were beginning to move about in large groups more often than they had ever done in the old days. Worst of all, the adult males were becoming increasingly aggressive. When we first offered the chimps bananas the males seldom fought over their food, . . . now . . . there was a great deal more fighting than ever before . . . (Goodall 1971:143).

The possibility that human interference was a main cause of the unusual behavior of the Gombe chimps was the subject of an excellent, but generally ignored book by Margaret Power (1991). Wrangham and Peterson (1996:19) footnote this book, but as with many other controversies, they essentially ignore its findings, stating that yes, chimpanzee violence might have been unnatural behavior if it weren't for the evidence of similar behavior occurring since 1977 and "elsewhere in Africa" (1996:19).

Further Evidence

What is this evidence from elsewhere in Africa? Wrangham and Peterson provide only four brief examples, none of which is very convincing:

(1) Between 1979–1982, the Gombe group extended its range to the south and conflict with a southern group, Kalande, was suspected. In 1982, a "raiding" party of males reached Goodall's camp. The authors state: "Some of these raids may have been lethal" (1996:19). However, Goodall describes this "raid" as follows: One female "was chased by a Kalande male and mildly attacked. . . . Her four-year-old son . . . encountered a second male—but was only sniffed" (1986:516). Although Wrangham and Peterson imply that these encounters were similar to those between 1974–77, no violence was actually witnessed. The authors also refer to the discovery of the dead body of Humphrey; what they do not mention is Humphrey's age of 35 and that wild chimps rarely live past 33 years!

(2) From 1970 to 1982, six adult males from one community in the Japanese study site of Mahale disappeared, one by one over this 12 year period.

None of the animals were observed being attacked or killed, and one was sighted later roaming as a solitary male (Nishida et al. 1985:287–289).

(3) In another site in West Africa, Wrangham and Peterson report that Boesch and Boesch believe "that violent aggression among the chimpanzees is as important as it is in Gombe" (1986:20). However, in the paper referred to, the Boesches simply state that encounters by neighboring chimpanzee communities are more common in their site than in Gombe (one per month vs. 1 every 4 months). There is no mention of violence during these encounters.

(4) At a site that Wrangham began studying in 1984, an adult male was found dead in 1991. Wrangham states: "In the second week of August, Ruizoni was killed. No human saw the big fight" (Wrangham & Peterson 1996:20). Wrangham gives us no indication of what has occurred at this site over the last 6 years.

In fact, this is the total amount of evidence of warfare and male-male killing among chimpanzees after 37 years of research!! The data for infanticide and rape among chimpanzees is even less impressive. In fact, data are so sparse for these behaviors among chimps that Wrangham and Peterson are forced to use examples from the other great apes, gorillas and orangutans. However, just as for killing among chimpanzees, both the evidence and the interpretations are suspect and controversial.

Can We Escape Our Genes?

What if Wrangham and Peterson are correct and we and our chimp cousins are inherently sinners? Are we doomed to be violent forever because this pattern is fixed within our genetic code?

After 5 million years of human evolution and 120,000 or so years of *Homo sapiens* existence, is there a way to rid ourselves of our inborn evils?

What does it do for us, then, to know the behavior of our closest relatives? Chimpanzees and bonobos are an extraordinary pair. One, I suggest shows us some of the worst aspects of our

past and our present; the other shows an escape from it. . . . Denial of our demons won't make them go away. But even if we're driven to accepting the evidence of a grisly past, we're not forced into thinking it condemns us to an unchanged future (Wrangham 1997:110).

In other words, we can learn how to behave by watching bonobos. But, if we can change our inherited behavior so simply, why haven't we been able to do this before *Demonic Males* enlightened us? Surely, there are variations in the amounts of violence in different human cultures and individuals. If we have the capacity and plasticity to change by learning from example, then our behavior is determined by socialization practices and by our cultural histories and not by our nature! This is true whether the examples come from benevolent bonobos or conscientious objectors.

Conclusion

The theory presented by Wrangham and Peterson, although it also includes chimpanzees as our murdering cousins, is very similar to "man the hunter" theories proposed in the past. It also does not differ greatly from early European and Christian beliefs about human ethics and morality. We are forced to ask:

Are these theories generated by good scientific fact, or are they just "good to think" because they reflect, reinforce, and reiterate our traditional cultural beliefs, our morality and our ethics? Is the theory generated by the data, or are the data manipulated to fit preconceived notions of human morality and ethics?

Since the data in support of these theories have been weak, and yet the stories created have been extremely similar, I am forced to believe that "Man the Hunter" is a myth, that humans are not necessarily prone to violence and aggression, but that this belief will continue to reappear in future writings on human nature. Meanwhile, primatologists must continue their field research, marshaling the actual evidence needed to answer many of the questions raised in Wrangham and Peterson's volume.

References Cited:

Ardrey, Robert. 1961. *African Genesis: A Personal Investigation into Animal Origins and Nature of Man.* Atheneum.
_____. *The Territorial Imperative.* Atheneum, 1966.
Brain, C. K. 1981. *The Hunted or the Hunter? An Introduction to African Cave Taphonomy.* Univ. of Chicago.
Dart, Raymond. 1953. "The Predatory Transition from Ape to Man." *International Anthropological and Linguistic Review* 1:201–217.
Darwin, Charles. 1874. *The Descent of Man and Selection in Relation to Sex.* 2nd ed. The Henneberry Co.
Cartmill, Matt. 1997. "Hunting Hypothesis of Human Origins." In *History of Physical Anthropology: An Encyclopedia,* ed. F. Spencer, pp. 508–512. Garland.
_____. 1993. *A View to a Death in the Morning: Hunting and Nature Through History.* Harvard Univ.
Goodall, Jane. 1986. *The Chimpanzees of Gombe: Patterns of Behavior.* Belknap.
_____. 1971. *In the Shadow of Man.* Houghton Mifflin.
Goodall, Jane. 1968. "The Behavior of Free-Living Chimpanzees in the Gombe Stream Reserve." *Animal Behavior Monographs* 1:165–311.
Nishida, T., Hiraiwa-Hasegawa, M., and Takahtat, Y. "Group Extinction and Female Transfer in Wild Chimpanzees in the Mahali Nation Park, Tanzania." *Zeitschrift für Tierpsychologie* 67:281–301.
Power, Margaret. *1991. The Egalitarians: Human and Chimpanzee: An Anthropological View of Social Organization.* Cambridge University.
Reynolds, V. and Reynolds, F. 1965. "Chimpanzees of Budongo Forest," In *Primate Behavior: Field Studies of Monkeys and Apes,* ed. I. DeVore, pp. 368–424. Holt, Rinehart, and Winston.
Suguyama, Y. 1972. "Social Characteristics and Socialization of Wild Chimpanzees." In *Primate Socialization,* ed. F. E. Poirier, pp. 145–163. Random House.
Sussman, R. W., ed. *1997 The Biological Basis of Human Behavior.* Simon and Schuster.
Sussman, R. W. 1995. "The Nature of Human Universals." *Reviews in Anthropology* 24:1–11.
Washburn, S. L. and Lancaster, C. K. 1968. "The Evolution of Hunting." In *Man the Hunter,* eds. R. B. Lee and I. DeVore, pp. 293–303. Aldine.
Wilson, E. O. 1997. "Sociobiology: A New Approach to Understanding the Basis of Human Nature." *New Scientist* 70(1976):342–345. (Reprinted in R. W. Sussman, 1997.)
_____. 1975. Sociobiology: *The New Synthesis.* Cambridge: Harvard University.
Wrangham, R. W. 1995. "Ape, Culture, and Missing Links." *Symbols* (Spring):2–9, 20. (Reprinted in R. W. Sussman, 1997.)
Wrangham, Richard and Peterson, Dale. 1996. *Demonic Males: Apes and the Origins of Human Violence.* Houghton Mifflin.

Further Reading

Bock, Kenneth. 1980. *Human Nature and History: A Response to Sociobiology.* Columbia University.
Gould, Stephen J. 1996. *Mismeasure of Man.* W. W. Norton.

Dr. Darwin

With a nod to evolution's god, physicians are looking at illness through the lens of natural selection to find out why we get sick and what we can do about it.

Lori Oliwenstein

Lori Oliwenstein, a former DISCOVER senior editor, is now a freelance journalist based in Los Angeles.

PAUL EWALD KNEW FROM THE BEGINNING that the Ebola virus outbreak in Zaire would fizzle out. On May 26, after eight days in which only six new cases were reported, that fizzle became official. The World Health Organization announced it would no longer need to update the Ebola figures daily (though sporadic cases continued to be reported until June 20).

The virus had held Zaire's Bandundu Province in its deadly grip for weeks, infecting some 300 people and killing 80 percent of them. Most of those infected hailed from the town of Kikwit. It was all just as Ewald predicted. "When the Ebola outbreak occurred," he recalls, "I said, as I have before, these things are going to pop up, they're going to smolder, you'll have a bad outbreak of maybe 100 or 200 people in a hospital, maybe you'll have the outbreak slip into another isolated community, but then it will peter out on its own."

Ewald is no soothsayer. He's an evolutionary biologist at Amherst College in Massachusetts and perhaps the world's leading expert on how infectious diseases—and the organisms that cause them—evolve. He's also a force behind what some are touting as the next great medical revolution: the application of

> *"If you look at it from an evolutionary point of view, you can sort out the 95 percent of disease organisms that aren't a major threat from the 5 percent that are."*

Darwin's theory of natural selection to the understanding of human diseases.

A Darwinian view can shed some light on how Ebola moves from human to human once it has entered the population. (Between human outbreaks, the virus resides in some as yet unknown living reservoir.) A pathogen can survive in a population, explains Ewald, only if it can easily transmit its progeny from one host to another. One way to do this is to take a long time to disable a host, giving him plenty of time to come into contact with other potential victims. Ebola, however, kills quickly, usually in less than a week. Another way is to survive for a long time outside the human body, so that the pathogen can wait for new hosts to find it. But the Ebola strains encountered thus far are destroyed almost at once by sunlight, and

even if no rays reach them, they tend to lose their infectiousness outside the human body within a day. "If you look at it from an evolutionary point of view, you can sort out the 95 percent of disease organisms that aren't a major threat from the 5 percent that are," says Ewald. "Ebola really isn't one of those 5 percent."

The earliest suggestion of a Darwinian approach to medicine came in 1980, when George Williams, an evolutionary biologist at the State University of New York at Stony Brook, read an article in which Ewald discussed using Darwinian theory to illuminate the origins of certain symptoms of infectious disease—things like fever, low iron counts, diarrhea. Ewald's approach struck a chord in Williams. Twenty-three years earlier he had written a paper proposing an evolutionary framework for senescence, or aging. "Way back in the 1950s I didn't worry about the practical aspects of senescence, the medical aspects," Williams notes. "I was pretty young then." Now, however, he sat up and took notice.

While Williams was discovering Ewald's work, Randolph Nesse was discovering Williams's. Nesse, a psychiatrist and a founder of the University of Michigan Evolution and Human Behavior Program, was exploring his own interest in the aging process, and he and Williams soon got together. "He had wanted to find a physician to work with on medical problems," says Nesse, "and I had long wanted to find an evolutionary biologist, so it was a very natural match for us." Their collaboration led to a 1991 article that most researchers say signaled the real birth of the field.

NESSE AND WILLIAMS DEfine Darwinian medicine as the hunt for evolutionary explanations of vulnerabilities to disease. It can, as Ewald noted, be a way to interpret the body's defenses, to try to figure out, say, the reasons we feel pain or get runny noses when we have a cold, and to determine what we should—or shouldn't—be doing about those defenses. For instance, Darwinian researchers like physiologist Matthew Kluger of the Lovelace Institute in Albuquerque now say that a moderate rise in body temperature is more than just a symptom of disease; it's an evolutionary adaptation the body uses to fight infection by making itself inhospitable to invading microbes. It would seem, then, that if you lower the fever, you may prolong the infection. Yet no one is ready to say whether we should toss out our aspirin bottles. "I would love to see a dozen proper studies of whether it's wise to bring fever down when someone has influenza," says Nesse. "It's never been done, and it's just astounding that it's never been done."

Diarrhea is another common symptom of disease, one that's sometimes the result of a pathogen's manipulating your body for its own good purposes, but it may also be a defense mechanism mounted by your body. Cholera bacteria, for example, once they invade the human body, induce diarrhea by producing toxins that make the intestine's cells leaky. The resultant diarrhea then both flushes competing beneficial bacteria from the gut and gives the cholera bacteria a ride into the world, so that they can find another hapless victim. In the case of cholera, then, it seems clear that stopping the diarrhea can only do good.

But the diarrhea that results from an invasion of shigella bacteria—which cause various forms of dysentery—seems to be more an intestinal defense than a bacterial offense. The infection causes the muscles surrounding the gut to contract

more frequently, apparently in an attempt to flush out the bacteria as quickly as possible. Studies done more than a decade ago showed that using drugs like Lomotil to decrease the gut's contractions and cut down the diarrheal output actually prolong infection. On the other hand, the ingredients in over-the-counter preparations like Pepto Bismol, which don't affect how frequently the gut contracts, can be used to stem the diarrheal flow without prolonging infection.

Seattle biologist Margie Profet points to menstruation as another "symptom" that may be more properly viewed as an evolutionary defense. As Profet points out, there must be a good reason for the body to engage in such costly activities as shedding the uterine lining and letting blood flow away. That reason, she claims, is to rid the uterus of any organisms that might arrive with sperm in the seminal fluid. If an egg is fertilized, infection may be worth risking. But if there is no fertilized egg, says Profet, the body defends itself by ejecting the uterine cells, which might have been infected. Similarly, Profet has theorized that morning sickness during pregnancy causes the mother to avoid foods that might contain chemicals harmful to a developing fetus. If she's right, blocking that nausea with drugs could result in higher miscarriage rates or more birth defects.

DARWINIAN MEDICINE ISN'T simply about which symptoms to treat and which to ignore. It's a way to understand microbes—which, because they evolve so much more quickly than we do, will probably always beat us unless we figure out how to harness their evolutionary power for our own benefit. It's also a way to realize how disease-causing genes that persist in the population are often selected for, not against, in the long run.

Sickle-cell anemia is a classic case of how evolution tallies costs and benefits. Some years ago, researchers discovered that people with one copy of the sickle-cell gene are better able to resist the protozoans that cause malaria than are people with no copies of the gene. People with two copies of the gene may die, but in malaria-plagued regions such as

tropical Africa, their numbers will be more than made up for by the offspring left by the disease-resistant kin.

Cystic fibrosis may also persist through such genetic logic. Animal studies indicate that individuals with just one copy of the cystic fibrosis gene may be more resistant to the effects of the cholera bacterium. As is the case with malaria and sickle-cell, cholera is much more prevalent than cystic fibrosis; since there are many more people with a single, resistance-conferring copy of the gene than with a disease-causing double dose, the gene is stably passed from generation to generation.

"With our power to do gene manipulations, there will be temptations to find genes that do things like cause aging, and get rid of them," says Nesse. "If we're sure about everything a gene does, that's fine. But an evolutionary approach cautions us not to go too fast, and to expect that every gene might well have some benefit as well as costs, and maybe some quite unrelated benefit."

"I used to hunt saber-toothed tigers all the time, thousands of years ago. Now I sit in front of a computer and don't get exercise, so I've changed my body chemistry."

Darwinian medicine can also help us understand the problems encountered in the New Age by a body designed for the Stone Age. As evolutionary psychologist Charles Crawford of Simon Fraser University in Burnaby, British Columbia, put it: "I used to hunt saber-toothed tigers all the time, thousands of years ago. I got lots of exercise and all that sort of stuff. Now I sit in front of a computer, and all I do is play with a mouse, and I don't get exercise. So I've changed my body biochemistry in all sorts of unknown ways, and it could affect me in all sorts of ways, and we have no idea what they are."

Radiologist Boyd Eaton of Emory University and his colleagues believe such biochemical changes are behind today's breast cancer epidemic. While it's impossible to study a Stone Ager's biochemistry, there are still groups of hunter-gathers around—such as the San of Africa—who make admirable stand-ins. A foraging life-style, notes Eaton, also means a life-style in which menstruation begins later, the first child is born earlier, there are more children altogether, they are breast-fed for years rather than months, and menopause comes somewhat earlier. Overall, he says, American women today probably experience 3.5 times more menstrual cycles than our ancestors did 10,000 years ago. During each cycle a woman's body is flooded with the hormone estrogen, and breast cancer, as research has found, is very much estrogen related. The more frequently the breasts are exposed to the hormone, the greater the chance that a tumor will take seed.

Depending on which data you choose, women today are somewhere between 10 and 100 times more likely to be stricken with breast cancer than our ancestors were. Eaton's proposed solutions are pretty radical, but he hopes people will at least entertain them; they include delaying puberty with hormones and using hormones to create pseudo-pregnancies, which offer a woman the biochemical advantages of pregnancy at an early age without requiring her to bear a child.

In general, Darwinian medicine tells us that the organs and systems that make up our bodies result not from the pursuit of perfection but from millions of years of evolutionary compromises designed to get the greatest reproductive benefit at the lowest cost. We walk upright with a spine that evolved while we scampered on four limbs; balancing on two legs leaves our hands free, but we'll probably always suffer some back pain as well.

"What's really different is that up to now people have used evolutionary theory to try to explain why things work, why they're normal," explains Nesse. "The twist—and I don't know if it's simple or profound—is to say we're try-

ing to understand the abnormal, the vulnerability to disease. We're trying to understand why natural selection has not made the body better, why natural selection has left the body with vulnerabilities. For every single disease, there is an answer to that question. And for very few of them is the answer very clear yet."

One reason those answers aren't yet clear is that few physicians or medical researchers have done much serious surveying from Darwin's viewpoint. In many cases, that's because evolutionary theories are hard to test. There's no way to watch human evolution in progress—at best it works on a time scale involving hundreds of thousands of years. "Darwinian medicine is mostly a guessing game about how we think evolution worked in the past on humans, what it designed for us," say evolutionary biologist James Bull of the University of Texas at Austin. "It's almost impossible to test ideas that we evolved to respond to this or that kind of environment. You can make educated guesses, but no one's going to go out and do an experiment to show that yes, in fact humans will evolve this way under these environmental conditions."

Yet some say that these experiments can, should, and will be done. Howard Howland, a sensory physiologist at Cornell, is setting up just such an evolutionary experiment, hoping to interfere with the myopia, or nearsightedness, that afflicts a full quarter of all Americans. Myopia is thought to be the result of a delicate feedback loop that tries to keep images focused on the eye's retina. There's not much room for error: if the length of your eyeball is off by just a tenth of a millimeter, your vision will be blurry. Research has shown that when the eye perceives an image as fuzzy, it compensates by altering its length.

This loop obviously has a genetic component, notes Howland, but what drives it is the environment. During the Stone Age, when we were chasing buffalo in the field, the images we saw were usually sharp and clear. But with modern civilization came a lot of close work. When your eye focuses on something nearby, the lens has to bend, and since bending that lens is hard work, you do

as little bending as you can get away with. That's why, whether you're conscious of it or not, near objects tend to be a bit blurry. "Blurry image?" says the eye. "Time to grow." And the more it grows, the fuzzier those buffalo get. Myopia seems to be a disease of industrial society.

To prevent that disease, Howland suggests going back to the Stone Age—or at least convincing people's eyes that that's where they are. If you give folks with normal vision glasses that make their eyes think they're looking at an object in the distance when they're really looking at one nearby, he says, you'll avoid the whole feedback loop in the first place. "The military academies induct young men and women with twenty-twenty vision who then go through four years of college and are trained to fly an airplane or do some difficult visual task. But because they do so much reading, they come out the other end nearsighted, no longer eligible to do what they were hired to do," Howland notes. "I think these folks would very much like not to become nearsighted in the course of their studies." He hopes to be putting glasses on them within a year.

THE NUMBING PACE OF EVOlution is a much smaller problem for researchers interested in how the bugs that plague us do their dirty work. Bacteria are present in such large numbers (one person can carry around more pathogens than there are people on the planet) and evolve so quickly (a single bacterium can reproduce a million times in one human lifetime) that experiments we couldn't imagine in humans can be carried out in microbes in mere weeks. We might even, says Ewald, be able to use evolutionary theory to tame the human immunodeficiency virus.

"HIV is mutating so quickly that surely we're going to have plenty of sources of mutants that are mild as well as severe," he notes. "So now the question is, which of the variants will win?" As in the case of Ebola, he says, it will all come down to how well the virus manages to get from one person to another.

"If there's a great potential for sexual transmission to new partners, then the viruses that reproduce quickly will spread," Ewald says. "And since they're reproducing in a cell type that's critical for the well-being of the host—the helper T cell—then that cell type will be decimated, and the host is likely to suffer from it." On the other hand, if you lower the rate of transmission—through abstinence, monogamy, condom use— then the more severe strains might well die out before they have a chance to be passed very far. "The real question," says Ewald, "is, exactly how mild can you make this virus as a result of reducing the rate at which it could be transmitted to new partners, and how long will it take for this change to occur?" There are already strains of HIV in Senegal with such low virulence, he points out, that most people infected will die of old age. "We don't have all the answers. But I think we're going to be living with this virus for a long time, and if we have to live with it, let's live with a really mild virus instead of a severe virus."

Though condoms and monogamy are not a particularly radical treatment, that they might be used not only to stave off the virus but to tame it is a radical notion—and one that some researchers find suspect. "If it becomes too virulent, it will end up cutting off its own transmission by killing its host too quickly," notes James Bull. "But the speculation is that people transmit HIV primarily within one to five months of infection, when they spike a high level of virus in the blood. So with HIV, the main period of transmission occurs a few months into the infection, and yet the virulence—the death from it—occurs years later. The major stage of transmission is decoupled from the virulence." So unless the pro-

tective measures are carried out by everyone, all the time, we won't stop most instances of transmission; after all, most people don't even know they're infected when they pass the virus on.

But Ewald thinks these protective measures are worth a shot. After all, he says, pathogen taming has occurred in the past. The forms of dysentery we encounter in the United States are quite mild because our purified water supplies have cut off the main route of transmission for virulent strains of the bacteria. Not only did hygienic changes reduce the number of cases, they selected for the milder shigella organisms, those that leave their victim well enough to get out and about. Diphtheria is another case in point. When the diphtheria vaccine was invented, it targeted only the most severe form of diphtheria toxin, though for economic rather than evolutionary reasons. Over the years, however, that choice has weeded out the most virulent strains of diphtheria, selecting for the ones that

---

"We did with diphtheria what we did with wolves. We took an organism that caused harm, and unknowingly, we domesticated it into an organism that protects us."

---

cause few or no symptoms. Today those weaker strains act like another level of vaccine to protect us against new, virulent strains.

"You're doing to these organisms what we did to wolves," says Ewald.

"Wolves were dangerous to us, we domesticated them into dogs, and then they helped us, they warned us against the wolves that were out there ready to take our babies. And by doing that, we've essentially turned what was a harmful organism into a helpful organism. That's the same thing we did with diphtheria; we took an organism that was causing harm, and without knowing it, we domesticated it into an organism that is protecting us against harmful ones."

Putting together a new scientific discipline—and getting it recognized—is in itself an evolutionary process. Though Williams and Neese say there are hundreds of researchers working (whether they know it or not) within this newly built framework, they realize the field is still in its infancy. It may take some time before *Darwinian medicine* is a household term. Nesse tells how the editor of a prominent medical journal, when asked about the field, replied, "Darwinian medicine? I haven't heard of it, so it can't be very important."

But Darwinian medicine's critics don't deny the field's legitimacy; they point mostly to its lack of hard-and fast answers, its lack of clear clinical guidelines. "I think this idea will eventually establish itself as a basic science for medicine, " answers Nesse. "What did people say, for instance, to the biochemists back in 1900 as they were playing out the Krebs cycle? People would say, 'So what does biochemistry really have to do with medicine? What can you cure now that you couldn't before you knew about the Krebs cycle?' And the biochemists could only say, 'Well, gee, we're not sure, but we know what we're doing is answering important scientific questions, and eventually this will be useful.' And I think exactly the same applies here."

Wonders

How Fast Is Technology Evolving?

By W. Brian Arthur

W. BRIAN ARTHUR is Citibank Professor at the Santa Fe Institute in New Mexico. Philip Morrison returns to "Wonders" next month.

My grandfather for some reason, wore a hat to meals. Some evenings—also hatted—he would play the fiddle. He was born in Ireland in 1874, and he lived to see, in his long life, satellites, computers, jet airplanes and the Apollo space program. He went from a world where illiterate people footed their way on dirt roads, where one-room schools had peat fires in the corner, where stories were told at night in shadows and candlelight, to a world of motor cars and electricity and telephones and radio and x-ray machines and television. He never left Ireland, although late in life he wanted to go to England in an airplane to experience flying. But in his lifetime—one lifetime—he witnessed all these birthings of technology.

Let's imagine speeding up biological evolution in history by a factor of 10 million.

It is young, this new technology. It is recent. It has come fast. So fast, in fact, that speed of evolution is regarded as a signature of technology itself. But how fast? How quickly does technology evolve? It is hard to clock something as ill defined as technology's speed of evolution. But we can ask how fast we would have to speed up the natural, biological evolution of life on our planet to make it roughly match some particular technology's rate of change.

Let's imagine speeding up biological evolution in history by a factor of 10 million. This would mean that instead of life starting around 3,600 million years ago, in our fast-forwarded world the first, crude blue-green algae appear 360 years ago, about the year 1640. Multicellular organisms arise in Jane Austen's time, about 1810 or so, and the great Cambrian explosion that produced the ancestors of most of today's creatures happens in the early 1930s, the Depression era. Dinosaurs show up in the late 1960s, then lumber through the 1970s and into the 1980s. Birds and mammals appear in the mid-1970s but do not come fully into their own until the 1990s. Humankind emerges only in the past year or two—and as *Homo sapiens* only in the past month.

Now let's lay this alongside a technology whose speed we want to measure—calculating machinery, say. We'll put it on the same timeline, but evolving at its actual rate. Early calculating machines—abacuses—trail back, of course, into antiquity. But the modern era of mechanical devices starts in the years surrounding the 1640s, when the first addition, subtraction and multiplication machines of Wilhelm Schickard, Blaise Pascal and Gottfried Wilhelm Leibniz begin to appear. These were rudimentary, perhaps, but early computational life nonetheless. The first successful multicellular devices (machines that use multiple instructions) are the Jacquard looms of Jane Austen's time. Calculators and difference engines of varying ingenuity arise and vanish throughout the 1800s. But not until the 1930s—the Cambrian time on our parallel scale—is there a true explosion. It's then that calculating machines become electrical, the government goes statistical, and accounting becomes mechanized. The 1960s see the arrival of large mainframe computers, our parallel to the dinosaurs, and their dominance lasts through the 1970s and 1980s. Personal computers show up, like birds and mam-

DUSAN PETRICIC

mals in the mid-1970s, but do not take hold until the late 1980s and early 1990s.

What then corresponds to humankind, evolution's most peculiar creation to date? My answer is the Internet or, more specifically, its offshoot, the World Wide Web. The Web? Well, what counts about the Web is not its technology. That's still primitive. What counts is that the Web provides access to the stored memories, the stored experiences of others. And that's what is also particular to humans: our ability not just to think and experience but to store our thoughts and experiences and share them with others as needed, in an interactive culture. What gives us power as humans is not our minds but the ability to share our minds, the ability to compute in parallel. And it's this sharing—this parallelism—that gives the Web its power. Like humans, the Web is new, although its roots are not. And its impact is barely two years old.

This correspondence between biology and technology is striking. And naturally, it's not perfect. Why should it be? This is fun, after all—more whimsy than science. But if we accept this correspondence, crude as it is, it tells us that technology is evolving at roughly 10 million times the speed of natural evolution. Hurricane speed. Warp speed.

From what I've said, it would seem that all the interesting things in technology or biology have occurred recently. But this is just appearance. In biological evolution, it is not the species markers that count but rather the new principles that are "discovered" at rare intervals. The miracles are not dinosaurs or mammals or humans but are the "inventions" of nucleotide-protein coding, cellular compartmentation, multicelled organisms with differentiated cells, networks of on-off regulatory genes. So it is with technology. The miracles are not computers or the Net; they are the original ideas that human reckoning can be rendered into movements of cogs and sprockets, that sequences of instructions can be used to weave silk patterns, that networks of electrical on-off switches can be used to pinpoint the zeroes of the Riemann zeta function. The miracles are these new principles, and they arrive infrequently. Evolution merely kludges them together to make new species or new machines in continually novel ways.

If technology is indeed evolving at something like 10 million times biology's rate, perhaps this is too fast. Perhaps we are careening into the future in a bobsled with no controls. Or being rocketed into orbit with no reentry possible. This is frightening, maybe. Until we realize that we use all the complicated, sleek, metallic, interwired, souped-up gizmos at our disposal for simple, primate social purposes. We use jet planes to come home to our loved ones at Thanksgiving. We use the Net to hang out with others in chat rooms and to exchange e-mail. We use quadraphonic-sound movies to tell ourselves stories in the dark about other people's lives. We use high-tech sports cars to preen, and attract mates. For all its glitz and swagger, technology, and the whole interactive revved-up economy that goes with it, is merely an outer casing for our inner selves. And these inner selves, these primate souls of ours with their ancient social ways, change slowly. Or not at all.

My grandfather died in 1968, the year before human beings landed on the moon. He never did realize his ambition to fly in an airplane. At 90, they told him he was too old. The world of his childhood no longer exists. It has all changed. Our world is changing, too, and rapidly. And yet nothing really is changing. For some of us at least, even lovers of technology like me, this is a comfort.

and, 9, 11, 14; of hominids, 100–105, 111–113, 114–117, 118–122
founder effect, 19
friendship, among East African baboons, 38–42; between human–gorilla, 43–49

Galileo, 8, 9
ganglioside, 18–21
Gardner, Trixie and Allen, 52–53
Garner, Karen J., 204
Gaucher's disease, 20–21
genes: biological factors of, 28–31; diseases and, 17–22, 23–27
genetic: cloning and, 28–31; drift of, 16, 19; races and, 186, 187
genocide, among the Interahamwe guerrillas, 201
Goodall, Jane, 35, 44, 55, 57, 59, 95; on minds of chimpanzees, 50–54
gorillas, African, 43–49
Gould, John, 11
gravitation, 9
great chain of being, 8, 9
Grim, Clarence, 26, 27
Guinier, Lani, 177–178

hair, race and, 183
Haldane, J. B. S, 15
hanta viruses, 210
Hanuman langur monkeys, infanticide of, 81–85
Haraway, Donna, 80
height, plasticity and, 190–193
heterozygotes, 18, 20–21
hexosaminidase, 18–20
Hill, Andrew, 102–103, 105, 113
Hill, Kim, 84
HIV (human immunodeficiency virus), 210; natural selection and, 221–222; poverty and, 206–208. See also AIDS
hominids: art of, 136–138; dating fossils of, 144–148; language of, 139–143; as scavengers, 118–121
Homo erectus, 100, 114, 116–117, 121, 126–129, 130, 145, 154, 183
Homo ergaster, 128–129
Homo habilis, 120, 128, 130, 145
Homo heidelbergensis, 122, 123
Homo sapiens, 106, 117, 130, 139, 145, 153, 177–179, 181, 183–184
homozygotes, 18–22
Hooker, Joseph, 12
hormones, 29
Howell, Clark F., 111, 117
Hrdy, Sarah Blaffer, 77–80, 81–85, 93
Hubbard, Ruth, 78
Hublin, Jean-Jacques, 149, 160
humans: evolution of bipedalism and, 106–110; as hunters, 118–121; male aggression against females, in, 95–97

Humphrey, Nicholas, 35–37
Hunt, Kevin, 105
hunter-gatherers, 119, 211, 216
hunting, cooperative, by chimpanzees, 55–58, 59–61
Hutton, James, 9
hypertension, blacks and, 23–27
"hypothetico-deductive" method, 13

inbreeding depression, 76
infanticide, among Hanuman langur monkeys, 81–85
infants, adult males and primates, 40–41
Interahamwe guerrillas, war against Tutsi minority, 201
isolation, evolution and, 16

Jablonski, Nina, 105
Jantz, Richard L., 170, 171–172
Java Man. See Homo erectus
jealousy, among baboons, 41–42
Jelderks, John, 171, 175
Jews, Tay-Sachs disease and, 182
Johanson, Donald C., 107
Johnson, Floyd, 168–169
Jones, Rhys, 147

Kachigan, Sam Kash, 77, 80
Kalinia, Jan, 203
Kennedy, John F., 197–199
Kennewick Man, 168–176
Kingston, John, 102–103
Klein, Richard, 130

language training, and development of chimpanzees, 50–54, 62–66
larynx, 140
Lasker, Gabriel, 190
Leakey, Louis, 51, 52
Leakey, Mary, 106–107
Leakey, Meave, 103; on evolution of human bipedalism, 106–110
Legionnaire's disease, 209, 210
Levantine paradox, 153, 156
Lieberman, Philip, 140
Linneaus, 177–178, 181
Lou Gehrig's disease, 192–193
Lovejoy, Owen, 105, 112
Lucy, 68, 103–105, 107, 112, 122–123, 129, 152
Lyell, Charles, 10–13
Lyme disease, 210
lysosomes, 18

Machiavellian intelligence, in primates, 34–37
mad cow disease, 210
malaria, 23, 182
Malthus, Thomas, 11–13
Man-Eating Myth, The (Arens), 163

manipulation hypothesis, of Hrdy, 79
mate-recognition system, 155–156
Mayans, height and, 192
Mayr, Ernst, 15, 155, 204
Mediterranean race, 178
Mendel, Gregor, 14, 15
menstrual taboos, 86
menstruation: natural selection and, 220; women's curse, among the Dogon people, 86–90
Milton, Katherine, 74, 75
mind, of chimpanzees, 67–69
Minthorn, Armand, 170–171, 174
miscegenation law, 177
missing link, 126
Mittermeier, Russell, 72–73, 76
Mongoloid race, 180–181, 183–185, 186, 188, 198
monkeys: cooperative hunting of colobus, by chimpanzees, 59–61; sexual behavior of muriqui, 72–76. See also primates
Morgan, Thomas Hunt, 15, 16
moulding, training method, for chimpanzees, 62
muriqui monkeys, sexual behavior of, 72–76
mutations, 15, 16

N

Native American Graves Protection and Repatriation Act (NAGPRA), Kennewick Man and, 168–176
natural selection: Darwin and, 8–16; genetic diseases and, 17–22, 23–27, 219–222; races and, 182, 184. See also evolution
natural theology, 9
natural-fertility population, 86, 88
Neanderthal people, 131–135; communication of, 139–143; cannibalism of, 163–165
Negroid race, 181, 183, 185, 186, 188, 198
Nesse, Randolph, 219–220, 221
neurotransmitters, 29
Niemann-Pick disease, 20–21
Nishida, Toshisada, 36
Nordic race, 178
North America, Kennewick Man of, 168–176
noses, races and, 183, 186

O

Ogbu, John, 188
Omran, Abdel R., 210–211
Origin of Species, The (Darwin), 8–16
Osborn, Henry Fairfield, 101
Owen, Richard, 14
Owsley, Douglas, 169–172, 175

P

Packer, Craig, 82–83
paradigm, of fossils, 100–105
Parkinson's disease, 192–193
parsimony, principle of, 69

Paterson, Ann, 137–138
Patterson, Bryan, 106
Patterson, Hugh, 155
Peking Man. *See Homo erectus*
Pereira, Michael, 82
Peterson, Dale, 214, 216–218
physical traits, between Neanderthal and hominids, 149–150
Pictet, Francois, 14
pigeonholing devices, 204
Pijoan, Carmen, 164
plagues, 209–213
plasticity, 190–193
Plooij, Frans, 36
polygynous species, 75, 78, 201
polygyny, among the Dogon people, 87
Pope, Alexander, 8
Pope, Geoffrey, 129
potassium-argon dating, of fossils, 127
Potts, Rick, 115, 118, 129
poverty, HIV and, 206–208
primates: cooperative hunting by, 55–58, 59–61; friendship among, 38–42, 43–49; Machiavellian intelligence, 34–37. *See also* bonobo chimpanzees; chimpanzees; gorillas; hominids; monkeys; sex
Principles of Geology (Lyell), 10
Profet, Margie, 89, 220
progesterone, 87
protein, DNA composition as, 29
protohominids, 103, 150
Proxmire, William, 197
Pusey, Anne, 82–83
Pygmies, height and, 191–192
pygmy chimpanzees. *See* bonobo chimpanzees

Quammen, David, 44
Quarterly Review of Biology, The (Profet), 89–90

R

race: culture and, 177–179, 186–189; evolution and, 180–185; Kennewick Man and, 168–175
radiocarbon dating, 145, 148, 152, 160
Ray, John, 9, 13
Redmond, Ian, 46, 49
Reed, Kaye, 104
Rensch, Bernhard, 15
Reznikoff, légor, 138
Rice, Patricia, 137–138
Ryder, Oliver A., 204

S

savanna hypothesis, 100–105
scanning electron microscope (SEM), 118, 119

scavengers, early hominids as, 118–121
scent glands, race and, 184
Schaller, George, 44
Schepartz, Lynne, 151–152
Schneider, Alan, 171, 175
Schwarcz, Henry, 147–148
Sedgwick, Adam, 14
self-concept, in chimpanzees, 53
"selfish gene theory," 216–217
sex: among baboons, 38–42; among bonobo chimpanzees, 91–94, 95–96; human male aggression against females and, 95–97; muriqui monkeys and, 72–76; myth of coy female and, 77–80
sexual dimorphism, in human evolution, 122–123
sexual selection, 75–76, 77
sickle-cell anemia, 18, 23, 182, 187, 220
Simpson, George Gaylord, 15
sinus prints, 199
skin color, race and, 182–183, 186, 187
Skinner, B. F., 67
slavery, genetic adaptation of blacks to, 23–27
smallpox, 209, 212
Snow, Clyde C., 196–200
social manipulation, 35
sociobiology, and human evolution, 215–216
Solutrean, culture, 172–173
sperm competition, 75–76
Stanford, Dennis, 170–171, 174, 176
stature, race and, 183, 186, 192
Stebbins, G. Ledyard, 15
Stewart, T. Dale, 133
Strassmann, Beverly I., 86–90
Strier, Karen, 72–76
survival of the fittest, 13
Swisher, Carl, 126–128, 130

taste, sense of, and race, 184
Tattersall, Ian, 128, 129, 155
Tay-Sachs disease, 17–22, 25; screening programs of, 22
teaching, race and, 186–189
teaching syntax, 63
technology, evolution of, 223–224
Teleki, Geza, 60
Temerlin, Jane and Maury, 50
theraputic hypothesis, of Hrdy, 79
thermaluminescence, 145–148, 152–153, 154
throwbacks, 13
Tobias, Phillip, 100, 104–105
tools, use of: by hominids, 126, 130, 172; by other primates, 52, 55–58, 59–61
travel, international, infectious diseases and, 212–213

Trinkaus, Erik, 159; on Neanderthal people, 122, 131–135, 154
Trivers, Robert, 82
tuberculosis, 209, 211, 213
Turner, Christy G., 162
turn-over pulse hypothesis, of Vrba, 102
Tutsi, height and, 191

uniformitarianism, 10, 13
United States, height in, 191–192
Upper Paleolithic transition, 144, 150
urine, race and, 184

Valladas, Helene, 146, 152, 153
Valle, Celio, 72
van Lawick, Hugo, 48
van Schaik, Carel, 83–84
Vandermeersch, Bernard, 152–153
violence, human nature and, 214–218
violent behavior, among Hanuman langur monkeys, 81
Vrba, Elisabeth, 101–103

Walker, Alan, 104, 111–112, 115, 117, 121; on evolution of human bipedalism, 106–110
Wallace, Alfred Russel, 11
Washoe, 52–54, 62–63
Watson, James D., 8
Wesley, John, 8
Wheeler, Peter, 103, 105
White, Francis, 92
White, Tim D., 104, 109, 112–113, 163
Whiten, Andrew, 34–37
Williams, George, 219–220
Wilson, E. O., 82; sociobiology and, 215–216
Wilson, Margo, 84
Wilson, Thomas, 25–27
Wood, Bernard, 112, 116, 122, 130
Wrangham, Richard, 74–75, 214, 216–218
Wright, Chauncey, 14
Wright, Sewall, 15

Y

Yellen, John, 144, 147
Yerkes, Robert, 52

Z

Zane, Daria, 175
Zihlman, Andrienne, 94
Zilhao, Jao, 157–161

AE Article Review Form

We encourage you to photocopy and use this page as a tool to assess how the articles in **Annual Editions** expand on the information in your textbook. By reflecting on the articles you will gain enhanced text information. You can also access this useful form on a product's book support Web site at **http://www.dushkin.com/online/.**

NAME: _____   DATE: _____

TITLE AND NUMBER OF ARTICLE: _____

BRIEFLY STATE THE MAIN IDEA OF THIS ARTICLE: _____

LIST THREE IMPORTANT FACTS THAT THE AUTHOR USES TO SUPPORT THE MAIN IDEA:

WHAT INFORMATION OR IDEAS DISCUSSED IN THIS ARTICLE ARE ALSO DISCUSSED IN YOUR TEXTBOOK OR OTHER READINGS THAT YOU HAVE DONE? LIST THE TEXTBOOK CHAPTERS AND PAGE NUMBERS:

LIST ANY EXAMPLES OF BIAS OR FAULTY REASONING THAT YOU FOUND IN THE ARTICLE:

LIST ANY NEW TERMS/CONCEPTS THAT WERE DISCUSSED IN THE ARTICLE, AND WRITE A SHORT DEFINITION:

ANNUAL EDITIONS revisions depend on two major opinion sources: one is our Advisory Board, listed in the front of this volume, which works with us in scanning the thousands of articles published in the public press each year; the other is you—the person actually using the book. Please help us and the users of the next edition by completing the prepaid article rating form on this page and returning it to us. Thank you for your help!

ANNUAL EDITIONS: Physical Anthropology 00/01

ARTICLE RATING FORM

Here is an opportunity for you to have direct input into the next revision of this volume. We would like you to rate each of the 44 articles listed below, using the following scale:

1. **Excellent: should definitely be retained**
2. **Above average: should probably be retained**
3. **Below average: should probably be deleted**
4. **Poor: should definitely be deleted**

Your ratings will play a vital part in the next revision. So please mail this prepaid form to us just as soon as you complete it. Thanks for your help!

RATING

ARTICLE

1. The Growth of Evolutionary Science
2. Curse and Blessing of the Ghetto
3. The Saltshaker's Curse
4. A Gene for Nothing
5. Machiavellian Monkeys
6. What Are Friends For?
7. Dian Fossey and Digit
8. The Mind of the Chimpanzee
9. Dim Forest, Bright Chimps
10. To Catch a Colobus
11. Language Training of Apes
12. Are We in Anthropodenial?
13. These Are Real Swinging Primates
14. The Myth of the Coy Female
15. First, Kill the Babies
16. A Woman's Curse?
17. What's Love Got to Do with It?
18. Apes of Wrath
19. Sunset on the Savanna
20. Early Hominid Fossils from Africa
21. A New Human Ancestor?
22. Asian Hominids Grow Older
23. Scavenger Hunt

RATING

ARTICLE

24. New Clues to the History of Male and Female Size Differences
25. *Erectus* Rising
26. Hard Times among the Neanderthals
27. Old Masters
28. The Gift of Gab
29. The Dating Game
30. The Neanderthal Peace
31. Learning to Love Neanderthals
32. Archaeologists Rediscover Cannibals
33. The Lost Man
34. Black, White, Other
35. Racial Odyssey
36. Culture, Not Race, Explains Human Diversity
37. The Tall and the Short of It
38. Profile of an Anthropologist: No Bone Unturned
39. Gorilla Warfare
40. HIV 1998: The Global Picture
41. The Viral Superhighway
42. Exploring Our Basic Human Nature: Are Humans Inherently Violent?
43. Dr. Darwin
44. Wonders: How Fast Is Technology Evolving?

(Continued on next page)

We Want Your Advice

ABOUT YOU

Name

Date

Are you a teacher? ☐   A student? ☐

Your school's name

Department

Address   City   State   Zip

School telephone #

YOUR COMMENTS ARE IMPORTANT TO US !

Please fill in the following information:
For which course did you use this book?

Did you use a text with this *ANNUAL EDITION*? ☐ yes ☐ no
What was the title of the text?

What are your general reactions to the *Annual Editions* concept?

Have you read any particular articles recently that you think should be included in the next edition?

Are there any articles you feel should be replaced in the next edition? Why?

Are there any World Wide Web sites you feel should be included in the next edition? Please annotate.

May we contact you for editorial input? ☐ yes ☐ no
May we quote your comments? ☐ yes ☐ no